ONE HUNDRED HARVESTS

Research Branch
Agriculture Canada
1886–1986

T.H. Anstey

Research Branch
Agriculture Canada

Historical Series No. 27
1986

Available in Canada through

Authorized Bookstore Agents
and other bookstores

or by mail from

Canadian Government Publishing Centre
Supply and Services Canada
Ottawa, Canada, K1A 0S9

Catalogue No. A54-2/27E
ISBN 0-660-12036-4

Canada: $14.95
Other countries: $17.95

Price subject to change without notice

Canadian Cataloguing in Publication Data

Anstey, T.H.

One hundred harvests : Research Branch,
Agriculture Canada, 1886-1986

(Historical series / Research Branch,
Agriculture Canada, ISSN 0829-7541 ; no. 27)

Issued also in French under title: Cent moissons.
Includes bibliographical references and indexes.
Cat. no. A54-2/27E
ISBN 0-660-12036-4

1. Canada. Agriculture Canada. Research Branch—
History. 2. Agricultural experiment stations—
Canada—History. I. Canada. Agriculture Canada.
Research Branch. II. Title. III. Series:
Historical series (Canada. Agriculture Canada
Research Branch) ; no. 27

S542.C32R47 1986   354.710082'33042   C86-099200-4

Staff editor: Sheilah V. Balchin
Designer: Carl Halchuk

Research Branch • Agriculture Canada
1886–1986

Dr. W. Saunders

Dr. J.H. Grisdale

Dr. C.G. Hewitt

Dr. E.S. Archibald

Dr. A. Gibson

Dr. J.M. Swaine

Dr. K.W. Neatby

Dr. E.S. Hopkins

Dr. C.H. Goulden

Dr. R. Glen

Dr. J.A. Anderson

Dr. J.C. Woodward

Dr. B.B. Migicovsky

Dr. E.J. LeRoux

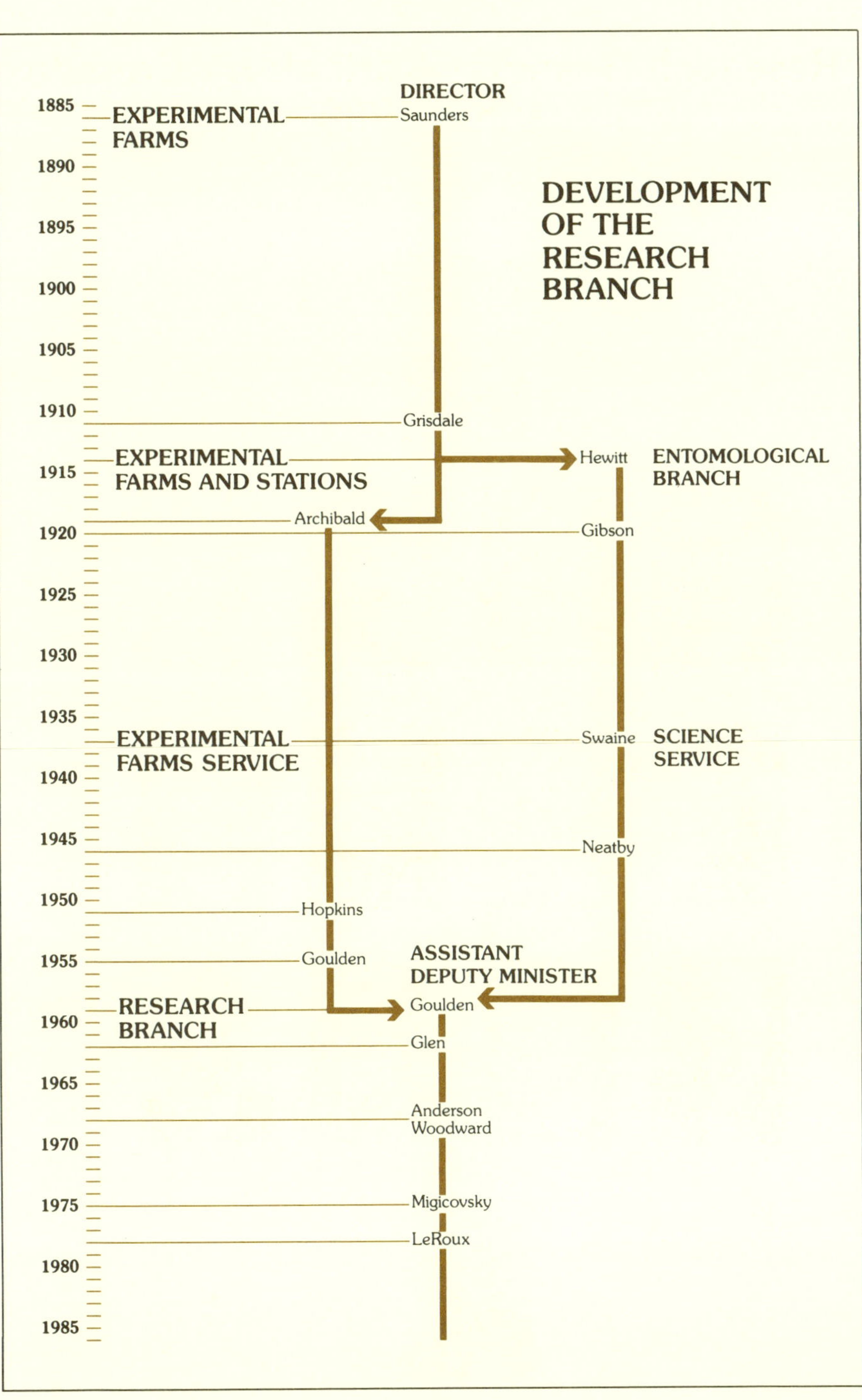

DEVELOPMENT OF THE RESEARCH BRANCH
DIRECTOR
EXPERIMENTAL FARMS
1885
Saunders
1890
1895
1900
1905
1910
Grisdale
EXPERIMENTAL FARMS AND STATIONS
1915
Hewitt
ENTOMOLOGICAL BRANCH
Archibald
1920
Gibson
1925
1930
1935
EXPERIMENTAL FARMS SERVICE
Swaine
SCIENCE SERVICE
1940
1945
Neatby
1950
Hopkins
Goulden
ASSISTANT DEPUTY MINISTER
1955
RESEARCH BRANCH
Goulden
1960
Glen
1965
Anderson
Woodward
1970
Migicovsky
1975
LeRoux
1980
1985

Contents

x

Part Three—Extramural activities

Appendixes

Indexes

Foreword

This book is about Canada and its development, about brave and dedicated men and women, and their successes in the face of adversity. It is about motivators of agriculture—federal ministers and their research teams of scientists and support staff—who have brought a wealth of technology to Canadian farmers and the agri-food industry. *One Hundred Harvests* has been written to help celebrate a century of agricultural research into soil management, crop and animal production, protection, utilization, food quality, and food processing. It recounts the ways in which scientific research is organized and operates—how it serves as the driving force behind a viable agri-food industry in this land of ice and snow.

Three years ago when plans to celebrate the 100th anniversary of the passage of The Experimental Farm Station Act were being formulated, I asked Dr. T.H. Anstey, a senior member of the Research Branch Executive, to prepare this history. Documenting the development of the branch and selecting samples from the thousands of activities of this national research organization would be a singularly important undertaking. Dr. Anstey's book is an excellent response to my request.

The Research Branch with its units stretching from St. John's West, Newfoundland, to Saanichton, British Columbia, has helped to unite the country by cooperating with farmers and their federations, by advising food processors and their distributors, by working with provincial governments and their agriculture departments, and by being partners with universities and their faculties of agriculture. The branch derives particular satisfaction from the assistance it provided to Manitoba and the North-West Territories during their early settlement.

In 1941, Tom Anstey began his career with the Experimental Farm, Agassiz, British Columbia, as a summer student working in horticulture. Following military service overseas on loan to the British 6th Airborne Division during World War II, he returned to Agassiz in 1946 as Assistant Superintendent (Horticulture) to Mr. W.H. Hicks. In 1953 he was appointed Superintendent, Experimental Station, Summerland, British Columbia. Following a 1-year exchange with Dr. C.J. Bishop, Superintendent, Kentville, Nova Scotia, in 1958–1959, he became the first director, Research Station, Lethbridge, Alberta, where he managed a smooth amalgamation of experimental farm and science service staffs. Dr. Anstey moved to Ottawa in 1969 as Assistant Director-General (Western), assuming responsibility for the operation of all stations from Winnipeg, Manitoba, to Saanichton, British Columbia. Ten years later, when the branch regionalized, Dr. Anstey joined my office as Special Assistant and Senior Adviser, International Research and Development. In this capacity he provided support to me and, as well, developed the Research Branch's first fully organized international program of cooperation, with ties to all major agricultural countries.

Dr. Anstey made significant research contributions by breeding strawberries and Italian green sprouting broccoli at Agassiz. He advanced tree fruit production in the Okanagan Valley at Summerland, and promoted and taught the use of electronic data processing while director at Lethbridge. He is a Past President and Fellow of the Agricultural Institute of Canada, and an Honorary Life Member of the Canadian Society of Horticultural Science. He is also a Vice-President (Honoraire), International Commission on Irrigation and Drainage, and holds a Public Service of Canada Merit Award for his work with that Commission.

I could have found no one more dedicated to the composition of the history of the Research Branch than Dr. Anstey. He has devoted all his energies to the preparation of this unique statement on Canadian agricultural research. His extensive knowledge of and keen interest in Canadian agriculture and its supporting scientific research, acquired over 45 years, have ideally equipped him to prepare a thorough account of the development of the Research Branch, its accomplishments, and its impact on Canadian agriculture.

xiii

Ottawa
January 1986

E.J. LeRoux
Assistant Deputy Minister
(Research)

Preface

The first agricultural research in Canada was done at Sainte-Anne-de-la-Pocatière, Quebec, in 1859, followed by that of the Ontario Veterinary College in 1863, and the Ontario School of Agriculture in 1873. The Research Branch of the Canada Department of Agriculture was started in 1886 as the Experimental Farms System. The complex of industry, university, provincial, and federal agricultural research that has grown from these beginnings serves a wide Canadian population. In a fundamental way it serves more Canadians than any other single research organization, since every citizen requires the most nutritious food obtainable in order to achieve both physical and mental health and growth potential.

Among nations of the world, Canada is youthful at 119 years of age; its past, just around the corner. Several Canadian institutions that had their beginnings shortly after Confederation are celebrating their centenaries during the last quarter of the twentieth century, well aged within the North American framework. Research Branch is proud to be part of this venerable company.

The celebration of the centenary of an institution is an appropriate time to compile its history; in this case a history which parallels that of the country and government. Indeed, more than that, it is at the central core of Canada's development. Without the Research Branch and its predecessor, the Experimental Farms System, the Great Plains of Canada would not have become so quickly one of the major bread baskets of the world nor would Canada likely have achieved the positive balance of trade that it holds and enjoys today.

The development of a country is dependent upon the rational and wise use of its renewable natural resources. For us this means our forests, our fisheries, our water, and our agriculture. The major role of the Research Branch has been to assure that Canada reaches its agricultural potential, but it has also made contributions in the development of forests and in the proper use of water.

The history of an institution, and particularly one that is required to generate ideas and solve problems, is essentially the history of its people. Throughout this treatise attention has been paid to the people (regrettably space permits attention to only a few people) who have made unique contributions to the functioning and progress of the Research Branch. It is hoped that a sufficiently large sample has been selected so that the reader can at least catch a glimpse of people of imagination, character, and devotion. Those most concerned were the ministers of agriculture who convinced Parliament of the original need for Experimental Farms and who continued to support them; the deputy ministers and their staffs who, in the face of other pressing problems, recognized the social and financial returns that would accrue to agriculture and to Canada when research was successful; the farmers who had the courage to try new varieties and techniques, and who made first-rate suggestions for the solutions of many problems; the colleagues in provincial departments and in university faculties and colleges of agriculture who freely shared experiences; the peers at agricultural research stations in the older countries of the world who were constructively critical of new theories and iconoclastic ideas; the support staffs at experimental farms, stations, laboratories, and institutes, upon whose hands and eyes scientists have learned to depend; and the scientists and officers in charge whose efforts are frequently crowned with success and who shoulder the blame when solutions are elusive.

Research stations and experimental farms, or "The Farm," as they are often still called in many of the communities where they are to be found across Canada, are different things to different people. To farmers and extension agrologists they are a source of information and consultation; to local gardeners they are a place to go for friendly advice regarding their problem weeds or insects; to an area's business community they are a reasonably large and steady source of customers; to convention bureaus they are a delightful attraction in apple blossom or chrysanthemum seasons; to school children they are an extension of their science class or a place to picnic and hike on Saturdays; and to many—scientists, technicians, typists, and artisans, they are a pleasant place of employment.

This history is an exoteric one, modified for those outside the inner circle. Although it will be useful to staff of the Research Branch in quickly reviewing the backgrounds of various subjects, it is not intended to be a definitive history either of the development of the organization or of the research accomplished by its staff. To have prepared such a history would have required much more than the 2 years available to me. It is my hope that some readers may be spurred on by certain parts of the book to search more deeply for themselves and build upon what I have begun.

During the hours spent in the Main Library of the Department of Agriculture, the Central Registry of the Research Branch, and the Public Archives of Canada, I have been extended innumerable courtesies and assistance in tracing elusive documents. Three of the photographs included in this publication were kindly provided by the Public Archives of Canada. They are the portraits of Dr. W. Saunders (PA-136872); Dr. A. Gibson (PA-140402); and Dr. K.W. Neatby (PA-139544). I have drawn freely from published reports of the Research Branch and its antecendents, but have made little direct reference to these papers, as they are not readily available to most readers.

Many people have been of great help and a source of strength throughout the preparation of the manuscript: Dr. E.J. LeRoux, Assistant Deputy Minister (Research), who first suggested I prepare this history, has continued to reassure and counsel; Drs. J.W. Morrison, J.J. Cartier, and R.L. Halstead suffered through reading first drafts of the text; Drs. R. Glen, J.A. Anderson H.K.C.A. Rasmussen, J.E. Andrews, and A.A. Guitard, and Mr. B.H. Whittle, each in his own way made a valued contribution; directors general of each region, directors of each research station, and literally dozens of research scientists have given freely of their time by supplying backgound papers and checking early manuscripts; Mr. J.P.F. Darisse kindly read the French text; Miss B.P. Jack gave valuable technical assistance during the research phase; Mrs. S.V. Balchin and Mr. D. Sabourin provided expert and thoughtful advice in editing the English and French texts; and Mr. C. Halchuk kindly selected the typeface and designed the format of the book. To each I extend my sincere thanks.

Finally, thanks to my wife, Wynne, who has made numerous suggestions and contributed much careful reading. There have been pleasant discussions of shared experiences touched on in the history. She has encouraged and assisted me throughout the writing.

Across the road from the building on the Central Experimental Farm in which much of the research for this history was compiled stands the Dominion Observatory, built in 1902. It is on the Central Experimental Farm but not part of it, for its operation is the responsibility of the Department of Energy, Mines and Resources with which the Department of Agriculture and particularly parts of the Research Branch cooperate closely. A plaque outside the Dominion Observatory commemorates Sir Sandford Fleming, the Canadian engineer who was associated with the survey and construction of the Intercolonial and the Canadian Pacific Railways, who designed the first Canadian postage stamp, and who was an early proponent of the idea of standard time.

Elsewhere on the Central Experimental Farm are similar plaques honoring other Canadians, including Macoun, Fletcher, Carling, Saunders (William and Sir Charles), and Neatby. They took an untamed land and made it fruitful; they took a new idea and made it great. Eventually another book will chronicle the second century; can we begin to imagine what its pages may say?

Ottawa, Ont.   T.H. Anstey
1985

PART I
THE METAMORPHOSIS

Research Branch
Agriculture Canada
1886–1986

Chapter 1
The Experimental Farm Station Act

Long before the white man occupied any portion of North America and while buffalo still roamed the abundant grasslands of the western plains, there were many Indian tribes such as the Hurons, the Tobaccos, and the Neutrals engaged in primitive agriculture (34). Indians of North America grew a variety of crops including maize, beans, squash, sunflowers, tomatoes, and peppers.

With the arrival and settlement of Europeans in North America, farming in areas later to be known as Nova Scotia, Prince Edward Island, and New Brunswick on the eastern seaboard, and as Lower Canada and Upper Canada in the St. Lawrence Valley, and on the Pacific coast had been practiced for more than a century before Confederation in 1867. In Prince Edward Island the basic industry was agriculture. The main industries for Nova Scotia and New Brunswick were lumbering, shipbuilding, and fishing, although agricultural production gradually increased to meet the demands of the other sectors. In Lower Canada (now Quebec) agriculture started with the arrival of Robert Giffard and 100 colonists in 1634, although Louis Hébert had farmed at Quebec City since 1617. About 30 years later, Jean Talon organized the seignorial system of land tenure, which assured the continuation of landlord–tenant agriculture. After 200 years, almost all the easily accessible arable land had been possessed. Agriculture was encouraged and practiced intensively along the St. Lawrence River. In Upper Canada (now Ontario), as early as 1802, exports of wheat and flour were over a million bushels (27 000 tonnes). Nearly all its arable land was in use well before Confederation. On the west coast, agriculture was limited primarily to the lower Fraser Valley and Vancouver Island, where it served the needs of local populations. Immigrants found that in these colonized regions cultivation methods that had been used in Europe were generally applicable, because each region had a relatively humid climate similar to that from which the population had emigrated.

The District of Assiniboia, containing the Red River Settlement, formally became part of Canada in 1870 when it was named Manitoba. It was largely self-sufficient in food production, although frequent plagues of grasshoppers in the summer and early frosts in the fall severely depleted the winter food supplies. In any event, until there was rail connection with either the east or the west coast, Manitoba had little opportunity of exporting surplus production except to the United States.

In 1884 settlers from the east started to move in colonist cars on the partially completed Canadian Pacific Railway to the "free" land in Manitoba and the North-West Territories, later to become Saskatchewan and Alberta, but named "God's Country" by the novelist James Oliver Curwood. They found a country with a dry and hostile climate quite unlike the one they had known in Ontario, Quebec, the Maritimes, or Europe (5). It was their unfamiliarity with the soil and the climate of this new land, and their inadequate farming methods that made the establishment of experimental farms an immediate necessity.

Prior to Confederation in 1867 and the passing of the British North America (B.N.A.) Act when the British Colonies north of the United States of America joined to form the Dominion of Canada, there was a Bureau of Agriculture under a Minister in the Province of Canada (formerly Upper and Lower Canada, see the B.N.A. Act c.6). The Act recognized the need for both federal and provincial authorities in agriculture and made provision for such under Section 95. This section states that "... the Parliament of Canada may ... make laws in relation to Agriculture ... in all or any Provinces; ... any law of the Legislature of a Province relative to Agriculture ... shall have effect in ... the Province as long ... as it is not repugnant to any Act of the Parliament of Canada." Section 93 states that "... each Province ... may exclusively make laws in relation to education ...." As a result of these two sections, provincial governments have generally reserved the right to education in agriculture, which includes agricultural extension, and the federal government has assumed the responsibility for research in agriculture, although not to the exclusion of similar activities on the part of the provinces.

The Department of Agriculture for Canada was organized under the Department of Agriculture Act, which was passed by Parliament and given royal assent on 22 May 1868. In addition to agriculture, the Minister and the department had other wide national responsibilities including immigration, public health, censuses and statistics, patents, copyrights, and trademarks. Agriculture was supposedly the prime responsibility of the department and the Minister, the Honourable J.C. Chapais, and his deputy, Dr. J.C. Taché, lost no time in presenting important Bills to the House of Commons for the protection and improvement of Canadian agriculture. One of the first Bills was an "Act Respecting Contagious Diseases of Animals," passed in 1869. This Act gave the Chief Veterinary Inspector, Prof. Duncan McEachran, who was also Dean of Medicine at McGill University (6), authority to prevent the introduction of animal diseases into Canada. Today's legislation is known as the Animal Disease and Protection Act of 1977 and it continues to help keep Canada free from the most dangerous animal diseases.

In 1883, the Honourable J.H. Pope, Minister of Agriculture, was concerned about the large collection of seeds that had been returned from the Philadelphia Exhibition in an infested state (12). He therefore appointed Mr. James Fletcher to the post of Entomologist for the department. (In 1884 Fletcher signed himself "Honorary Entomologist." The term "Dominion Entomologist" was not used until 1910.) Fletcher was asked to advise on appropriate action to prevent imported insects from becoming serious problems in Canada. At that time, Fletcher worked in the Parliamentary Library as an accountant, but he was also an avid amateur entomologist.

The North-West Territories were acquired from the Hudson's Bay Company by the federal government the year following Confederation. Settlers gradually moved into these great expanses of arable land, but the department had no way of helping the new arrivals decide which crops to grow or how to grow them. Parliament was concerned about this situation and on 30 January 1884 it established a Select Committee of the House of Commons to determine the needs for the improvement of Canadian agriculture.

Mr. G.A. Gigault, M.P., was appointed to chair the Select Committee. The Committee's first action was to send a questionnaire to 1500 Canadians whose names were suggested by Committee members; 385 replies were received, some in great detail. The questionnaire contained five specific questions and these, together with the responses, are given in Table 1.1.

Table 1.1 Questionnaire prepared by the Gigault Committee

Question Are you in favor of —	Number of respondents		
	For	Against	No answer
Establishing an experimental farm?	278	64	43
Appointing an entomologist?	198	117	70
Establishing a central bureau?	256	62	67
Establishing a section devoted to statistics?	211	74	100
Publishing handbooks, reports, and bulletins?	255	48	82

The Committee suggested the establishment of an experimental farm, to be set up as a garden, where varieties of foreign grain, trees, and fertilizers could be tested. It also suggested that samples of seeds and plants be distributed throughout the Dominion by such an experimental farm. In addition to an experimental farm, the Committee envisioned a central bureau with the objective of collecting information on all matters relating to agriculture. This central bureau would have skilled staff to offer advice, conduct experiments, and note improvements effected in other countries that might be introduced into the Dominion for the benefit of agriculture.

During eight meetings of the Select Committee, 14 witnesses were called. Among the witnesses were Prof. Penhallow, Department of Botany, McGill College; Prof. William Brown, Ontario Agricultural College; Mr. A.M. Ross, Commissioner of Agriculture for Ontario; Mr. Charles Gibb, fruit grower, Abbotsford, Quebec; and Mr. James Fletcher, Honorary Entomologist, Department of Agriculture.

On 21 March 1884, the Committee reported its findings to Parliament. The recommendations of the Select Committee were that the federal government should (1) establish a central bureau of agriculture, and (2) establish experimental farms that would operate in conjunction with the proposed central bureau. The bureau would be part of the Department of Agriculture and have the following objectives:
1. Introduce plants, determine the comparative value of fertilizers, test seeds for purity and vitality, and test the health of plants and animals.
2. Investigate methods of controlling insects and diseases of plants and animals.
3. Study the qualities of breeds of animals, how to protect them from parasites and diseases, and how best to feed them.

4. Gather useful statistical information.

5. Publish informative bulletins on the foregoing subjects.

The idea of having experimental farms for the express purpose of providing current technology to farming communities was relatively new in North America. The first organized agricultural experimental station in the world was sponsored by Sir John Lawes at Harpenden, just north of London, England, in 1843. It was called Rothamsted, and is frequently referred to as the "mother of experimental stations." Nine years later in Germany, farmers banded together in order to test plants and animals on one farm rather than have individuals conduct tests on their own farms. When the work grew beyond their means, the German farmers requested help from their government, which applauded their initiative and encouraged others toward group experimentation. Within 30 years, that is by 1882, there were 80 experimental stations in Germany. By 1878, France had established 43 experimental stations, which were said to be as important to the farming community as physicians were to sick people and lawyers were in litigious matters. The first agriculture experiment station in the United States was organized in Michigan in 1857. Five years later the government of the United States passed the Morrill Act, which established land grant colleges. Not until 1887, however, was the Hatch Act passed; this Act granted lands for the development of state experiment stations.

Parliament appointed Prof. William Saunders of Northwestern University, London, Ontario, to make further and detailed studies into the practicality of establishing experimental farms in Canada. The appointment was dated 2 November 1885. Saunders worked quickly, for on 20 February 1886, he reported (35) to the Minister, the Honourable John Carling, his findings on 33 states in the United States and on four provinces in Canada. He also gave a summary of the situation in the Dominion Department of Agriculture and referred to the organizations in Europe. Saunders concluded that the benefits derived from agricultural teaching colleges in America under the Morrill Act did not warrant the cost of establishing such institutions in Canada at that time. He did report, however, that agricultural experiment stations were of very great service in supplying information and stimulating progress in agriculture at a comparatively nominal cost. Because agriculture lay at the foundation of Canadian prosperity, Saunders recommended that any reasonable expenditure in the development of agricultural experiment stations should be incurred without delay, leaving the matter of agricultural education in colleges for future consideration.

From Saunders' report we learn that a start had already been made in Canada with respect to education and experimentation in agriculture. Prince Edward Island farmers established an Agricultural Society in 1827 (8), which was formed to import superior blood lines of several kinds of livestock and which eventually resulted in the development of a government stock farm for the same purpose. New Brunswick followed suit, and formed a government stock farm that provided stud and young stock to help farmers upgrade their own herds. Actually, what is believed to be the first agricultural society organized in Canada was the Society for Promoting Agriculture, established in Nova Scotia in November 1789 (1). In 1859, a school of agriculture was started by the Reverend Abbé Pilote at Sainte-Anne-de-la-Pocatière, Quebec. In 1912, this school became affiliated with Laval University in Quebec City. Reaman (34) recounts that Prof. George Buckland, who emigrated from Scotland in 1851, gave lectures at the University of Toronto on the science and art of agriculture. In 1863, Dr. Andrew Smith founded the Ontario Veterinary College. At the instigation of John Carling, then Commissioner of Agriculture for Ontario, the Ontario School of Agriculture was organized in 1873 and conducted some agricultural experimental work; it moved to Guelph in 1874.

On 22 April 1886, the Honourable (later Sir) John Carling, by now Minister of Agriculture for Canada, moved that the House of Commons "resolve itself into Committee of the Whole ... to consider the following resolution ...." The resolution empowered the government to establish experimental farm stations. The motion for the House to go into Committee was approved, but the matter was delayed until the Minister had resumed his duties following a short illness. On 30 April 1886, the Minister reopened the subject, reviewed the work of the Select Committee chaired by Mr. G.A. Gigault, M.P., and reminded the House that Prof. Saunders had been appointed to visit different agricultural experimental farm stations in the United States, and to enquire into the workings of similar institutions in England, Germany, Russia, and France to determine the amount of land each experimental farm needed, their annual expenditures, and the results of experimental work. His report (35) also included observations regarding institutions in Belgium, Ireland, Austria, Hungary, Italy, and Japan. The Minister then said that it was the intention of the government to establish a principal, or central, experimental farm for Ontario and Quebec jointly, in the vicinity of Ottawa, and four other farms in different parts of Canada. According to the proposed Act (see Appendix III), of these four other farms one was to be located in Prince Edward Island, Nova Scotia, or New Brunswick, one in Manitoba, one in the North-West Territories, and one in British Columbia.

Of the 16 Members of Parliament who spoke, none from either side of the House opposed the resolution. One must recall that at the time nearly half the men, and hence voters, in Canada derived their livelihood from the soil. Some Members from the opposition suggested that the appropriate provincial governments should establish such farms because they would be more familiar with the types of problems that required solving in their particular areas. They would also be able to supervise the activities of the experimental farms better than the federal government could from Ottawa. One Member thought markets were needed, not experiments to learn how to grow more produce. This Member stated that "our farmers as a body are intelligent and know precisely what to do in their business ... our farmers are raising too much [produce] and are not paid enough for what they raise."

The remaining speakers were enthusiastically in support of the resolution. Several said that there was nothing before the House during that session as important as the proposal brought by the Minister of Agriculture. Many speakers identified problems that required attention. The most frequently raised one was that of determining the best varieties of crops to grow under each of the climatic conditions selected for the five experimental farms. Such information would make it unnecessary for each farmer to experiment on his own and risk losing some of a year's harvest when part of an experiment was a failure. Another subject of interest was the planting of trees both for shelter and for timber. Those who knew the climatic conditions in the North-West Territories claimed that large stands of trees planted in Manitoba and the Territories would bring substantial benefits to the settlers by providing shelter and timber, and might even soften the severe climate. Others were concerned about fencing and its cost, saying that the experimental farm stations should test various kinds of bushes for their hedge-making capabilities. The matter of manures, chemical fertilizers, pastures, crop rotations, feeding values of different kinds of crops, and harvesting times for grain were all subjects for lively comment in Parliament. Probably the most significant proposal, however, was that Canada should produce a spring wheat that would ripen before the fall frosts in Manitoba and the North-West Territories. In addition, this spring wheat should be of superior baking quality. One Member wisely noted that since the experimental farm stations were to replace individual experimentation, the results farmers were given by such stations should be dependable.

The second reading was on 7 May 1886. This time, only eight Members of Parliament spoke in addition to the Minister of Agriculture. Questions related to details on how the various experimental farm stations would be set up, where they would be located, the cost to establish them, and the cost to maintain them. The Minister read the conclusions of the Gigault Committee into the Commons Debates. These conclusions, together with Saunders' recommendations, form the basis of Bill 124, The Experimental Farm Station Act (see Appendix III). Some Members suggested again that provincial governments assume the responsibility for these stations or that individual farmers be supplied with the necessary seed and that one farmer in each constituency, recommended by the local Member of Parliament, perform experiments as planned by a central staff in Ottawa. None of these suggestions was adopted, however, and the House agreed to the second reading. The final reading and passage of the Bill, under Prime Minister, the Right Honourable Sir John A. Macdonald, was on 11 May 1886. The House prorogued on 2 June 1886, and Bill 124, together with many others, was given royal assent and became law.

The Experimental Farm Station Act in force today is the same prescient Act that was passed in 1886. It was so well conceived that only minor amendments have had to be made to provide for experimental farm stations in new provinces as they were formed, and to make some administrative adjustments. The Act has served agriculture and Canada well. It gives ministers the freedom to organize according to changing needs, but it charges their officials and scientists with solving problems as these emerge. Initially, the Act met the requirements of an agriculture dependent upon horsepower, then, without change, adjusted to the mechanical evolution, and now provides for the high technology era. Twelve deputy ministers have assured that spending estimates are prepared each year in support of Experimental Farms. Twenty-one ministers have justified them in Parliament. Today, after going through several reorganizations, the Experimental Farms System is known as the Research Branch of the Canada Department of Agriculture. The names used (frequently preceded by "Dominion"), with the dates, are as follows:

1886–1899 Experimental Farms or Experimental Farms System,
1900–1910 Experimental Farms Branch,
1911–1937 Experimental Farms and Stations,
1938–1958 Experimental Farms Service, and
1959–   Research Branch.

Chapter 2
The First Five Experimental Farms
1886–1905

Experimental farm superintendents were by no means the first to experiment with farming in Canada. Many farmers in Ontario had developed productive enterprises by trial and error prior to 1886. MacEwan (29) describes the trials and experiments with different crops and machinery that were conducted by Archibald Wright at Winnipeg, as early as 1882. He was the first farmer in the west to grow sweetclover for the tannin in its seed, he grew all kinds of grain, he tried to grow sugarcane and peanuts, and he imported purebred Holstein dairy cows from Minnesota.

Much care and good judgment led to the selection of a central experimental farm site at Ottawa, Ontario, and branch experimental farms located at Nappan, Nova Scotia, Brandon, Manitoba, Indian Head, North-West Territories, and Agassiz, British Columbia. The sites were chosen by Dr. William Saunders, Director of the new Experimental Farms. An advanced agricultural industry developed and spread around each establishment as farmers recognized the benefits accruing from the new technology now available to them.

Dr. William Saunders was appointed in October 1886. In 1848, when he was 12 years old, he and his family had moved to London, Ontario, from Devon, England. His scientific bent was evident at an early age; he apprenticed as a druggist and opened his own pharmacy in 1855 when he was only 19 years old. Saunders was also an avid gardener, and because insects attacked the plants he grew, he became interested in entomology. He helped organize the Entomological Society of Canada in 1863. As a result of his interest in and knowledge of gardening and entomology, he was made a Fellow of the American Society for the Advancement of Science in 1874, and of the Royal Society of Canada in 1881. His remarkable book, *Insects injurious to fruits*, was published in 1883 and remained the primary reference in that field for many years.

Saunders described to the Minister of Agriculture in some detail the property selected for the Central Experimental Farm and its value for agricultural research. The land was located on the south side of the road leading to Merivale, just west of the City of Ottawa in Nepean Township. On 25 June 1908, this portion of Merivale Road was named Carling Avenue (Ottawa By-law 2777). Among the characteristics of the property listed by Saunders were the following:

a. proximity to the boundary of the provinces of Quebec and Ontario in the Township of Nepean;

b. ease of approach by road, rail, and water;

c. height above both the Rideau and Ottawa rivers such that drainage to each river occurs; and

11

d. variety of soils ranging from heavy clay to sandy loam, with most being good-
    quality dark sandy loam or friable clay loam, making it admirable for experi-
    mental work.

The 188 ha were purchased from 14 people, the largest blocks coming from
J.R. Booth in 1886, V. Hallat & J. Reid, and T. Stackpole in 1887, and Mary
Fellows and J. Warnock in 1888. The small fields had been divided by fences
well-packed with rocks, which later were used for the base of roads within the
farm. Under the careful supervision of the first farm foreman, John Fixter,
between 4000 and 5000 stumps were removed with dynamite and second-
growth trees uprooted and burned. More than 6 miles (approximately 10 km) of
tile and box drains were laid, some through blasted rock. The farm was fenced
and considerable grading done to prevent snow drifting. By December 1887, the
whole farm had been brought under the plow and made ready for a crop in the
following year, even though title to the Fellows and Warnock properties was not
obtained until 1888.

Saunders emphasized the value of having the Central Experimental Farm
on almost virgin soil in order that experiments with fertilizers on various crops
would not be affected by previous applications of manures and different cropping treatments. By the end of the 1st year, experiments with barnyard manure in
varying stages of decomposition and with several forms of phosphate and
nitrogen fertilizers were planned. A topographical map that identified proposed
roads, buildings, shelter belts, and an arboretum was completed. A temporary
office and seed-testing house was built early in the 1st year, as well as a
temporary laboratory for the chemist. Barns for horses and other livestock were
under construction by the end of 1887. Even at this early date, Saunders
visualized an agricultural museum where visiting farmers would have "the
opportunity of comparing the different varieties [which can be grown] in different parts of the Dominion of Canada."

Agriculture Minister Carling took a keen interest in the development of the
Central Farm and frequently visited the new property on Merivale Road. In
1887, the Minister took up residence at 236 Metcalfe Street. This house was
occupied by two other cabinet ministers, Sir Alexander Campbell in 1883 and Sir
Louis Henry Davis in 1897. It is now owned by the Chelsea Club and the City of
Ottawa has designated it a heritage building.

Residences were built for the director and for several officers and foremen
on the Central Experimental Farm. This became common practice not only in
Ottawa but on branch farms as well. Because of a lack of suitable commercial
accommodation near branch farm locations, a "government room" with linen
was provided in each superintendent's residence for the use of the director and
other officials when they made their annual tours. Usually the bed was a brass
four-poster. The large porcelain washbasins, pitchers, and washstands that were
provided in each one of these rooms are today held by some research stations as
treasured mementos. The wife of each superintendent brought out her best
linen, china, and silver on which to serve specially prepared meals featuring the

produce of the experimental farm and the region. Each of the first four experimental farms was near a Canadian Pacific Railway station. The superintendent usually met the director at the station and drove him to the farm in a democrat or a carriage harnessed to a sleek pair of bay trotters.

During his 1st year (see, in particular, the *Report of the Director for 1892*) Saunders made three trips to the Maritime Provinces and two trips to Manitoba, the North-West Territories, and British Columbia. Indeed, the day following his appointment, Saunders left for the Maritime Provinces. He traveled continuously for the next 3 months, inspecting all the farms that had been offered for sale as experimental farms. Travel was by train, probably in a lower berth. Roomettes were not available until much later and Saunders certainly did not have his private rail car.

Superintendents were appointed for the branch farms in the Maritimes, Manitoba, and the North-West Territories. All three (William M. Blair, S.A. Bedford, and Angus Mackay) were brought to Ottawa in 1887 to help with the work of establishing the Central Farm and to give each superintendent the opportunity of becoming familiar with the way in which Saunders wished the work of the Experimental Farms to proceed.

Mr. William Blair was born at North River, near Truro, Nova Scotia, in 1836. He farmed near his hometown and was the first person to supply milk to Halifax by rail. He was active in both local and national politics, having been elected twice to Parliament. He was Colonel, 78th Highland Regiment, from 1880 to 1888. The Nova Scotia Agricultural College, Truro, was established in large measure through his urging. In the spring of 1888, he moved to the Experimental Farm, Nappan, Nova Scotia, as superintendent.

Mr. Bedford was 12 years old in 1863 when he came with his family from Sussex, England, to Goderich, Ontario. In 1877, at the age of 26, he homesteaded near Darlingford, Manitoba. He was employed by land companies to examine properties and guide new settlers to prospective homesteads. He himself then farmed in the Moose Mountain District of the North-West Territories and was elected to its first Legislative Assembly. In 1888, he arrived in Brandon, Manitoba, as the first superintendent of that experimental farm.

Mr. Angus Mackay was born in 1840 in Pickering Township, Ontario. In 1882, together with three companions, Mackay moved to Indian Head, North-West Territories. Each one took up a homestead and operated the four properties cooperatively. After his appointment as superintendent for the branch farm in the North-West Territories, Mackay spent part of 1887–1888 in Ottawa with Saunders. In the spring of 1888 he returned to Indian Head with the assigned task of selecting suitable land and developing an experimental farm.

It was not until July 1889 that Mr. Thomas A. Sharpe was appointed superintendent of the experimental farm at Agassiz, British Columbia, where he assumed his duties on the following 19 September. Sharpe was born near Kingston, Ontario, in 1847. He moved to southern Manitoba as a young man, where he raised cattle on unsurveyed land. He had one of the first registered herds of Shorthorn in that new province.

Under these four pioneering superintendents, inspired by the drive of their director, William Saunders, the experimental farms developed rapidly and set the course for agriculture in Canada. All four were amateurs in the sense that none was trained in technical agriculture. Each one, however, was keenly interested in advancing agriculture throughout Canada and had a concern for soil, crops, and livestock. Also, each one was brilliant in his own way and took the responsibility to act in the best interests of Canadian farmers.

Correspondence built rapidly. In 1889, only 3 years after the experimental farms were authorized, the Central Farm received nearly 7000 enquiries from farmers and responded with 5400 letters, 41 500 pamphlets, and 3700 packages of seed. The following year, more than 17 000 letters were received and nearly 20 000 responses mailed. By 1894, 15 000 farmers were being supplied with samples of improved varieties of seed, most of which were cereals. The director visited each experimental farm annually and attended meetings of farmers in each province. He promoted the planting of hardwood trees, such as hickory, ash, elm, oak, beech, walnut, and cherry, noting in particular that these species were absent from the forests of British Columbia.

On 20 August 1890, when Saunders was visiting Indian Head, the temperature dropped to 27°F (− 3°C). This caused severe damage to grain that had been sown late in the spring. It confirmed to Saunders the need, previously expressed by Members in the House debate on 22 April 1886, for early maturing varieties of cereals, which would permit earlier harvests and lengthen the harvesting period. The Indian Head experience had a far-reaching effect on the development of wheat varieties in Canada, as we shall see. When Saunders visited Lethbridge in the southwestern part of the North-West Territories, he was impressed with the "energy and industry" of the settlers.

He also noted the value and beauty of several experimental gardens of the Canadian Pacific Railway. These gardens had been supplied with seed and young plants from the Central Experimental Farm and became part of the national plant testing system. For many years thereafter, the Canadian Pacific Railway continued to encourage landscaping around its railway stations, and offered annual awards to those stationmasters whose gardens were judged to be outstanding.

At Agassiz, by the fall of 1890, Superintendent Sharpe had cleared and brought under cultivation approximately 50 acres (20 ha) of land, with a similar amount underbrushed and ready to log off. The mild climate and heavy rainfall, unique in Canada, were conducive to the planting of more than 600 varieties of fruits and 400 varieties of ornamental trees and shrubs during the 1st year of operation. Several of the exotic ornamental species planted in the arboretum could survive in no other province. Today, they are beautiful mature trees approaching their 100th anniversary. In addition, Sharpe had conducted experiments with varieties of grain, corn, root crops, and potatoes, all of which impressed Saunders and caused him to remark that Agassiz would soon become "one of the most attractive places on the continent."

One of the first purchases at experimental farms was that of a team or two of heavy draft horses and one or two lighter animals, as indicated in Table 2.1. Each farm kept a different breed of draft horse, and many superintendents took great pride in having their best mares and stallions shown at local, national, and even international exhibitions. The policy was to have the farm stallions available for breeding purposes, with the objective of improving the quality of draft horses throughout Canada.

Table 2.1 Horses on experimental farms 1886–1890

Farm	Number of horses	Comments
Central	12	$42.50 was also spent for hired horses.
Nappan	6	These were 4- and 5-year-olds bought in P.E.I., plus one pair of oxen in 1889.
Brandon	11	These were black and bay mares and horses.
Indian Head	11	Some of these were bought in Toronto.
Agassiz	6	These were four heavy and two light horses bought in Toronto.

In 1891, the 5th year, a large dairy barn was built on the Central Experimental Farm, but it was completely destroyed by fire in 1913. Fortunately, the livestock were saved, but all equipment, machinery, and a large amount of feed was lost, for the fire occurred in October when the mows were full. The following year, a new barn was raised on the original site and still stands. It currently houses the Showcase Dairy Herd and the agricultural section of the Museum of Science and Technology. The seed-testing building, constructed in 1889, was used effectively. By 1891, farmers sent 2957 samples to be tested, double the number submitted the previous year. The average vitality of the samples received was 85.6 percent. Those that germinated poorly were discarded; thus the tests saved farmers the cost of sowing defective seed and reaping poor harvests.

By 1893, experiments on soil fertility were well established and Saunders had determined the amounts of nutrients removed from the soil by various cereal crops. For instance, a 25-bushel (680-kg) crop of wheat removed 40 pounds (18 kg) of nitrogen, 18 pounds (8 kg) of phosphoric acid, and 19 pounds (8.6 kg) of potash. Similar data were developed for barley, oats, and many of the root crops such as turnips, mangels, carrots, and sugarbeets. From experiments conducted on the experimental farms, Saunders concluded that grain gave a better return to farmers if it was fed to steers, dairy cattle, or swine than if it was sold as grain at the prevailing prices. He noted that the elements taken from the land by grain were largely returned through the application of manure to the fields.

In 1895, Saunders reported that Canada had too few experimental farms which were "too widely separated to fully represent all the different climates and other conditions affecting agriculture throughout the Dominion." As a result, he enlisted the cooperation of farmers to test varieties of grain. During the year, he received applications from more that 31 000 farmers for samples of seed but regretted that only 26 000 3-pound (1-kg) samples were available for distribution, and for new varieties of cereals, only 1-pound (0.5-kg) samples were supplied. Several of the new varieties proved to be better than those then under general cultivation.

In the same year, Saunders acknowledged the donation of a large number of trees and shrubs from Russia, England, Jamaica, Japan, California, and Minnesota. In addition, the Geological Survey of Canada collected and forwarded to the Central Experimental Farm seed from remote areas of Canada. Prof. John Macoun, who had resurveyed the Palliser Triangle in the 1870s and had demonstrated the importance of summer rainfall to the growth of plants on the Great Plains, was the botanist for the survey.

Branch experimental farms started to expand their staff in 1896. Mr. William M. Blair, superintendent of the Nappan farm, resigned and Mr. G.W. Forrest replaced him but served for only 1 year. Mr. R. Robertson followed in 1898 and remained for 15 years. In addition, the first assistant superintendent in horticulture, Mr. W. Saxby Blair, son of William, was appointed to Nappan in 1896. He wrote his own report directly to Saunders in Ottawa as "the work done in the Horticultural Division of the Experimental Farm for the Maritime Provinces." It became accepted practice for divisions (see Chapter 3) at the Central Experimental Farm to authorize and oversee the research work done at the branch farms and stations. This development throughout experimental farms and the other organizations that followed was to be felt until 1959, when planning and reporting became the responsibility of stations, following only general guidelines developed by national concensus.

The year of 1896 was the 10th anniversary of the establishment of experimental farms and Saunders reviewed their accomplishments. Although there had been rivalry among districts across Canada to obtain experimental farms, Saunders said there had been no "adverse criticism worthy of attention" and he therefore concluded that the sites chosen were supported by the Canadian farming population.

Among the practical results obtained from experimental farms during the first decade, Saunders emphasized that the recommendations for maintaining soil fertility and renewing cropping capabilities of land were of prime importance. In addition, farmers should value new information on subjects such as handling manure, deciding on fertilizers, seedbed preparation, and seeding. Of equal importance was the testing of various species and varieties of cereals and forage crops throughout the country. Plants from all parts of the world with similar climatic conditions to Canada were tested, and several new species as well as many new varieties were introduced during the first decade. To improve the chances of success, crossbreeding between varieties exhibiting superior characteristics was undertaken with the objective of producing offspring better than either parent. In order to comply with farmers' requests, more than 35 000 samples of seeds were distributed in 1896 by the five experimental farms to all parts of Canada.

The experiments done with fodder crops and ensilage showed how dairy and beef cattle could be properly fed during the winter months. The results of these experiments greatly stimulated both the dairy and beef industries, making winter activities of farmers more profitable than before. In addition, Saunders recorded that the dairy industry was now producing a better and more profitable grade of butter. The value of exported cheese increased from $7.3 million in 1886 to $17.6 million in 1898 as the result of research done by the experimental farms. Similar increases in the export of cattle and pork products were recorded.

Particular attention was paid to the 50 000 people who had settled farmland in the North-West. Extremes of climate required many experiments on the culture of hardy plants. Low precipitation brought about new methods of cultivating soil to maximize moisture retention. Saunders said that the experimental farms were "due the credit" for introducing awnless bromegrass, *Bromus inermis*, to the Canadian northwest and demonstrating its resistance to drought, its tolerance of low temperatures, and its usefulness as both a pasture and a hay crop. It is still the only species of bromegrass used today, and many varieties have been developed and licensed since 1896.

The chief of the horticulture division gave thought to the health and happiness of Canadians by encouraging farmers to cultivate fruits, vegetables, and ornamental plants. He ensured that each experimental farm grew a good selection of all imported material. The planting of trees and shrubs for shelter and ornament transformed a sometimes bleak landscape to one of comfort and beauty.

Each year brought an increased response from the public for information. Controls for insects and plant diseases were frequently requested. Noxious weeds received attention. The analysis of feeds and seeds, the relative nutritional value of different manures and various chemical fertilizers, and the determination of the quality of well water were among the services rendered by the chief of the chemistry division.

Forty varieties of spring wheat were compared at Brandon and Indian Head. Indian Head experienced some leaf rust, and at Brandon, rust was particularly severe, resulting in premature ripening, weak straw, shrunken heads, reduced yield, and a light-weight sample. The days to maturity ranged among varieties from 126 to 104 days at Brandon, and from 118 to 108 days at Indian Head. Hard Red Calcutta wheat from India was one of the parents used in the wheat-breeding program and it was among the earliest maturing varieties available.

Sharpe, superintendent at Agassiz, reported a marked increase in visitors during the 10th year. On one day early in August the British Columbia Fruit Growers' Association and a Vancouver newspaper group visited the Farm. In total, there were more than 1000 visitors that year, the Provincial Deputy Minister of Agriculture, Mr. J.R. Anderson, being among them.

Saunders calculated the average yield of oats, barley, and wheat on the Central Experimental Farm during the years 1896 through 1898 and compared these yields with the 3 years from 1889 to 1891. The increased yields were outstanding, being 72 percent for oats, 38 percent for barley, and 33 percent for wheat. He attributed the increases to a moderate use of fertilizer, the plowing under of green manure crops, an improved seedbed preparation, the early sowing of seed, and the selection of better varieties. He also calculated that the increase of 1 bushel/acre in the yield of each of these three crops would add nearly a million dollars to the earnings of farmers in Ontario alone. Similar results of crop improvement and animal production brought about by better methods of management and the use of superior varieties of plants were reported from the four other experimental farms. Of particular note was the work of Angus Mackay at Indian Head. He demonstrated the value of summerfallow by early summer plowing followed by several harrowings to destroy weeds and thus conserve moisture for the next year's crop.

As early as 1860, the best lands in the St. Lawrence Valley, the Ottawa Valley, and the Great Lakes areas were taken up by settlers and from 1880 to 1900 farming developed into a highly profitable enterprise, with many products being exported. Experimental farms' new and improved varieties and methods were of immediate use in developing a solid agri-food industry in Eastern Canada. By 1900, the farming population in the Maritimes, Quebec, and Ontario was starting to decline in real terms. It had dropped from 3.5 million in 1871 to just under 0.5 million in 1901. In Manitoba, however, the rural population increased from 12 000 to 32 000 during the same period. The establishment of experimental farms preceded the majority of settlers to the North-West Territories by only a few years.

East or west, Saunders had the well-being of the Canadian farmer uppermost in his mind. He was determined to improve the ability of the primary producer to make a better return on his investment in land and labor and to improve the conditions under which farming families lived. He and his staff established a remarkable agricultural base during the final 15 years of the nineteenth century.

Chapter 3
Development of Divisions
1886–1913

In 1886, Dr. William Saunders organized divisions based upon five scientific disciplines, namely agriculture (field crops and animal husbandry), cereals, chemistry, entomology and botany, and horticulture. The divisions were staffed by only the divisional chief until 1890, when an assistant was appointed to the Chemistry Division. There were no divisional representatives at branch farms until Mr. Wm. Saxby Blair was appointed as horticulturist at Nappan in 1896. The system of divisional control from Ottawa headquarters continued until 1959, when the Research Branch was organized.

Chiefs of divisions and later their staffs, as they were appointed, had a marked influence upon the conduct of research at branch experimental farms. The superintendent at each farm or station was responsible to the director in Ottawa for the general activities of his farm and for the expenditure of and accounting for annual funds voted each year by Parliament. In the early years, chiefs of divisions planned research projects for branch farms to execute. Divisional officers received experimental data for analysis, drew appropriate conclusions, and often prepared bulletins, reports, and scientific papers based on their conclusions. The reason for their doing so was that the divisions were better equipped than the branch farms and therefore could recruit more formally educated and skilled personnel. It was not until 1903, however, that any divisional chief other than Mr. Frank T. Shutt, in chemistry, held an earned degree. All had been keen self-educated amateurs turned professional.

In November 1886, the year Saunders was appointed director, he selected Mr. W.W. Hilborn as horticulturist. The following year, Mr. James Fletcher was confirmed as entomologist and botanist and Mr. Frank T. Shutt, a university graduate, as chemist. Saunders, in addition to being Director of the Experimental Farms System, also assumed responsibility for heading the divisions of agriculture and of cereals. The first *Annual Report of the Director*, as called for in Section 10 of The Experimental Farm Station Act, was dated 31 December 1887. It included reports from each of his three colleagues, Messrs. Hilborn, Fletcher, and Shutt.

ENTOMOLOGY AND BOTANY DIVISION

The Entomology and Botany Division had its beginning before The Experimental Farm Station Act was passed in 1886. The Minister of Agriculture, the Honourable J.H. Pope, and his deputy, Dr. J.C. Taché, realized that farmers and fruit growers were suffering damage to their crops from insect attack. The Reverend C.J.S. Bethune, an avid amateur entomologist and student at Trinity

College, Toronto, met with William Saunders and eight other people interested in entomology, on 16 April 1862 and formed the Entomological Society of Canada, which formally came into being on 26 September 1863 (12). The government of Ontario recognized that members of the Society could be useful to the farming community of that province and therefore gave them an annual grant to assist them in their studies. When James Fletcher, an accountant in the library of the House of Commons, was appointed honorary entomologist to the Department of Agriculture on 1 June 1883, it was the second formal recognition by a Canadian government of the value of well-informed entomologists. Following confirmation of his appointment[1] Fletcher prepared his first report in December 1884. He subsequently prepared a report each year, with the exception of 1886, until his death in 1908.

As Chief, Entomology and Botany Division, Fletcher continued his detailed reports on injurious insects, described ways in which they might be controlled, and added information on progress made toward developing a national arboretum. Tent caterpillars, codling moths, and plum curculios were among those insects that did the greatest damage to orchards and even though satisfactory remedies were available commercially they were not generally used by orchardists. Insects attacking cereals and field crops were of particular concern to Fletcher, because he realized the financial losses farmers might suffer from such an infestation. He took special care when explaining the life history of each insect commented upon in his reports, which he circulated widely among the farm press and interested farmers. Gibson (12) has referred to Fletcher's reports as containing a vast amount of information on economic entomology. Insects affecting forest trees, particularly pine, also received his attention.

In 1893, as the result of several petitions to the Minister, Fletcher began experiments with honey bees. Ten swarms of common black bees were purchased and set up in an apiary near a house on the Central Experimental Farm where Mr. John Fixter, the farm foreman, lived. Fletcher's experiments included different methods of wintering bees and different seeding dates for buckwheat as pasture for the bees. He was able to supply a continuous flow of nectar from 18 July until the first frost of 14 September, using only four buckwheat seeding dates. Italian queen bees were introduced during the first season and these gradually replaced the black bees. Fixter managed the apiary until 1906, when he resigned to accept a position with Macdonald College at Sainte-Anne-de-Bellevue, Quebec.

Mr. J.A. Guignard, B.A., was appointed in 1895 to assist Fletcher. Guignard took responsibility for the herbarium, to which Fletcher had presented his

---

[1]There is some confusion concerning the date upon which Fletcher was actually appointed Chief, Entomology and Botany Division. Estey (10) cites an Order-in-Council of 18 July 1886, recommending that Fletcher be appointed to the position. However, Saunders was not appointed director until October 1886 and the auditor-general's reports show the first payment of $1500 to Fletcher by Experimental Farms was in 1887. Indeed, he also received $16 in the same year from the library of the House of Commons. The previous year the library paid him $1450. Saunders records Fletcher's appointment as 1 July 1887, which is presumably the date upon which Fletcher actually moved to the Central Experimental Farm.

personal collection of over 3000 mounted specimens. In addition, Guignard dealt with some species of insects. In 1899, Fletcher obtained the assistance of Mr. Arthur Gibson. Gibson was born in Toronto in 1875, where he received his early education. Like Fletcher, Gibson was an enthusiastic amateur entomologist. He joined the department to assist Fletcher with the entomological side of the division's responsibilities. Gibson remained with the department until his retirement from the position of Dominion Entomologist in 1942.

Fletcher was instrumental in starting the Ottawa Field-Naturalists' Club in 1879, and wrote many papers for its journal, *The Ottawa Naturalist*. He took an intense interest in all phases of agriculture, and was the first member on the Board of Directors of the Central Canada Exhibition from the Experimental Farm. He organized the arboretum, which gave him great pleasure, setting aside 65 acres (26 ha) on which to plant trees, shrubs, and other plants of the Dominion. By 1889, Fletcher had collected 210 species from 12 families of woody plants. He relinquished his responsibilities for the arboretum to the Horticulture Division in 1895. The numerous people and institutions who donated to the arboretum included Mr. C. Gibb, Abbotsford, Quebec, who gave a collection of hardy plants from Russia; the Arnold Arboretum, Boston, Massachusetts; the Imperial College of Agriculture, Tokio [sic], Japan; and the Royal Botanic Gardens, Kew, England. Fletcher also became interested in grasses, both for forage and for turf. He seeded named varieties as well as native grasses from across Canada in search of material that would grow vigorously under severe weather conditions.

On 8 November 1908, at age 56, Dr. James Fletcher died, following a brief illness. He had come to Canada from England when he was 22 years old and served Canada well for 34 years. Fletcher was highly respected by his colleagues and many friends throughout his adopted country and the world. Queen's University conferred the degree of LL.D. on him in 1906. He was elected Fellow of the Royal Society of Canada and served as its secretary. He was a vice-president of the Entomological Society of Ontario and a founding member of the Ottawa Field-Naturalists' Club. His accomplishments are recorded by Bethune in the *Canadian Entomologist* (4) and by Estey in the *Canadian Journal of Plant Pathology* (10). In his honor a monument in the form of a drinking fountain has been erected on the Central Experimental Farm by the Ottawa Field-Naturalists' Club.

It was no easy task to replace Fletcher. The work of the division had grown such that in 1909, Saunders divided it into two: entomology and botany. Dr. C. Gordon Hewitt, a zoologist, was recruited from the University of Manchester, England, to head the Entomology Division. By 1914, this dynamic man had developed it from a unit comprising only himself, an assistant, and a stenographer, into the Entomological Branch, independent of the Experimental Farms. His professional interests were broad and included wildlife conservation as well as control of insects affecting crops.

Dr. H.T. Güssow, from the British Museum, was hired as Chief, Botany Division. He received his botanical training in Breslau, Leipsig, and Berlin,

Germany, before accepting a position with the British Museum in 1903. He too, was a vigorous and farsighted individual. With Hewitt, he drafted The Destructive Insect and Pest Act, which replaced The San Jose Scale Act of 1899. The first Act prohibited the importation into Canada of host plants from countries where San Jose scale was known to exist. The new Act of 1910 had much wider powers and, with slight modification, is the one under which Canada protects itself today from unwanted insects and pests. Drs. Hewitt and Güssow were to have a marked and lasting effect upon the development of Experimental Farms.

HORTICULTURE DIVISION

The Chief, Horticulture Division, Mr. W.W. Hilborn, also collected plant material, but unlike that in the arboretum, his plants were selected because of their potential for commercial fruit production. By 1888, Hilborn had collected apples and crab apples from Russia, pears from northern Europe, plums from Russia and northern Europe, and cherries, peaches, and apricots from China. The collection also included grapes, currants, gooseberries, raspberries, blackberries, and strawberries, totaling well over a thousand named varieties and several hundred unnamed seedlings. It is interesting to note that most of the raspberry seedlings were obtained from Prof. William Saunders, who had grown them from seed when he lived in London. Among the material were many hybrids between black and red varieties, which produced fruits of large size and fine quality during their first growing season. Hilborn died in office during 1889.

Mr. John Craig was appointed to succeed Hilborn later in the same year. The division extended its testing of fruits and vegetables, made extensive distributions of forest and fruit trees, determined and tested formulae for controlling apple scab, and continued to plant out several hundred seedlings of each of the small fruits. By the 10th year of operation, in 1896, Craig was able to report the results of variety trials on most kinds of fruits, giving firm recommendations as to where each could and could not be grown. He investigated the keeping qualities of apples and pears, the various methods of pruning grapes for optimum yields, the suitability of Russian mulberries for producing fruit in Canada, and the use of various legumes and legumes mixed with cereals or grasses for cover crops in orchards. For this he recommended alfalfa, mammoth red clover, and alsike clover with orchardgrass, in descending order of suitability. He investigated the processing and preservation of fruit, and selected the best varieties of both fruits and vegetables for canning.

Craig left his position in 1898 to accept the professorship of horticulture at Cornell University, and Mr. W.T. Macoun became Chief, Horticulture Division. Macoun had been Saunders' assistant and foreman of forestry since 1889. He therefore had a full knowledge of Saunders' system of operation on the Central Farm. Macoun was born in Belleville, Ontario, in 1869. Among his many awards was an honorary doctor of science degree from Acadia University.

Macoun had a lasting influence upon the division. He was personally familiar with growing conditions throughout Canada, for he had traveled with his father, Prof. John Macoun, on botanical exploratory trips in Quebec and Western Canada. Mr. W.T. Macoun was instrumental in establishing arboreta of native and introduced species of trees and shrubs at each of the experimental farms in order to develop a broad list of material suitable for planting on Canadian farm homesteads. By 1899, blossoming dates of 68 varieties of apples had been recorded. This information had been gathered with the help of 48 observers distributed from Salt Spring Island on the west coast to Prince Edward Island in the east. On the basis of these data, Macoun classified varieties into early, medium, and late blooming groups so that apple growers might plant those within a group close to one another in order to optimize pollination. Additional lawns were seeded in the arboretum at the Central Experimental Farm, which then contained over 3000 species and varieties of trees and shrubs.

In 1900, Canada sent an attractive display of grain and fruit to the Exposition Universelle in Paris. Some of the samples came from commercial growers, but by far the largest portion was supplied by the various experimental farms. Brandon, Indian Head, and Agassiz sent samples of grain, both threshed and in the sheaf. Nappan and the Central Farm provided fruit for the horticulture display. Saunders went to Europe that summer and attended meetings of the British Association for the Advancement of Science in London and the Pomological Congress in Paris, where he presented papers.

The results of an interesting experiment that had been started in 1890 were reported in 1905. The experiment compared yields of apples from different trees of the same variety. Over a 5-year period the yield variation ranged as follows: among Wealthy trees from 39 gallons to 103 gallons of apples per tree, among McMahan White trees from 143 to 476 gallons, and among McIntosh Red trees from 168 to 373 gallons. No explanation for the large differences was given (current knowledge would indicate a virus infection in the low-yielding trees), but Macoun propagated more trees from both the high and low yielders to determine if the characteristics persisted.

CHEMISTRY DIVISION

Mr. Frank T. Shutt was appointed chemist in August 1887. Shutt came to Canada with his family from England in 1870. He became interested in chemistry while he was an assistant to the Public Analyst in Toronto. In 1885, he graduated as a chemist from the University of Toronto where, as a demonstrator, he instructed the sons of Dr. Saunders.

Clause 7, items (d) and (e) of The Experimental Farm Station Act charged officers of each farm station with the duty of analyzing natural and artificial fertilizers and conducting experiments to test their comparative value. They were also required to examine the composition and digestibility of food for domestic animals. Saunders, being a druggist, knew he had a well-educated chemist in Shutt, one who could perform as required. Shutt developed a system of soil and fertilizer analyses, of testing various fertilizers under actual field conditions, and of cooperating with officers of other divisions to further the ends of his chemical work. He was honored with several awards for his thorough and practical work, including a special prize from the American Society of Agronomy, and the Sir Joseph Flavelle Medal from the Royal Society of Canada; he was also made a Commander of the British Empire.

At the end of his first 5 months with the Experimental Farms, Shutt made a trip with Saunders through the eastern United States to examine chemistry laboratories and to determine the most modern apparatus for analytical work. He saw fume cabinets, balance rooms, filter pumps, Kjeldahl apparatus for nitrogen determinations, photography rooms, and special rooms for gas analyses and combustion work. Upon returning to Ottawa, he drew up plans for a chemistry laboratory and supervised its construction. While the laboratory was being built, he obtained a small room in the Russell House Block on Sparks Street and proceeded with his chemical work there. Analytical work done during the year included two samples of water from Manitoba, seven samples from the City of Ottawa, and one sample of marl (a natural mixture of lime, clay, and sand) from Ottawa. He also started a study of the composition of wheat to determine the influence that variety and climate had upon the quality of its flour. The two varieties used were Red Fyfe and the newly imported Ladoga from Russia.

By June 1889, construction of the chemistry laboratory was complete. Shutt said that "the early months of the year [were] occupied in the personal supervision of the manufacture of the interior fittings [for the laboratory]." Equipment, which arrived from Germany, was installed, and Shutt moved from his temporary laboratory to the more commodious accommodation on the Central Experimental Farm. He noted that the new laboratory was so up to date that chemists visiting from other parts of Canada and from the United States copied the plans for their own use.

During the remainder of the year Shutt analyzed many types of mud from Prince Edward Island. He found that swamp mud was high in nitrogen but in a form not immediately available to plants. However, if it was composted with manure, wood ash, or lime, the nitrogen became available and the mud was useful as a top dressing. Other types of mud such as river and oyster were examined, but most of them were found to be of little practical use as additives to upland soils.

The work accomplished during Shutt's first year of operation in the new laboratory was remarkable. During this period, he hired an assistant, Mr. Adolph Lehmann, B.S.A., from Guelph Agricultural College. Analyses included samples of foundation beehive comb, soil and mud, well water, fodder, potatoes, sugarbeets, milk, and apple leaves, and samples of wheat to determine the affect of treating plants with copper sulfate to destroy smut. In addition to the various chemical analyses, Shutt wrote a short explanation of what the data meant. He did this for each farmer who submitted a sample.

In the 1890s paris green (arsenic oxide) was used as a post-blossom spray to control codling moth on apples. Canada exported large volumes of apples to Great Britain and, in 1891, the British press circulated a report that Canadian apples contained a small quantity of arsenic and therefore were poisonous.

Concerned about the poor press, Shutt analyzed apples that had been sprayed with paris green. He found no trace of arsenic, even though he used the most delicate analytical method then available, which could detect one part in 50 000. This is the first known instance of the Experimental Farms becoming involved in a nontariff trade barrier. Many more were to follow.

The number of samples analyzed in the chemistry laboratory continued to increase. They came from Shutt's own experimental plots, from other divisions, and from farmers. Shutt continually pleaded for more laboratory help as the numbers grew, but it was not until 1899 that a second assistant chemist was appointed.

A serious fire occurred in the special chemistry laboratory at about 6:00 p.m. on 6 July 1896. It was the result of someone accidentally breaking a flask containing boiling sulfuric acid that was being used to determine nitrogen in an organic substance. Even at that hour there were still many workers on the farm, and they rushed to control the fire. However, one man was seriously burned, the special laboratory was gutted, and many records and samples were lost. A special appropriation was made and within a few weeks the laboratory was back in operation. The construction of a new chemistry building was started shortly afterward and occupied in 1899. This handsome one and a half storey stone and brick building still stands.

Shutt involved himself with activities which, by today's standards, were not related to chemistry. However, there was no one else to do many of the experiments that were to prove of great value to Canadian farmers. In 1897, for instance, he inoculated red clover seed or the soil in which the clover was to be planted with "nitragin," a culture of bacteria that grows on the roots of legumes and converts atmospheric nitrogen to organic nitrogen. He produced sufficient tuberculin for veterinarians to test over 3500 adult cows for the presence of tuberculin bacteria. He ran extensive experiments to learn how best to preserve manures against nutritional loss and he showed that manure which was properly handled and protected had about twice the value of unprotected manures. During the next few years, he did extensive work on the milling and baking qualities of hard red spring wheat, on feeding concentrates to livestock, on the formulation of insecticides and fungicides, and on the optimum use for agricultural wastes from starch factories, tobacco kilns, and flour mills.

POULTRY DIVISION

In 1888, Mr. A.G. Gilbert, a successful local poultryman, was appointed to manage the poultry work on the Central Experimental Farm. Although Gilbert was a journalist by profession, he had studied the raising of chickens. He put his journalistic skills to good use by publishing many bulletins on raising poultry and by corresponding extensively with poultrymen across Canada.

When the Poultry Division was established, eggs of different strains from 19 breeds were obtained from Canadian, English, and American breeders. Gilbert outlined a program to cross strains within breeds and to make crosses between breeds; thus was started a long series of poultry-breeding projects, the results of which have been of inestimable benefit to Canada and the world. The second report Gilbert prepared for Saunders, in 1889, was actually a short course for those who wished to raise chickens on a commercial basis. He dealt with topics such as the breeds most suitable for various climatic conditions, the types of poultry house that should be constructed, and the best feeds for laying hens.

Gilbert was particularly concerned about improving egg production during the winter months. At that time of the year production was low, prices were high, and profits could be maximized. As expected, good management within the poultry house and proper feeding proved to be the key. He prepared detailed plans of the poultry house at the Central Experimental Farm in order that farmers might copy or adapt them. He conducted experiments in holding fresh eggs at different temperatures to maintain quality. He paid close attention to rations that would produce the best flavor of egg. He determined that the Plymouth Rock breed best met the meat production requirement. He ran feeding trials with meat birds, took them to the killing floor, and packaged them to be assessed on the British market.

At the request of the House of Commons Agricultural and Colonization Committee, he ran an experiment in 1897 to determine how much profit a farmer could make raising only 50 hens. The flock produced 4773 eggs during the year and the return on these eggs was $78.69. In addition, some eggs were sold for hatching and some cockerels were sold for meat, giving a total return of $139.19. Feed cost $45.26 for the year, leaving a profit of $93.93 or nearly $2 per bird per year, which is equivalent to about $13.90 per bird in 1980 dollars.

Incubators for hatching chicks were generally available during the last few years of the nineteenth century. Gilbert tried several of the new oil-heated machines but without too much success. The object of using incubators was to hatch a large number of chicks at the same time, in order to have a flock of uniform age. Gilbert believed that suitable techniques would be developed, making the use of incubators acceptable, and that the high loss of chicks following incubator hatching could have been caused by poor stock rather than by faulty methods of incubation. In addition to chickens, Gilbert introduced turkeys, ducks, and geese to his program. He did no experimental work with them but carefully recorded how they were raised and tried to improve upon the methods.

Experiments became more complicated at the turn of the century, and the Dominion Chemist (Shutt), started to work with Gilbert on the preservation of eggs and the fattening of chickens. To preserve eggs, Shutt found that limewater was as effective as and much cheaper than waterglass (magnesium sulfate). Experiments to fatten birds were extensive. He used many different rations on eight breeds and a cross between Plymouth Rock and Brahma. He compared fattening in a pen versus fattening in a crate. He was able to add more than 0.5 lb (0.2 kg) per week to the weight of a bird over a 4- to 6-week period. At the same time, Gilbert started trap-nesting pens of 12 birds from each of several purebred and crossbred hens. The largest number of eggs produced by any one hen in

12 months was by a Plymouth Rock, which laid 172 eggs. The second largest number was 145 eggs, and the third largest number was 128 eggs. Most of the birds, whether Plymouth Rock or White Leghorn, laid fewer than 100 eggs[1] in the year.

CEREAL DIVISION

Dr. William Saunders himself kept the responsibility for experimental work with agriculture, which included cereals, forage, and livestock. Saunders first ran variety trials of wheat, barley, and oats. By 1890, he seeded two varieties of each kind of cereal on six successive weeks, starting with the 3rd week in April and finishing the last week in May. Statistics were not needed to show that early seeding produced a much larger crop than later ones for all varieties. He emphasized that repeated tests would be needed to produce reliable averages. He did, however, calculate that if farmers in Ontario delayed seeding by as little as 2 weeks, the loss of income would amount to $1.5 million for barley alone— and barley was selling for only fifty cents per bushel!

In 1888, Saunders started hybridizing wheat, using several different varieties as parents. His stated objectives were to produce varieties that were early ripening and of high quality. By 1896, 10 years after The Experimental Farm Station Act was passed, he had 15 new varieties under test. However, in addition to these detailed and time-consuming experiments, he ran variety trials with oats and barley, and tests for the prevention of smut in oats. By 1900, he reported experiments to produce high oat yields. He planted crops of grain, flax, beans, corn, and millet one year, and the next year planted oats on the same plots. As might be expected, the highest yield followed horsebeans but not soja [sic] beans. The second highest yield followed grain. The lowest yield followed millet. The oat crop following horsebeans was 50 percent larger than the one following millet.

Saunders continued to distribute samples of grain for farmers to grow. In 1899, he sent 2858 samples of seed, each sample sufficient to plant one tenth of an acre (0.04 ha), to those who were most interested in variety tests. In addition, over 29 000 smaller samples were distributed to encourage other farmers to select the best and most productive sorts of seed. During this period, Saunders gave much attention to testing the vitality of seeds. In 1901, he reported that 2384 farmers had sent seed samples of 19 kinds of crop to be tested for germination. Wheat, oats, and barley accounted for more than 90 percent of them.

By 1902, the extent of the variety tests had increased to include 268 varieties, 118 of them being spring wheat. These were grown in one fortieth-acre (0.16-ha) plots, with care being taken to use soil that was as uniform as possible in both fertility and texture. Nevertheless, there was no replication and the results were undoubtedly confounded with differences in soil.

27

---

[1]Today's methods of rearing and breeding, produce 300 eggs on average from each bird in a flock.

The responsibilities of the director had become such that by 1903, Saunders found it necessary to reliquish the duties of Chief, Cereal Division. Since 1887, he had visited all the experimental farms each summer at times suitable for observing crop responses to the different climatic zones. He took two of his five sons on these summer trips and they helped him with field notes and with making crosses between appropriate varieties of grain. Charles, the middle son, was in his early twenties when he started the trips with his father (33). He proved to be dextrous with his hands and meticulous in his observation and note-taking. Although he was a graduate chemist, Charles Saunders also had a passion for music and became a teacher of voice, adding to the cultural life of early Toronto.

William Saunders appointed Charles to the position of Experimentalist, Cereal Division, in 1903, and in 1905, Charles became chief of the division. Charles threw himself wholeheartedly into his work, first by carefully testing and reselecting from the new varieties introduced by his father, then by developing milling and baking tests suitable for small samples of grain. During the field selection period in late July and early August he told Prof. M.A. Carleton of Washington, D.C., in a letter dated 9 July 1906, that he could not give the professor as much time on a visit as he would like because he, Charles Saunders, was in the field from 5:00 a.m. to 8:00 p.m. By such dedication, Charles proved worthy of the confidence his father had placed in him—he produced Marquis hard red spring wheat, thereby making Canada a world leader in the bread–wheat export trade.

AGRICULTURE DIVISION

William Saunders' prodigious energy enabled him to also take the responsibility for the Agriculture Division. The work of this division dealt with research on all field crops except cereals and with all livestock except poultry.

During the first few years, a comparison of all available varieties of corn, sugarbeets, root crops, and potatoes was made. Plots covering 6 acres (2.4 ha) were planted with root crops. It is true that William Saunders had Fixter, the farm foreman, to help him from the beginning, and in 1889 he hired W.T. Macoun to help with seeding, managing, and note-taking on the plots. However, Saunders shouldered the main responsibility. It was a heavy load. In 1889, he bought the first cattle for the Central Experimental Farm totaling 54 head, consisting of three dairy breeds and two beef breeds. A number of bulls were bought of each breed in order that one could be supplied to each of the branch experimental farms. With the exception of two bulls and five heifers that were selected from a high-grade herd in Syracuse, New York, all the stock came from among the best herds in Canada. The intention was that surplus stock would be made available to Canadian farmers, and Saunders wanted them to have the best.

In 1890, Saunders relinquished his responsibilities for the Agriculture Division and appointed Mr. J.W. Robertson to the post. At the same time, Robertson was named Dominion Dairy Commissioner. He was born in Scotland in 1857 and farmed in Ontario for 11 years before being appointed, in 1886, as professor of dairying at the Ontario Agricultural College. Although his position as dairy commissioner took more of Robertson's time than his position as agriculturist, he served the Experimental Farms well. During his first year he had a dairy building and a piggery constructed. He bought three breeds of swine, and added to the cattle herd. He asked Saunders for a sheep building to complete the facilities on livestock work. The following year, Robertson conducted the first feeding trial for fattening steers. Three pairs of steers were each fed a different ration as follows: hay, roots, and meal; corn ensilage and meal; and hay, roots, corn ensilage, and meal. In summary those fed corn ensilage and meal gained more weight, and although they ate more feed per day, the cost of the feed was between four and seven cents per day less than the other two rations.

As Dominion Dairy Commissioner, Robertson established 19 experimental dairy stations to study the manufacture of cheese and butter. He spent more than half his time visiting these stations, lecturing at conventions and meetings of dairymen, and arranging cooperative programs with provincial departments of agriculture. In 1891, he attended 49 conventions, speaking from two to five times at each.

Because of his interest in livestock, Robertson spent most of his time on the Central Experimental Farm with stock and less with crops, at least in the first few years. He worked with mixtures of corn and fodder plants to find one which would make a well-balanced ration and cost less than cereals or concentrated by-products. He decided upon a mixture of corn, horsebeans, and sunflower mixed at the time of chopping and ensiling. The corn and beans were seeded together and the whole plants were harvested and chopped. Sunflowers were grown in a separate field and only the heads were used. He called this the Robertson Mixture for Ensilage, and in 1893, he sent samples of seed, together with a circular of directions, to more than 60 farmers for testing. He had already tested the butter made from milk produced by cows fed his ensilage mixture. The ensilage containing sunflower heads was judged to produce milk that made butter richer and more highly colored than that produced from corn ensilage. Farmers who tried the mixture reacted favorably. With this encouragement, Robertson continued the feeding trials, comparing his mixture with many others. In most instances, the Robertson Mixture produced beef more cheaply per unit of live weight than other mixtures. In one experiment he reported the cost to be half that of the comparison mixture.

Robertson resigned his position with the Experimental Farms in 1896 and went on to other challenges in agriculture. In 1904, he founded the Canadian Seed Growers' Association, became the first principal of Macdonald College, and chaired the 1909 Royal Commission on Technical Education.

Saunders resumed the responsibility for the Agriculture Division for the next 2 years. Then, in 1898, he appointed J.H. Grisdale to the post. Grisdale was the first university graduate in agriculture to be employed on the Central Experimental Farm. He was born at Sainte-Marthe, Quebec, spent 2 years at the Ontario Agricultural College, then attended Iowa State College, from which he graduated in 1898. He received an honorary doctor of science degree from Laval

University in 1918. Grisdale entered into the pattern set by Messrs. Saunders and Robertson with both livestock and crops. In addition, he set 200 acres (80 ha) of the Central Experimental Farm aside to be managed as closely as possible to a commercial farm of that size. This farm was divided into five lots of 40 acres (16 ha) each. On these, he established 5-year rotations of peas seeded down to clover, row crops of corn and roots, and cereals seeded to grass, hay, and pasture. He also tested new varieties of cereals and other crops such as sorghum and rape in 1899. Neither of the two varieties of sorghum succeeded, but the rape was judged to be a good succulent forage for either sheep or swine.

Over the next few years, Grisdale made comparative feeding trials with horses, dehorned steers, swine, and sheep. He experimented to determine optimum times to milk dairy cattle. In all trials, he paid close attention to costs of inputs and returns from sales of stock. The results were judged mainly upon the net returns from each treatment. He established a number of rotations to examine different depths of plowing, and some rotations with and without legumes.

In 1912, the Agriculture Division was divided into the Field Husbandry Division, which remained with Grisdale, the Forage Division, headed by Dr. M.O. Malte, and the Animal Husbandry Division, headed by Dr. E.S. Archibald. More will be said about these three men in later chapters.

TOBACCO DIVISION

The first mention of this division was in 1908. Mr. F. Charlan, reporting as Chief, Tobacco Division, did so directly to the Minister until 1912, when the division joined Experimental Farms. Work at the Quebec locations of Saint-Césaire and Saint-Jacques began in 1908 under Mr. O. Chevalier. When Experimental Farms was given the responsibility for the tobacco work in 1912, Saint-Césaire was closed because it was too difficult to get to, and Farnham was opened instead under Chevalier. Harrow, Ontario, was established in 1908 as a tobacco station, with Mr. W.A. Barnet in charge.

In 1910, an important change occurred affecting the whole system. Director Saunders delegated the responsibility for supervising and inspecting the research work on branch experimental farms and stations to the chiefs of divisions. For more than 20 years he had done both tasks himself. This formalized the national responsibility of each chief, and their titles were altered to reflect the change—they became "Dominion" Cerealist, Horticulturist, or Chemist. These titles held for the next 40 years until, in 1950, "Chief" was again used.

By 1912, all divisions but two had been established. Their histories, including amalgamations and separations into other divisions and services, will be chronicled in subsequent chapters as the years unfold. The pattern established by Saunders of having all superintendents reporting directly to the director on the operation of experimental farms has continued. As other organizations were spawned, different ways of managing were developed. However, the decision

Saunders made in 1910 to delegate responsibility for supervising and inspecting the research work on branch experimental farms and stations to division chiefs prevailed for the next half century. The system was efficient because it avoided duplication of effort, productive because it could marshal substantial resources, and catholic in its use of scientific disciplines.

Chapter 4
Development of Experimental Stations
1906–1913

The western world was becoming industrialized and Europe was placing heavy demands on North America for raw materials to feed her factories and people. Most of the land south of the 49th parallel had been occupied, leaving the nearly empty plains of the Canadian prairies as the land of promise. There were many Canadians in industrial Ontario, many Americans, and many western Europeans who wanted fresh air and fresh opportunity. There were also thousands of peasants in eastern Europe who could now dream the hitherto impossible dream of owning their own land. Some settlers had moved into Manitoba and a few into the North-West Territories, but it had not been the rush anticipated by either the government of Canada or the Canadian Pacific Railway.

There had been four Prime Ministers since Sir John A. Macdonald died in 1891 and Sir Wilfrid Laurier's election in 1896. The flood gates opened during 1896, Laurier's first year in office. Sir Clifford Sifton, the energetic Minister of the Interior from Winnipeg, published pamphlets in 20 languages. He advertised in 7000 American newspapers, and sent agents to Europe, all with the same message: come and work the free homesteads in Western Canada. Transportation was thirty dollars from Liverpool, England, to St. John, New Brunswick, and a colonist rail car from St. John to Winnipeg, Manitoba, jammed with people, cost six dollars per head (25).

Saskatchewan and Alberta became provinces in 1905, each with its own government, and cities within each province began developing at strategic locations. Manitoba, Saskatchewan, and Alberta grew from 390 000 people in 1901 to 1 322 000 in 1911. Some of the new settlers were experienced farmers, but many had no idea of how or when to prepare the land, sow the seed, or harvest the crops. Saunders, who by now had visited the Great Plains many times, pleaded for more experimental farms.

The decision as to where to establish the next experimental farm, to be designated a station, was not taken casually. Johnston (24) in his history of the Research Station, Lethbridge, Alberta, recounts that in 1901 the Alberta Railway and Irrigation Company sent for the superintendent of the Experiment Station, University of Wyoming, Laramie, to operate a Model Farm at Lethbridge. Mr. W.H. Fairfield arrived, and introduced nitrogen-fixing bacteria to the prairie soils in order that alfalfa and other legume crops could be grown successfully. By October 1903, the editor of the Lethbridge newspaper was writing supportive articles regarding the benefits of the Model Farm to the agriculture of the community. Other communities in Alberta also recognized the advantages, and petitions to establish government experimental stations were received from

Macleod, Claresholm, Medicine Hat, Calgary, Red Deer, Lacombe, and Edmonton.

On 6 April 1906, Senator L.G. DeVeber announced to the Lethbridge Board of Trade that Lethbridge had been chosen as the site for the Experimental Station in southern Alberta. In August 1906, Elliot T. Galt, general manager of the Alberta Railway and Irrigation Company, contributed again by donating 400 acres (160 ha) of land, half of which was irrigated, for an experimental station. As noted by Saunders, the industry of the people of southern Alberta, the construction of a sugarbeet factory in 1904, and the establishment of a Model Farm 5 years earlier were the deciding factors in the choice of Lethbridge. Fairfield was appointed superintendent of the new experimental station on 1 August 1906, and its development into a large multidiscipline organization was under way. Fairfield was the first person with a degree in agriculture to be appointed as a superintendent. Later in 1906, Mr. N. Wolverton at Brandon, Manitoba, became the second person.

In March 1907, the site for the second experimental station in Alberta was chosen some 350 km to the north of Lethbridge at Lacombe, halfway between Calgary and Edmonton. With considerable pressure from the Lacombe Board of Trade, the Honourable Sidney Arthur Fisher, Minister of Agriculture, eventually agreed to the purchase of developed land, at a costly $50/acre ($125/ha) from a private owner.

Mr. G.G.H. Hutton, a progressive local farmer, became its first superintendent and took possession of the property in March 1907, although, according to Fredeen (11) he had been chosen for the position on 22 July 1906, 10 days prior to the appointment of Fairfield in Lethbridge. Hutton was born in Ontario, graduated from the School of Agriculture, Guelph, in 1900 and farmed in the Lacombe area until his appointment to the experimental station. In 1919, he resigned his position to become superintendent of agricultural and animal industry for the Canadian Pacific Railway.

Gold had been discovered in the Klondike area of the Yukon in 1896, with peak production in 1903. Saunders wanted to find out what the agricultural potential was there. On 18 April 1905, therefore, he shipped grain, grass seed, and potatoes from Ottawa to Dawson City, but because the rivers were frozen the mail did not reach its destination until early June. However, some of the samples were left at White Horse [sic] about the middle of May, and Superintendent A.E. Snyder of the Royal North-West Mounted Police distributed them to seven people, probably at different locations. Some recipients reported excellent results, whereas others had frost in June or August. It was agreed that the seed should have been made available in early May. However, the exercise was encouraging enough that further samples were supplied in 1906, with more satisfactory results this time.

The third Alberta experimental station had its beginnings in 1907 at Fort Vermilion, about 350 miles (560 km) due north of Edmonton, on the lower reaches of the Peace River. Farming was already well established in that area, for the Hudson's Bay Company had installed a roller flour mill capable of processing

25 000 bushels (68 tonnes) of wheat annually. Mr. F.S. Lawrence of Fort Vermilion, visited Ottawa in April 1907, and took a supply of seeds and trees from the Central Experimental Farm to Edmonton by train. Leaving Edmonton on 1 May 1907, he drove a team of horses and a wagon 400 miles (650 km) to Peace River Crossing, which he reached on 17 May. There he spent a day organizing a raft for the last 300 miles (480 km) to Fort Vermilion down the Peace River. Rafting the river took him 3 days and 5 hours, without stopping. Seeding of the grain samples was complete 6 days later on 27 May. The seeds grew well, but a frost on 30 August damaged the kernels of all varieties of wheat, oats, and barley. Lawrence resigned during the winter and Mr. Robert Jones, a local farmer, was hired to continue the experiments.

In 1909, two other experimental stations were opened, one at Charlotte-town, Prince Edward Island, and the other at Rosthern, Saskatchewan. Mr. J. Artemus Clark, the first superintendent of Charlottetown, took possession of part of the property in August 1909. Clark was born in Prince Edward Island, graduated from the Ontario Agricultural College, Guelph, then worked for the Dominion Seed Commissioner in Ottawa for 2 years. Six parcels of land were bought by the provincial government and leased to the Canada Department of Agriculture. Interesting among them, from an historical point of view, was the Pope property, on which sits a fine wooden mansion known as Ravenwood House. It was built and occupied in 1824 by the Honourable William Johnston, Attorney General for Prince Edward Island. When Johnston died in 1828, his son-in-law, William Forgan, a lawyer and member of the government moved into Ravenwood House. The Honourable J.C. Pope, first Premier of Prince Edward Island after Confederation, resided in the house from 1873. Most superinten-dents of the experimental station at Charlottetown and directors of that research station have lived in this historic residence during their tenure of office. Today it is the home of the current director of the Charlottetown Research Station. Raven-wood House is one of the attractions of Charlottetown and has been visited by the royal family on three occasions when they were guests of the province.

The Rosthern property was bought in 1908 and Mr. W.A. Munro was appointed its superintendent in March of the following year. Munro was born in Ontario, graduated from Queen's University, then from the Ontario Agricultural College, Guelph. He was an agriculture extension officer in both Alberta and Ontario prior to moving to Rosthern. Munro, and later Mr. F.V. Hutton, made important contributions to the agricultural development of Saskatchewan north of Saskatoon. They did considerable work with cereals, forages, horticultural crops, dairy cattle, beef cattle, sheep, and swine. A station was opened in 1911 at Scott, west of Saskatoon, within the Palliser Triangle. Shortly after, the Dominion Forage Crop Laboratory was set up in Saskatoon and a new station started at Melfort. Rosthern was then considered superfluous and was closed in 1940.

Starting in 1909, Saunders arranged to have small experimental plots or substations in parts of Canada that were not easily served by established experimental farms or stations. He contacted leading farmers, ranchers, and even some missionaries, asking them to seed test plots with varieties of vegetables, cereals, and grasses in order to learn the extent of Canada's farming possibilities. The first such arrangement was made with Mr. E.W. Calhoun, superintendent of the Harper Ranch, Kamloops, British Columbia. Calhoun reserved 10 acres (4 ha) on which he grew several varieties of cereals, potatoes, and apple trees to find the best cultural methods to use under dry farming conditions. A similar arrangement was made with Mr. Frank Moberley on his farm near Lake Abitibi in Quebec. As time went on, the experiments were extended to Lesser Slave Lake, Alberta; Forts Resolution, Smith, and Providence, MacKenzie District; and Minto Bridge, Yukon Territory.

The divisions of Entomology and of Botany and Plant Pathology as well as a few other divisions, opened laboratories at locations other than Ottawa. Sometimes they were at or near an experimental farm or station and sometimes they were in other government buildings such as a post office. Each had workers from only one division and rarely did they have much land for experimental work. Frequently, a greenhouse or an insectary was attached. The first such laboratory was entomological and opened by G.E. Sanders (not to be confused with Saunders) at Bridgetown, Nova Scotia, in 1910. Sanders was born in Roundhill, Annapolis County, and obtained his entomological education at the Ontario Agricultural College, Guelph, and at the University of Illinois. He worked first as a nursery inspector and then as an assistant to the Illinois State Entomologist. The laboratory was moved to Annapolis Royal in 1915 and then to Kentville in 1952.

The French-Canadian breed of horse is found primarily in the province of Quebec and was very popular in the 1700s when there were 5275 such horses recorded (3). By 1910, the Canada Department of Agriculture became involved with the breed, because only 969 of the 2528 French-Canadian horses inspected could meet the criteria to enter its registry. The Experimental Farms Branch was concerned that the breed might disappear, so a station was opened at Cap Rouge on 1 January 1911, in an effort to preserve both the French-Canadian horse and the French-Canadian cow. Cap Rouge is about 14 km west of Quebec City, on the north side of the St. Lawrence River. The purchased property was known as the Stadacona Farm and consisted of a solid block of 380 arpents (130 ha), of which 185 arpents (63 ha) were cultivated, the rest being in paddocks, buildings, steep hills, brush, and forest. It had three houses as well as a boarding house, eight farm buildings, and a complete line of equipment. There were three heavy teams and a driver but no French-Canadian horses or French-Canadian cattle.

The French-Canadian horse, now called the Canadian, was brought to Canada in June 1647 from northwestern France (2). It is medium in size and weighs between 900 and 1100 pounds (400 to 500 kg). It has a long, deep body with a well-rounded and heavily muscled rump; thus it is able to travel great distances without tiring. In 1886, it became a registered breed. The French-Canadian cow is a direct descendant of cows brought from Normandy and Brittany by Jacques Cartier in 1541. The breed is only a moderate producer of milk of about 4.4 percent butterfat.

Registered animals of both breeds were bought by the Cap Rouge Station. In 1920, property at Saint-Joachim, 40 km east of Quebec City, was leased for 20 years from the Quebec Seminary, specifically for breeding, experimental feeding, and management of Canadian horses. By that time, Cap Rouge and Saint-Joachim had 96 registered French-Canadian horses and 59 head of French-Canadian cattle. The Cap Rouge station later became heavily involved with research on horticultural crops because of the interest farmers had in supplying the Quebec City market.

Additional experimental stations on the Great Plains came none too soon. Precipitation in 1910 throughout much of Alberta and Saskatchewan was the lowest in the memory of the local people (30). That year, many crops were a complete failure and the International Dry-Farming Congress, held in Colorado Springs in 1911, attracted a great deal of Canadian attention. Previous congresses had been held in Billings, Montana (1909), and Spokane, Washington (1910). By 1911, the congress had 13 500 individual members in 50 countries. Canadians dominated the congress in Colorado Springs and therefore, even though Salt Lake City, Utah, and Prescott, Arizona, were leading contenders for the 1912 congress, both cities withdrew in favor of holding the next congress at Lethbridge. Fairfield was a member of the Congress Executive and in 1912 he was Chief, Jury of Awards. Each congress had a large display of farm machinery and held competitions for the choicest samples of grain and other farm produce. Seager Wheeler of Rosthern, Saskatchewan, won the wheat championship at the Colorado Springs Congress with a sample of Marquis, the new variety bred by Sir Charles Saunders and released by Experimental Farms. Marquis confirmed its superiority in 1912, when it took the championship at Lethbridge with a sample shown by Henry Holmes of Raymond, Alberta.

For some time, additional stations and laboratories were set up, one or more each year. The growth of Experimental Farms and Stations is reflected in the annual budgets and numbers of staff as summarized in Appendix IV. The sequence of establishing these stations is given in Appendix II and no attempt will be made here to follow the development of each unit throughout the remainder of this history, although note will be made of interesting situations as they occurred.

The last Experimental Farm Report prepared by Dr. William Saunders before he retired was that for the year 1910–1911. By this time Saunders was 75 years old, having been born at Crediton, Devon, on 16 June 1836. During that last year he showed no signs of slowing down, as he had paid his annual visit to each of the farms and stations. He left Ottawa on 1 May to visit Charlottetown, Nappan, Brandon, Indian Head, Rosthern, Scott, Lacombe, and Lethbridge. He was home again on the 12 June. Six weeks later, on 22 July, he was off again to visit the west when the crops were nearly ready to harvest, paying second visits to Brandon, Indian Head, Rosthern, and Lethbridge. In addition, he went into British Columbia to spend 2 days at Agassiz where "the crops were looking well." He returned to Ottawa a month later.

Regrettably, Dr. William Saunders, C.M.G., LL.D. (Queen's and Toronto), F.R.S.C., F.A.A.A.S., lived only 3 years following his 1911 retirement, which he took in London, Ontario. During his long life he had made an herculean effort for Canada and its agriculture. The William Saunders Building, erected on the site of his residence on the Central Experimental Farm was opened in June 1936, 100 years after his birth. Today, it houses the herbarium of the Biosystematics Research Institute, a fitting use for a building bearing the Saunders' name.

Joseph Hiram Grisdale succeeded Saunders shortly after the latter's retirement. He had been a member of Saunders' staff since 1899, when he was appointed head of the Agriculture Division, following Robertson's resignation. Grisdale retained his responsibility for the Agriculture Division during 1911, and the general format of reporting to the minister that had been established by Saunders was continued. In 1912, however, the report appeared in two sections. The first section, prepared by the director, was general and outlined the progress made by the branch over the year. The second, prepared by divisional chiefs, reported work done in each division, to which were added appropriate subject reports from each farm and station. For instance, Macoun reported on all the horticultural research done within the service. Grisdale was of the view that the new arrangement would be "more convenient and useful to the farming community than [was] the former system . . . ." The new reporting method more than doubled the size of the annual report which, for 1913, required two volumes.

Thomas A. Sharpe retired from the Experimental Farm, Agassiz, in 1911. He had bought a property at Salmon Arm, British Columbia, in 1905 and proceeded to clear it in preparation for farming upon retirement. He made arrangements with Grisdale to test various crops there. In 1912, he brought 30 acres (12 ha) under cultivation, on which he planted nearly 100 varieties of apples and 50 varieties of pears, cherries, plums, and berries. Sharpe continued reporting his results to the director through 1919. When Angus Mackay was 73, in 1913, he retired from the superintendency of the Experimental Farm, Indian Head. He was retained as Inspector, Western Experimental Farms (29) until his death on 10 June 1931, at the age of 91.

By the end of 1913, the branch had grown from 5 divisions and 5 experimental farms, to 10 divisions and 18 experimental farms or stations, plus 9 entomology or plant pathology laboratories. All divisions were centralized at Ottawa. There were six establishments in the Maritime Provinces, four in Quebec, four in Ontario, nine in the Prairie Provinces, and four in British Columbia. Saunders and Grisdale had exercised their usual care in selecting sites according to the needs of the agricultural industry. At the same time, working with successive ministers, they showed consideration when requests were received for attention to a particular locale.

Chapter 5
The War Years and into the Dry Period
1914–1936

By the time World War I began, in 1914, most of the divisions of Experimental Farms and Stations had been established and their chiefs were in place.

The first break in the Experimental Farms occurred in 1914 when C.G. Hewitt, the Dominion Entomologist, persuaded the department to place entomologists in a separate branch. The Honourable Martin Burrell, a fruit grower from British Columbia, was Minister of Agriculture then and Mr. G.F. O'Halloran was his deputy. O'Halloran was the son of a lawyer and had successfully practiced law himself in Montreal before being appointed deputy minister of agriculture and deputy commissioner of patents in June 1902. His expertise was clearly toward the legal rather than the agricultural component of the department. The report of the minister to the governor-general for 1914 includes the Entomology Division of Experimental Farms. The 1915 report makes no mention of entomology either as a division of Experimental Farms or as a branch on its own. The next report of the Minister for 1916 devotes four pages to the Entomological Branch but gives no explanation as to why the separation was made. The 1915 annual report of Director Grisdale of Experimental Farms included a section from the Bee Division, prepared by its chief, Mr. F.W.L. Sladen. Sladen addressed his report to Grisdale as the "first report of the Bee Division" and said it came into existence with the separation of the Division of Entomology from the Dominion Experimental Farms.

One possible explanation for the move is given in a short paper by Hewitt (20) to the *Canadian Entomologist*, which says that "the urgent need of legislation ... to prevent the introduction into Canada ... of serious insect pests and plant diseases" made it necessary to pass the Destructive Insect and Pest Act of 1910. The Entomology Division administered part of the Act, and because such administration was not the work of the Experimental Farms, the Entomology Division needed to be separated. By 1914, there were eight branch entomology laboratories from Bridgetown to Agassiz. Hewitt appointed three inspectors and six superintendents of fumigation. He became the first director of the branch on a footing equal to J.H. Grisdale, both reporting directly to the deputy minister. The Botany Division, under H.T. Güssow, continued to administer the plant disease part of the same Act while remaining within the Experimental Farms and Stations. He hired nearly 30 temporary employees to inspect potatoes, in order to help combat a powdery scab outbreak in 1914. Güssow administered his portion of the Destructive Insect and Pest Act within the Experimental Farms, whereas entomology, with fewer inspectors, was formed into a new branch.

The bee work, done under the supervision of Mr. F.W.L. Sladen, was organized into its own division. Sladen was an enthusiast. He established nine apiaries in 1913 and 1914, and three more between 1916 and 1919 as new experimental stations were added to the five originals (27). Although the prime objective was to evaluate honey production potential at each location, a few experiments on management and wintering of hives were conducted. Sladen himself analyzed much of the data drawn from these apiaries and learned a great deal about the relationship between honey flow and the flowering sequence of plants growing at various locations. He also initiated queen breeding experiments, using stock imported from Italy. Unfortunately, Sladen died of a heart attack in 1921, at the age of 45. Charles B. Gooderham, Sladen's assistant, became the new Dominion Apiarist.

Gooderham was a graduate of Macdonald College, Sainte-Anne-de-Bellevue, and had joined the Bee Division in 1917. He retained his position until retirement in 1949. Gooderham actively promoted the expansion of Canadian honey production from a mere 5 million pounds (2.3 million kg) in 1921, to nine times that quantity in 1964. He increased the number of experimental stations at which colonies were set up from 17 to 21. He used his public speaking ability to address interest groups across the country and arranged for the distribution of technical information, which resulted in more than 45 000 beekeepers in Canada by 1964. Gooderham added to Sladen's accomplishments by promoting research on the attractiveness of different plants to bees, the effect of insecticides on bees, the effectiveness of bees in pollinating commercial crops such as tree fruits, clovers, and vegetables for seed, as well as the control of diseases of bees, and methods of handling and processing honey.

In January 1915, the first conference of superintendents was held at the Central Experimental Farm. They discussed the current and proposed experimental work, revised the aims of the Experimental Farms and Stations now that the entomologists were in a separate branch, and endeavored to foster a spirit of cooperation among the various farms, stations, and laboratories. It was the forerunner of many such conferences held during the next 60 years.

The Dominion Chemist commented, in 1915, upon enlistments in the Armed Forces and the negative effects of consequent vacancies upon his programs. He had to defer some work because he could not hire qualified chemists to replace those who had joined the services. All other divisions, branch farms, and stations reported an increase in their activities in order to assist farmers in the production of food. Employees of the Experimental Farms Branch were not slow to enlist, for by 31 March 1916, Grisdale named 102 experimental farm people who were on active service.

By 1914 and 1915, Grisdale felt the new experimental stations were in good running order; therefore he added the divisions of Extension and Publicity, and Illustration Stations to publicize and demonstrate results of experiments. Prior to 1914, the research done on experimental farms represented a spectacular contribution to agricultural presentations at large international exhibitions. Superintendents of individual experimental farms, when asked by local agricultural fair associations, showed samples of their research. In 1914, the Central Experimental Farm staff prepared five complete exhibits, each of which toured a separate circuit of agricultural exhibitions throughout Canada. At each of the 18 exhibitions attended, the particular work of the experimental farm nearest to it was also displayed. These activities led to the formation of the Extension and Publicity Division in 1915, headed by Mr. J.F. Watson.

Mr. John Fixter, who had been the first foreman on the Central Experimental Farm and who later joined the staff of Macdonald College as farm manager, became the first Supervisor of the Illustration Station Division. Fixter, with the help of Grisdale and some superintendents, chose locations for 37 illustration stations during 1915 and 1916. They arranged with private farmers in Quebec, Saskatchewan, and Alberta to use experimentally proven methods of cultivation and crop rotation, and to grow the crops most suited to their own locale. Farmers were compensated for any possible loss the newer methods might cause, and they were paid for their time when working on those experimental plots that would give no financial return.

Rain when needed, reasonable freedom from pests, and ideal harvesting conditions produced heavy yields in nearly all parts of Canada in the banner year of 1915. About 60 percent of the 360 million bushels (9.8 million tonnes) of wheat produced in the three Prairie Provinces was Marquis, the first big year for the relatively new variety. Everything came together in 1915 to demonstrate the value of agricultural research. The investment of time and effort by Messrs. Carling, Gigault, and Saunders was paying high dividends. Earlier in the year, the Honourable Martin Burrell, Minister of Agriculture, had called for an all-out production effort from Canadian farmers to meet the needs of those engaged in the war in Europe. Canadian farmers, with the full backing of experimental farms, did not disappoint him. The euphoria, however, was short-lived. In 1916, a severe rust epidemic resulted in a loss of 100 million bushels (2.7 million tonnes) of wheat. Marquis showed little resistance to the fungus.

An interesting development occurred in 1916 when an experimental station was opened at La Ferme, near Lac Esprit in the Abitibi district of Quebec, on a piece of land reserved in 1914 for this purpose. It had been used in 1915 as a prisoner-of-war camp, when prisoners cleared 155 acres (63 ha) of dense woods and the government sold 2500 cords (9000 m^3) for pulp. In September 1916, Mr. P. Fortier took charge of the Experimental Station as foreman–manager, and by January 1917, the internment camp was removed and Experimental Farms resumed control. The property (1200 acres, 485 ha) was deeded to the federal government by the government of Quebec for a nominal sum, on condition it be used for an experimental station. A great variety of crops and livestock was grown on the station and much valuable information was obtained as to what would flourish in the region. La Ferme continued its operations until 1936.

At one time, before World War I, there was an important fiber flax industry in Canada. However, because of the scarcity and high price of labor (there being little labor-saving machinery), flax fiber could not be produced in Canada to compete cost-wise with that imported from Russia. The industry declined until, by 1914, it was reduced to fewer than 2000 acres (800 ha). The war, however, cut off supplies from eastern Europe and by 1917, flax again became an important crop, such that the Experimental Farms organized a Division of Economic Fibre Production, with Mr. G.G. Bramhill as its officer in charge. Bramhill established a small, but complete, experimental flax mill in Ottawa, with the most up-to-date machinery available. He hired a practical worker with 20 years' experience in flax growing and retting, and the manufacturing of linen. Experiments were undertaken to determine the areas in Canada suitable for growing fiber flax, the fertilizers needed, the extent to which flax reduced the fertility of the soil, and other factors related to successful fiber flax production. He considered the establishment of a paper or fiberboard mill to use waste material. He encouraged farmer groups and owners of flax mills to increase their acreage of fiber flax. Other experiments were started to determine the extent to which hemp could be economically grown in Canada for fiber purposes. Bramhill stayed with the division for only a year, and in 1918 Mr. R.J. Hutchinson succeeded him.

Hutchinson was the "practical worker" whom Bramhill had hired the previous year. He was a native of Northern Ireland, where he trained for 5 years at the Flax Spinning Company, Armagh, then in 1912 went to Stuttgart, Germany, and Bruges, Belgium, to learn about their methods of fiber production. In 1914, the government of Northern Ireland commissioned him to buy and grade flax fiber. He came to Canada in 1916 to take charge of the spinning department of a large flax firm that made gun strings, rifle pulls, rope, and halters. His lengthy term as chief of the division extended to 1952.

In 1918, Deputy Minister O'Halloran retired and the Honourable T.A. Crerar, Minister of Agriculture, appointed Director Grisdale as his deputy. Dr. E.S. Archibald, Dominion Animal Husbandman from 1912, replaced Grisdale as director in 1919. Archibald was born in Yarmouth, Nova Scotia, earned an arts degree from Acadia University and an agricultural degree from the University of Toronto. He lectured at the Nova Scotia Agricultural College for 4 years prior to joining experimental farms. Grisdale had retained the responsibility for the Field Husbandry Division during his tenure as director. Archibald followed his lead until he appointed Dr. E.S. Hopkins to the post in 1920.

Hopkins was born at Lindsay, Ontario, graduated from the Ontario Agricultural College in 1911, and received his master's degree in soils from Cornell University in 1915. He had been an agricultural representative in Peterborough County, Ontario, for 2 years while studying for his master's degree, following which he became an instructor at the Vermilion School of Agriculture, Alberta, for a period of 5 years. Before accepting the position of Dominion Field Husbandman he had been in charge of provincial soil investigations at the School of Agriculture, Olds, Alberta.

The economy of Canada strengthened following World War I. Although employment was easily obtained, the universities had only just begun to graduate students in agriculture. As a result, Archibald reported that in 1920 he could hire only 12 technically trained people but lost 14 through resignations. Demand for graduates from the private sector was brisk and salaries were better than those being offered by the federal government. The situation improved during the next 5 years, however, because Archibald was able to hire 78 technical staff and lost only 26, many of those retiring because they had completed their service. It was during this period that names such as Goulden, Neatby, Hilton, Conners, Craigie, Ripley, and Margaret Newton appeared on the lists of staff additions. The organization chart for Experimental Farms showed 13 divisions, 32 experimental farms and stations, and 8 branch plant pathology laboratories. The pathology laboratories were directly responsible to the Dominion Botanist, and two of the experimental stations where tobacco was the main crop were directly responsible to the Dominion Tobacco Husbandman.

Like Grisdale before him, Archibald changed the organization of his annual reports. During the war years, the size of each report was reduced to a brief outline of the work done. Following the war, Archibald returned to the method used by Saunders. He himself wrote a brief summary and overview, each division reported on its own subject, and each experimental farm or station prepared its own report. Archibald considered it to be a better service to farmers if they could have the results of the research done within their own climatic and soil zone. He was eager that timely, up-to-date information should reach farmers promptly. For this reason the Experimental Farms published a bulletin, *Seasonal Hints*, every 4 months and mailed it to 300 000 farmers. In addition, over 400 press articles were issued each year.

In most of Canada during 1920 the amount of rainfall was low in June and July. Below average crops were harvested except at one or two stations in Eastern Canada. Soil drifting was experienced on the Great Plains' farms and stations at Brandon, Indian Head, Scott, Rosthern, and Lethbridge. It was also commonly found on privately owned farms, particularly where dust mulch was used on summerfallow. Some farmers, however, had abandoned dust mulches and left their summerfallow with a lumpy surface. Others adopted a method of strip farming, using alternating strips 16 rods (80 m) wide of crop and of summerfallow running north and south at right angles to the prevailing westerly winds. Experimental farms and stations began testing the effectiveness of these and other methods of preparing seedbeds and handling summerfallow in ways that would minimize the loss of soil and moisture. The station at Swift Current, Saskatchewan, was located in the middle of the lowest rainfall area, and was just being organized specifically to study ways of adapting farming practices to these climatic conditions. The drought of 1920 was a foretaste of things to come.

The following year, 1921, saw the end of the Saunders name in the department, when Dr. Charles Saunders, the Dominion Cerealist, and the person directly responsible for introducing Marquis hard red spring wheat, retired due to ill health. The United Farmers' Guide of Gardenvale, Quebec, in its editorial of 1 February 1922, lamented the resignation of Charles Saunders, saying that it was not "for any reason that will give ... added glory to the Department of Agriculture at Ottawa." Saunders himself wrote in a paper, which was read at the convention of the Agricultural Societies of Manitoba, that he had "decided to give up agricultural research work altogether on account of the

43

discouragements of recent years which have at last exhausted my buoyancy and enthusiasm." He did not present the paper himself, however, because he was indisposed. H.L. Newman, who had been secretary-treasurer of the Canadian Seed Growers' Association since 1905, was hired in 1923 as Dominion Cerealist. Newman was a graduate of Guelph, had studied at Iowa State College, at Cambridge, England, and at the Svalof Plant Breeding Research Station, Sweden. He knew Saunders and respected the contributions he had made to Canadian agriculture. Newman was an excellent choice as Saunders' successor.

In 1920, Macoun, Dominion Horticulturist, was given additional responsibilities as a commissioner in the Ottawa Improvement Commission (OIC), a post he held until 1933, when he retired. In 1927, the OIC became the Federal District Commission, which in turn, became the National Capital Commission in 1959. During the period Macoun was a Commissioner, a number of OIC parks and driveways within Ottawa were planned and expanded, undoubtedly with the professional guidance of Macoun. They included The Driveway, Confederation Square, Island Park Drive, Rockcliffe Park, and Majors' Hill Park, all principal drives and parks in today's Ottawa.

The last division to be established within the Farms System, except for reorganizations, was the Bacteriology Division. In 1923, Dr. A. Grant Lochhead was recruited as the Dominion Bacteriologist. Lochhead was a graduate of McGill University and had studied in Germany for 2 years, where he was interned during World War I from 1914 to 1918. He lectured at Macdonald College in bacteriology, worked for Canadian Milk Products Ltd., then for the Malt Products Company of Canada, and finally returned to lecturing, this time at the University of Alberta. Lochhead remained as chief of the division until he retired in 1956. The intent when forming the Bacteriology Division was to provide other divisions and scientists outside Ottawa with a service similar to that provided by the Chemistry Division. Some bacteriological work, which was transferred to the new division, had been done by the Division of Botany.

Many people were concerned about the build-up of rust in Canadian wheat crops. On 25 June 1924, Dr. J.H. Grisdale, Deputy Minister of the Canada Department of Agriculture, and Dr. H.M. Tory, President of the National Council for Scientific and Industrial Research, commonly called the National Research Council of Canada, or NRC (40), met to arrange for a conference of experts to discuss the situation and formulate a plan to contain and control the spread of rust.

Dr. E.C. Stakman at the University of Minnesota had recently found that the fungus causing wheat stem rust included many distinct forms to which different varieties of wheat were susceptible in different ways. This opened up a whole new line of thinking, the significance of which is explained in Part II. The conference planned by Grisdale and Tory was held on 9–10 September 1924, with Drs. Hayes and Stakman, from Minnesota, in attendance. According to Johnson (23), this conference was a major landmark in the control of rust in Canadian wheats. The outcome was the formation of the Dominion Rust Research Laboratory on the campus of the Agricultural College at Winnipeg.

The interesting story of the cooperation between Experimental Farms and the National Research Council, the Agricultural College, Winnipeg, other universities in Canada, universities in the United States (particularly the University of Minnesota), and the divisions of Cereal and Botany is to be found in Johnson (23) and in Gridgeman (18).

It was not until 1924 that divisions other than Botany and Tobacco appointed technical staff outside Ottawa to take charge of their own areas of research. That year, Sydney E. Clarke was appointed assistant agrostologist in the Forage Division, and Leonard B. Thomson became field husbandman in the Division of Field Husbandry, both stationed at Swift Current and serving Saskatchewan and Alberta. Each would have a significant impact upon the development of Canadian agriculture, Clarke by his contribution to the use of native range in Saskatchewan, Alberta, and British Columbia, and Thomson as superintendent of the Swift Current Experimental Station and then as director of the Prairie Farm Rehabilitation Administration.

The Division of Chemistry worked closely with the other divisions. For instance, in 1925 Shutt made a comprehensive study of the effects of variety and environment upon the protein content of hard red spring wheat. He analyzed samples of the same varieties of wheat grown at 18 different locations from Charlottetown, Prince Edward Island, to Saanichton, British Columbia, and found that some varieties such as Marquis consistently had a higher protein content than other varieties. In like manner, those samples grown under dry conditions had higher gluten contents than those grown under moist conditions. Saunders had recognized this fact from his chewing studies at about the turn of the century, but because of inadequate analytical methods he was not able to measure the differences accurately. The Chemistry Division did cooperative work on soft pork, feeds for livestock and poultry with the Animal Husbandry Division, dehydrated fruits and vegetables with the Horticulture Division, insecticides and fungicides with the Entomological Branch and the Botany Division, and meat and dairy products with the Health of Animals Branch.

In 1919, the first poultry egg-laying contests were instigated, involving 1610 birds laying on average 122 eggs. Six years later, 13 contests were conducted, involving 4100 birds laying on average 172 eggs, an increase of 40 percent, or more than 6 percent per year. All birds laying 200 eggs, or more, were granted breeding registration, provided they met other criteria. Of the 1301 birds that laid the required number of eggs, only 666 were registered, the others being eliminated because they laid undersized eggs or because they failed to meet breed specifications. Without doubt, the egg-laying contests provided a stimulus for Canadian poultry breeders to improve the quality and performance of their flocks. In 1926, the Agassiz Experimental Farm received worldwide publicity resulting from its egg-laying contest, in which one bird, owned by the University of British Columbia, produced 351 eggs in 365 days, a world's record at that time.

Below average summer rainfall in some years caused poor grain yields in the Prairie Provinces. Research in 1926 at Swift Current and at Lethbridge, therefore, began to focus on improving the efficiency of trapping, conserving, and utilizing the small amount of moisture available. Types and varieties of crops were studied to learn which ones could withstand drought conditions and at the same time produce enough fodder for livestock.

From 1920 to 1930 the Botany and Plant Pathology Division, under Güssow, made great strides. Laboratories were established at Sainte-Anne-de-la-Pocatière, Winnipeg, Edmonton, Summerland, and Saanichton. A tribute should be paid here to a prominent agricultural family, some of whom were pathologists. Dr. William (Bill) Newton, who became officer in charge of the plant pathology laboratory at Saanichton, started his career as a plant pathologist at the Experimental Farm, Agassiz, in 1927. The next year he reported from the plant pathology laboratory on the campus of the University of British Columbia, Vancouver, where he worked with Mr. H.S. MacLeod on potato viruses. A third move in as many years brought him to Saanichton, where he studied different methods of preparing Bordeaux mixture (a fungicide), and investigated downy mildew in hops. He remained at Saanichton until retirement 30 years later, having finally found a spot where the fishing was to his liking. The Newton family produced five prominent agriculturists. Margaret Newton was a plant pathologist at the Dominion Rust Research Laboratory, Winnipeg. Dorothy Newton was a botanist; she later married Dr. W.E. Swales, an animal pathologist at the Canada Department of Agriculture Institute of Animal Parasitology, Sainte-Anne-de-Bellevue. John Newton was professor and head of the Soil Science Department at the University of Alberta. Robert Newton at one time was Assistant Dominion Cerealist in Ottawa. He then spent some years at the University of Alberta and the National Research Council of Canada. Returning to Alberta as Dean of the Faculty of Agriculture, he later became President of that University.

The Division of Botany was also responsible for the Certified Seed Potato program throughout Canada, taking its authority from the Destructive Insect and Pest Act. The potato inspection service within the Division of Botany started in a small way in 1915. At that time, there was a limited survey of potato fields in New Brunswick, Nova Scotia, and Prince Edward Island, followed by an inspection of tubers after harvest by scientists from the laboratories within each province. In 1916, the survey was continued and a number of fields were certified as being free from disease. It was from these fields that the first shipment of certified potato seed went to Bermuda. A limited survey was made in Quebec the same year. Each year an additional province was included. By 1920, Alberta potato growers had the benefit of the service. Up until 1920, Dr. Paul A. Murphy, officer in charge of the Charlottetown plant pathology laboratory, was in charge of the program, but upon his retirement, Güssow moved the management of the program to the Ottawa headquarters of the Division, making Mr. G. Partridge responsible. Growers of table stock potatoes found they obtained yields of up to 25 bushels (670 kg) more per acre when they planted certified seed than when they planted uncertified seed. Seed growers found they received higher prices and more easily gained overseas markets for their seed if it was certified. By 1920, 338 thousand bushels (9.2 thousand tonnes) were certified. Seven years later 2.5 million bushels (68 thousand tonnes) were certified, which required the inspection of 28 500 acres (11 500 ha). The activities in other areas of the Division of Botany and Plant Pathology increased in a like manner, as Güssow's annual report grew from 100 pages in 1921 to 250 pages in 1927.

The Central Experimental Farm obtained an additional 361 acres (146 ha) from the lumber baron, J.R. Booth, in 1929, in order to extend experimental plots. The Division of Forage Plants, for instance, increased its plantings of hybrid corn and sunflowers, many of which gave yields appreciably higher than standard commercial mixtures. In the same year, Illustration Stations had increased in number to 186 across Canada.

A small committee of the Privy Council of Canada on Scientific and Industrial Research met during the lunch hour of 25 April 1929. It had the task of deciding upon major policies with respect to the work of the National Research Council of Canada (NRC). Among the ministers present was the Honourable Richard W. Motherwell, Minister of Agriculture. The committee met several more times during the next 2 years and on 5 November 1931 considered the advisability of "centralizing the control of the research activities" of the federal government under NRC. Minutes of the meeting report that the consensus of the ministers present was that such centralization would be desirable. However, the then Minister of Agriculture, the Honourable Robert Weir, did not agree. He brought Grisdale, Archibald, Güssow, Hopkins, MacMillan, Newman, and Watson from the department to the next meeting. Dr. H.M. Tory, President of NRC, was also present and explained the way in which cooperative investigations were planned and directed by associate committees of NRC. He said that each cooperating laboratory was assigned specific problems from a complete program "laid down" by the committee; thus unnecessary duplication of work was avoided. Weir suggested it might be necessary to establish a parallel Agricultural Research Council to direct activities in agriculture, as was done in England. Tory denied the need because NRC already had the legislative authority and the facilities for performing such service. Grisdale acknowledged that certain committees of NRC had rendered valuable service to agriculture, but he was of the opinion that a similar service could have been done by the Department of Agriculture had NRC not undertaken the task.

The minutes report that "considerable discussion ensued" but that no consensus was reached as to how the "whole-hearted cooperation" of the officers of the Department of Agriculture could be secured. Since there already was good cooperation among the working scientists of the department and NRC, the problem was one of administrators trying to enlarge their portion of the limited resources available in the depth of the 1930s depression. The problem was resolved 3 years later at the 105th meeting of the Council of NRC on 11 December 1934. The NRC associate committees involving agriculture were reorganized as joint committees between the Department of Agriculture and NRC. Dr. J.M. Swaine, who earlier in the year had been appointed Director of Research for the department, agreed to serve on all joint committees. Barton had replaced Grisdale as deputy minister by this time.

Hot winds and low rainfall across the Prairie Provinces ushered in the 1930s. Archibald noted that although rainfall was average up to the end of May, a prolonged drought set in in June and crops deteriorated steadily. This resulted in higher grades for wheat but much lower yields. The spring of 1931 again brought powerful winds, soil drifting, and the need to reseed many areas in all three provinces. Crops in Eastern Canada were about normal, but in Western Canada the desert, which had started to form in 1929, was now firmly established in eastern Alberta and western Saskatchewan. Gray (17) points out that the disaster on the prairies during the 1930s was caused as much by low wheat prices due to a world depression as by low wheat yields due to a lack of moisture. Yields per acre dropped from an average of 23 bushels (625 kg) in 1928 to 9 bushels (240 kg) in 1931, going to a low of 3 bushels (80 kg) in 1937. At the same time, the returns per bushel to the farmer dropped from $1.25 in 1928 to $0.60 in 1931. Yields of 3 bushels per acre at almost any price is a disaster!

The desert did not form overnight. Scientists at Indian Head, Scott, Swift Current, and Lethbridge knew of the dangers caused by drifting soil. Fields from which the soil was blown lost their precious organic matter. In addition, fields onto which soil was blown were damaged either through the bombardment of their soils by small particles or by the piling up of drifts along fence lines. By 1930, Experimental Farm staffs had many, but not all, of the answers needed to fight back the desert. They knew that farmers must work their fields at the right time or not at all. They knew that some areas should be returned to grass. They knew that stubble from the previous year's crop and dead weeds from fallow land should be left on the soil surface to trap snow and prevent soil from blowing. They knew that when soil on a field started to blow, a farmer must take immediate action to prevent further damage, by ridging soil at right angles to the direction of the wind. All this information was gathered in 1934 by Messrs. Hopkins, Barnes, Palmer, and Chepil and published (21) in 5000 copies. At the same time, the Honourable Robert Weir, Minister of Agriculture for the federal department, introduced a bill to establish the Prairie Farm Rehabilitation Administration (PFRA). It had its final reading on 11 April 1935.

The story of PFRA, its development and achievements, has been told by others (38, 39). To begin with, however, PFRA was administered by Experimental Farms staff. Archibald was responsible for the expenditure of funds voted to control drifting soil, and he knew how to use them. Since 1929, with the start of the depression, his divisional chiefs and experimental farm superintendents had been squeezed for money. All did what they could with the funds available. Now, with the formation of PFRA and one million dollars per year to spend in helping farmers control their drifting soil and reclaim their degenerated fields, everyone felt a new confidence. The technical details of what they did are to be found in Part II of this history. It was a hectic 2 years for the stations concerned. In 1937, the Honourable James Garfield Gardiner, Minister of the Canada Department of Agriculture, proposed an amendment to the Act respecting PFRA, which strengthened the Administration and reduced the involvement of the Experimental Farms.

The Experimental Farms Branch celebrated its 50th anniversary in 1937. The last divisional chiefs appointed by Dr. William Saunders retired in 1933. The Dominion Chemist, Shutt, was 71 years old and the Dominion Horticulturist, Macoun, was aged 64. Shutt lived to attend the 1937 celebrations, and it was a happy occasion for him. In 1937, however, other forces were afoot which changed forever the nature of Experimental Farms.

49

Chapter 6
Formation of the Entomological Branch
1914–1936

We must retrace our steps to 1914 in order to follow the development of the Entomological Branch, which separated from Experimental Farms in that year. Entomology was the only division to separate as a special branch, although later, several other divisions joined the entomologists where they remained for just over 20 years until they reunited with Experimental Farms to form the Research Branch.

The birth of the Entomological Branch coincided with the outbreak of World War I and its relatively brief existence ended shortly before the advent of World War II. The years between the wars witnessed the expansion of commercial orchards and forest industries in the Maritimes, Quebec, Ontario, and British Columbia, and the extension of cereal acreages in the Prairie Provinces. The greater abundance of suitable host plants led directly to the proliferation of such pests as codling moths and various mites in fruits, spruce budworms and bark beetles in forests, and grasshoppers, cutworms, wheat stem sawflies, and wireworms in cereal and forage crops. Dusts and sprays had long been used in orchards and forests, whereas poisoned baits proved useful against grasshoppers and other insects in field crops. Insect controls throughout this period were mainly variations and modifications of established forest and farm practices.

When entomologists separated, Dr. C.G. Hewitt was Dominion Entomologist and at that time he became director of the new branch as well. He was 24 years old when he was hired by Director Saunders as head of the Division of Entomology of Experimental Farms. Hewitt earned his doctor of science (D.Sc.) degree from the University of Manchester at the age of 20 and lectured at that university until he came to Canada in September 1909. He immediately entered the scientific community, for he was elected vice president of the Ontario Entomological Society in 1910. As director of the new branch, he organized entomologists into four functional divisions: Field Crops and Garden Insects (1914), Forest Insects (1914), Foreign Pest Suppression (1919), and Systematic Entomology (1919). This foundation was subsequently augmented by a fifth division under his successor, Mr. Arthur Gibson, who added Stored Product Insect Investigations (1932).

In 1917, by Order-in-Council of 10 April, Hewitt was given additional duties as Consulting Zoologist. Under this title, he advised the government of Canada on matters relating to the protection of birds and mammals, and he also became Secretary to the Advisory Board on Wild Life Protection. He was pleased to announce the passage of the Migratory Birds' Convention Act in 1917. This Act protected all insectivorous birds and therefore was of distinct benefit to agriculture.

On 29 February 1920, Dr. Hewitt died of pneumonia, following a brief but severe attack of influenza. Although only 35 years old, he had become well recognized throughout Canada and the world as an outstanding economic entomologist, wildlife specialist, and administrator. In addition to his wide involvement in scientific societies, he contributed to civic organizations in Ottawa. He was succeeded by Mr. Arthur Gibson as Dominion Entomologist.

Gibson was born in Toronto in 1875. As a boy, he was an eager amateur entomologist. He joined the Experimental Farms System in 1899 as Fletcher's second assistant (Mr. J.A. Guignard, a clerk, was the first assistant). Gibson had a comprehensive knowledge of butterflies and published several papers on that subject in the reports of the Entomological Society of Ontario before 1899. He was elected president of the American Association of Economic Entomologists in 1926. In 1927, he was appointed Honorary Curator of Entomology for the National Museum of Canada, while continuing to represent the Department on the Advisory Board of Wild Life Control. His responsibility at the museum was to develop an educational exhibit of insects and their activities. In 1935, his outstanding service to entomology in Canada was recognized by Queen's University, which conferred upon him an honorary LL.D. degree.

FIELD CROPS AND GARDEN INSECTS

Arthur Gibson was appointed Chief, Field Crops and Garden Insects Division, upon its formation in 1914. As with the Experimental Farms, scientists in the Entomological Branch and particularly those in the Field Crops and Garden Insects Division gave special attention during the years of World War I to crop protection, so that farmers might maximize their food production. The demand from farmers for such help made it necessary for the branch to increase the number of scientists outside Ottawa. By 1919, there were 10 field laboratories, at least one in each province except Prince Edward Island. Frequently, they were placed at or near experimental farms or stations.

The importation of all corn fodder and various other corn products from certain parts of the United States was prohibited effective 19 May 1919, in an attempt to prevent the introduction of the European corn borer, an established pest in parts of the United States. Unfortunately, the borer was found for the first time in late August 1920 in Canada. Despite a concentrated effort to eradicate the pest, it continued to spread in southwestern Ontario, such that by 1925 Gibson reported its increase to be alarming. The story of its eventual control is told in a later chapter.

When Gibson became Dominion Entomologist following Hewitt's untimely death in 1920, Mr. R.C. Treherne moved to Ottawa from Vernon, British Columbia, to head the Field Crops and Garden Insects Division. Treherne came to Canada in 1905 when he was 19 years old, entering the Ontario Agricultural College that fall. Following graduation, he worked as a provincial nursery inspector and then as a field officer in New Brunswick for the Division of Entomology. In 1911, he opened the Entomological Laboratory at the Vineland

Station, and later that year took charge of the new entomological laboratory on the Experimental Farm, Agassiz. With the formation of the Entomological Branch in 1914, Treherne was made officer in charge for British Columbia and moved to Vernon. He died suddenly on 7 June 1923 at the age of 37, 3 years after moving to Ottawa from Vernon. In 1925, Mr. H.G. Crawford replaced him as chief of the division.

Crawford was born at Mahone Bay, Nova Scotia, and graduated from the Ontario Agricultural College in 1915, receiving his master's degree at the University of Illinois in 1917. During his summers as a student he worked for the Ontario provincial entomologist and at the Entomological Laboratory, Strathroy, Ontario. He lectured in entomology at Guelph for 2 years, then joined the Entomological Branch in 1920, quickly becoming the Canadian authority on the European corn borer. Crawford was known as a man of vigor, with buoyant spirits, keen perception, and penetrating humor.

The thirties were extraordinary times in the Prairie Provinces. Only the unparalleled problems of drought and soil drifting outranked the trauma that resulted from insect pests. Soil drifting forced changes in farm practices, which in turn made the control of insects more difficult. For example, strip farming greatly increased the frequency and area of margins between fields bearing crop and those remaining in stubble. This eased the access to crops for grasshoppers and sawflies migrating from adjoining stubble land, a principal source of infestations. The persistent drought added to the damage potential of the wireworm and the expansion of infestations of the pale western cutworm.

Prodigious contributions to means of reducing crop losses were made with minimal professional staffs through the leadership of men such as Arthur Kelsall (Annapolis Royal), J.D. Tothill (Fredericton), C.E. Petch (Hemmingford), W.A. Ross (Vineland), H.G.M. Crawford (Chatham), Norman Criddle (Treesbank), K.M. King (Saskatoon), H.L. Seamans (Lethbridge), and E. Hearle (Kamloops).

Criddle, a distinguished self-taught Canadian naturalist, artist, and friend of Dr. James Fletcher, was appointed by Hewitt in 1913. Messrs. Seamans and King were recruited from the Montana State Agricultural College at Bozeman in 1921 and 1922, respectively, by Arthur Gibson. All built their staffs slowly, exclusively from young Canadian entomologists. These neophytes were swiftly launched into important assignments and encouraged to take advanced degrees at reputable universities in the United States in order to develop a staff with a broader background. At the time, there were only two universities in Canada offering such degrees in entomology; thus, R.D. Bird and R.H. Handford (Treesbank), A.P. Arnason, R. Glen, and H. McDonald (Saskatoon), and C.W. Farstad, G.F. Manson, and R.W. Salt (Lethbridge) all moved on to assume greater responsibilities in the development of entomology and of agricultural research in Canada.

With only one entomological laboratory located in each of the Prairie Provinces, provincial boundaries set the general geographic responsibilities. Since the insects did not respect these boundaries, the three officers in charge, with the approval of the Dominion Entomologist, arranged a new alignment of authority covering the entire prairie region: Treesbank guided the early research on grasshoppers, Saskatoon assumed full responsibility for wireworm and red-backed cutworm investigations, and Lethbridge covered the wheat stem sawfly and pale western cutworm requirements. The program leaders received all-out support for their assigned fields from the staffs residing and working within their home provinces.

Extensive collaboration was maintained between the entomological laboratories, the experimental farms, and provincial departments of agriculture in the conduct of relevant research and control campaigns. Two examples will illustrate. J.G. Taggart, superintendent of the Experimental Station, Swift Current, provided up to 30 acres (12 ha) of land and the required tillage for a long-term cooperative project with K.M. King to study the effects of four different methods of summerfallow on wireworm abundance and resulting damage to wheat. At the height of the grasshopper outbreaks of 1933–1934, all members of the entomological laboratory at Saskatoon joined with all available staff from the Saskatchewan Department of Agriculture (such as agricultural representatives, and dairy and livestock workers) in operating the enormous baiting program in that province.

FOREST INSECTS

Hewitt appointed Dr. J.M. Swaine as chief of the Forest Insects Division in 1914. Swaine was a Nova Scotian and graduated from Cornell University in 1905, receiving his doctorate from the same university in 1916. He lectured at Macdonald College, Sainte-Anne-de-Bellevue, from 1907 to 1912, at which time he joined the Entomology Division of the Experimental Farms and Stations as forest entomologist.

Dr. John D. Tothill, appointed as field entomologist at Fredericton in 1911 to introduce parasites of gypsy and browntail moths, became the first head for natural control investigations (now called biological control). The gypsy moth, although not in Canada, was moving north in the United States and had reached Massachusetts. Swaine arranged with the U.S. Department of Agriculture to have some Canadian entomologists work at an American laboratory where parasites and predators of the gypsy and the browntail moths were raised for release in Nova Scotia, New Brunswick, and Quebec. The program was successful, for neither moth had become of economic importance in Canada until the late 1970s when several outbreaks of the gypsy moth did occur in the eastern townships of Quebec and in southwestern Ontario.

Forest entomologists became heavy users of aircraft, which proved invaluable in the survey of insect damage and the only practical method of applying control sprays. The use of aircraft started in 1919, when Eric Hearle (12) investigated mosquito problems in the Lower Fraser Valley from the air. Initially, aircraft from the Dominion Air Service were used. By 1923, numerous departments made such heavy demands upon the limited number of airplanes that an Interdepartmental Committee on Air Operations had to be established to coordinate this activity. The Associate Dominion Entomologist, Dr. J.M. Swaine, was the departmental representative.

Each year, additional species of insects attacked forests throughout Canada. In most instances, control measures involved the proper management of forests. A branch bulletin regarding the spruce budworm recommended, for instance, that mature balsam fir be removed as quickly as possible, and that in future, balsam stands be cut when young to prevent a buildup of the budworm. Other insects that were causing damage included the spruce beetle and larch sawfly in Eastern Canada, mountain pine beetle and cedartree borer in British Columbia, and tent caterpillars and various scale insects on shade trees throughout Canada.

During this period entomologists such as Dr. R.E. Balch, Dr. J.J. de Gryse, and Mr. Ralph Hopping were recruited. Balch came to Canada from England to take up farming. Following service in World War I, he received his education at the University of Toronto and the New York State College of Forestry. He took charge of the Forest Entomological Laboratory, Fredericton, in 1930.

De Gryse was born and educated in Belgium, moved to the United States, then was appointed entomologist to the Forest Nursery Station, Indian Head, in 1923, while the station was still under the Department of the Interior. In 1925, he transferred to Ottawa and in 1934 followed Swaine as Chief, Forest Insect Division.

The third entomologist, Hopping, also had his roots outside Canada. He was born in New York and in 1891 moved to California. He became knowledge-able about forest insects, specializing in the control of the western pine beetle. In 1919, Swaine hired Hopping as forest entomologist for British Columbia, with his office in Vernon.

The European spruce sawfly, first discovered in the Gaspé Peninsula in 1930, had spread to cover 5000 square miles (13 000 km^2) by 1934. The chief of the division, de Gryse, enlisted provincial governments and various forest industry associations, which collaborated in liberating a million parasites imported from England, since they were not native to Canada. The following year, although not part of the Forest Insect Division, Messrs. A.B. Baird and J.D. Tothill at the Belleville, Ontario, laboratory raised and released millions of parasites. Some species were judged to have established themselves but showed no evidence of controlling the spread of the sawfly at that time. More importantly, however, the introductions brought with them an insect virus that has contrib-uted to keeping the sawfly in check since then (see Chapter 18). The successes at the Entomological Laboratory, Belleville, resulted in an ultramodern laboratory being built in 1936, which provided for the controlled rearing of insect parasites and predators. Because of stringent quarantine safeguards to prevent the escape of foreign hosts upon which the insects were reared, none ever escaped, or if any did, they never became established in North America.

The division organized a forest insect intelligence service in 1931. To begin with, circulars on the principal forest insects were published by Messrs. Swaine, de Gryse, Balch, and E.B. Watson. Then, in 1936, the division asked leading

55

forestry firms and provincial forestry services to collect forest insects during the summer months and send them to the division. In return, the division published annually the results of the survey, and divisional officers gave courses in forest entomology to woods' managers, foresters, and rangers at a number of locations across Canada. The Canadian Forest Insect Survey, as it became known, gathered much new scientific data and fostered close cooperation between entomologists and forest managers throughout Canada.

FOREIGN PEST SUPPRESSION

Hewitt gave top priority in the new branch to the administration of the Insect and Pest Regulations of the Destructive Insect and Pest Act. Well he might, for the parliamentary appropriation to administer the Act was more than three times the appropriation allotted to the Entomological Branch. To begin with, in 1914, this function was managed by Gibson, Chief, Field Crops and Garden Insects Division.

The Division of Foreign Pest Suppression was spun off from the Field Crops and Garden Insects Division in 1919, in order to better administer the regulations under the Destructive Insect and Pest Act of 1910. Mr. L.S. McLaine became its chief. Although born in England, he graduated from the Massachusetts Agricultural College. From 1913 until his appointment as chief, he had been in charge of the browntail moth studies in Nova Scotia and New Brunswick.

Up until 1922, the administration of the Destructive Insect and Pest Act was handled by the Entomological Branch and the Division of Botany of Experimental Farms. In 1922, however, an Order-in-Council passed on 21 April set up the Destructive Insect and Pest Act Advisory Board for the purpose of administering the Regulations under the Act. The Board was chaired by the Dominion Entomologist, Mr. Arthur Gibson with Dr. E.S. Archibald, Director, Dominion Experimental Farms and Stations, as vice-chairman. Members of the Board were Deputy Minister Dr. J.H. Grisdale, Dominion Botanist Dr. H.T. Güssow, and Mr. L.S. McLaine, Chief, Division of Foreign Pest Suppression, as secretary. During its 1st year, the Board prepared two Ministerial Orders and three Orders-in-Council, all dealing with insect pests. One goal of the Advisory Board was to improve communications between the Entomological Branch, which was located on the sixth floor of the Confederation Building, and the Division of Botany and Plant Pathology, which had its offices and laboratories at the Central Experimental Farm 8 km distant. The Dominion Entomologist continued to be responsible for that portion of the regulations which applied to insects.

Funding for the branch came largely from monies voted for the Administration of the Destructive Insect and Pest Act. No data are available on the division of work between the Entomological Branch and the Division of Botany in the Experimental Farms Service; however, from circumstantial evidence it appears that entomology spent 60 percent and botany spent 40 percent of the funds. Thus, in 1930, when $570 000 was spent under the Destructive Insect and Pest Act, the Entomological Branch could have spent about $342 000 of the total. In the same year, the vote for the Entomological Branch itself was only $25 000, thus more than 90 percent of funding for the branch originated with its administration of the Act. Until the fiscal year 1934–1935, all funding for laboratories outside of Ottawa came from monies voted under the Destructive Insect and Pest Act (14).

In 1928, the Division of Foreign Pest Suppression formally organized the Plant Inspection Service. Although this seemed to be a new service, it was not a new function. The Entomological Branch, and the Division of Entomology of Experimental Farms before it, had inspected plant material entering Canada since the passage of the Destructive Insect and Pest Act in 1910. The entomologists had, of course, the full support and cooperation of the Division of Botany, which inspected plants for diseases. In 1928, Canada had 11 plant inspection stations at strategic ports of entry. The volume of nursery stock inspected increased from 900 000 plants in 1919 to 48 000 000 in 1929, an increase of over 50-fold in 10 years!

On 1 August 1933, the horticultural inspection service operated by the province of British Columbia was transferred to the Division of Foreign Pest Suppression in the Entomological Branch, thus aligning it with the rest of Canada. Until that time officers of the province had acted as collaborating inspectors of import and export shipments of plants on Canada's west coast. During the year, of the 9606 plants carried by passengers entering Canada, 719 were prohibited entry. Of the 344 912 plants imported through the mails, 54 were refused entry because of doubt regarding their freedom from disease and insects.

SYSTEMATIC ENTOMOLOGY

In 1916, the Parliament Buildings burned and parliamentarians moved to the National Museum. Because of the resulting congestion, in 1917 the entomological collections at the new Victoria Memorial Museum, which housed the National Museum, were transferred to the Entomological Branch in the Birks Building (19), thus bringing the entire national collection of insects together. Dr. J.H. McDunnough was appointed Chief, Systematic Entomology Division, and took charge of the collection in 1919. McDunnough was born in Toronto, graduated from Queen's University, then took his doctorate in Berlin, Germany. He was curator of Barnes' Lepidoptera Collection in Decatur, Illinois, for 9 years before joining the Entomological Branch. Until the appointment of McDunnough and the organization of a division to properly look after collecting and classifying insects, entomologists had shared the task as their time permitted and interests dictated. The new arrangement provided a foundation upon which to build a reliable reference collection of specimens.

Housing the national collection within the Department of Agriculture was a unique arrangement. It is one that has greatly benefited agriculture in the identification of confusing complexes of insect pest species with differing food plant preferences and behavioral patterns. It has immeasurably aided Canadian entomologists in their approach to control methods. Although many amateur and professional entomologists across Canada contributed to the growth of the collection, the high standards that led to its greatness were laid down by the first three appointees: McDunnough specializing in Lepidoptera (butterflies and moths); W.J. Brown specializing in Coleoptera (beetles); and G.S. Walley specializing in Diptera and Hymenoptera (flies and wasps).

By 1930, McDunnough and his staff were identifying up to 6000 specimens each year. They had built by far the best insect collection in Canada and one of the most important in North America. At that time there were 2100 steel drawers in 42 steel cabinets housing the collection. Today the Biosystematics Research Institute has 1200 cabinets for insects and 750 for plant material.

STORED PRODUCT INSECTS

The Department of Agriculture from its inception in 1868 until 1919 included public health among its many responsibilities. The Entomological Branch, therefore, had a concern for insects and other pests affecting households and public health as well as those contaminating stored products such as grains and flour.

The Division of Stored Product Insects was formally organized in May 1932 when Dr. E.H. Gray was appointed as its chief, with his headquarters in Winnipeg. Gray was born in Ottawa shortly before his family moved to Lethbridge, Alberta. After graduating from Montana State College, he joined the Entomological Laboratory, Lethbridge. He developed his interest in stored product insects at the University of Minnesota, where he received his master's degree prior to moving to Winnipeg.

The first major task of the new division, in cooperation with the Division of Foreign Pest Suppression, was to assure that no live insects entered Canada in the samples of grain destined for the 1933 World's Grain Exhibition to be held in Regina. None did. Following the Exhibition, Gray moved his laboratory and office to Ottawa and made significant contributions to the control of insects in cereals and flour being shipped across Canada and internationally.

We return now to the activities of the branch as a whole. Dr. J.H. Grisdale who had been deputy minister in the department since 1918 and had served under four different ministers of agriculture, retired in 1932 because of failing health. He was in his 62nd year. The Honourable Robert Weir, Minister of Agriculture from 1930 to 1935, appointed Dr. G.S. Barton, Dean of Agriculture at Macdonald College, Sainte-Anne-de-Bellevue, to succeed him. Barton was born in Vankleek Hill, Ontario, near Ottawa, graduated from the School of Agriculture, Guelph, in 1907, and immediately went to Macdonald College, where he lectured in animal husbandry. He became professor of Animal Husbandry in 1911 and Dean in 1925. His honorary doctor of science degree was received from Laval University in 1928, and his Award of Agricultural Merit (Commander) came from the government of Quebec in 1929.

After Barton had been deputy minister for 2 years he appointed Dr. J.M.

Swaine, the chief of the Division of Forest Entomology in the Entomological Branch, to a newly created position of Director of Research for the department. This appointment put Swaine in the hierarchy over Messrs. Gibson and Archibald, directors of the Entomological Branch and the Experimental Farms and Stations, respectively. On 15 July 1934, Archibald wrote to Barton asking him for a statement with respect to the relations between the Director of Research and the staff of the Experimental Farms. Archibald, in his nicely worded letter, assumed that the director "will be an officer of the department" who will assist branches to conduct research rather than "any alteration in jurisdiction." Five days later, Barton responded by advising Archibald that a letter was being sent to heads of all branches. That letter said that Swaine had been appointed under a recent Order-in-Council, that he would be on the administrative staff of the department, and that his duties called for him to pay special attention to research, to improve relations between workers within the department, to produce a more balanced research program, to improve the level of work attained, and to effect certain economies. The deputy minister also said it would be understood that Dr. Swaine might have "close contact and association with those ... actually engaged" in research.

The following May, in 1935, Swaine advised Barton that he was laying plans to reorganize research work in the department. He said that the objectives of the reorganization were to bring about greater efficiency in research and to advance the training of a number of the staff. He noted that, because of personalities and for political reasons, he found difficulty in eliminating minor experimental farms or stations or even reducing the work of larger ones. The passage of the Prairie Farm Rehabilitation Administration Act (PFRA) occurred simultaneously. PFRA was organized by the Experimental Farms under Archibald; thus rather than experimental farms and stations being eliminated or reduced their influence was being extended. Nothing seemed to come of Swaine's proposed reorganization until 1 April 1937, when Science Service was formed. It included the Entomological Branch, and the third phase in the development of the Research Branch was entered upon.

Chapter 7
The Development of Science Service
1937–1958

The severance of entomologists from Experimental Farms in 1914 caused a minor tremor compared with the split of 1937. Three of the 13 Experimental Farm Divisions whose scientists worked closely with colleagues in the Cereal, Forage, Horticulture, Poultry, Animal Husbandry, and other divisions were broken away and united with the Entomological Branch to form a new service called Science Service. Dr. Swaine's vision for the new service was perhaps an ivory tower approach to research in the Department of Agriculture. Young men and women returned to Canada in large numbers after World War II, and with their added momentum, the next 15 years became a period of unmatched expansion of both staff and facilities, a period when Science Service was able to build upon its solid base.

On 22 February 1937, Deputy Minister Barton conferred with the Civil Service Commission, winning approval for the establishment of a skeletal staff of new senior positions necessary for the departmental reorganization recommended by Dr. J.M. Swaine in 1935. The plan was approved, effective 1 April 1937, by Order-in-Council P.C. 6/834 of 16 April 1937. An unsigned memorandum, date-stamped 5 May 1937, filled in the new alignment of functions in the department. It refers to Science Service as having the "more strictly science services in the Department, which ... will be placed under the direction of a competent director trained in science." Dr. J.M. Swaine, the science adviser to the deputy minister, was chosen as that director.

Barton regrouped the department by bringing each function of a similar nature and purpose under one administrative head (15). Thus, he established four operating services: Production, Marketing, Experimental Farms, and Science, a reduction from nine to four, and changed the designation of each function from "branch" to "service." In addition, matters involving the whole department such as publicity and extension, the departmental library, and the Prairie Farm Rehabilitation Administration, were handled by the Administration Service under the direct control of the deputy minister.

Those functions formerly in the Entomological Branch and in Experimental Farms that dealt with production or marketing, were placed either in the Production Service or in the Marketing Service. As an example, the administration of the Destructive Insect and Pest Act was moved to Production Service. However, research on animal pathology, formerly in the Health of Animals Branch, was moved to Science Service. Research in agricultural economics, nevertheless, was retained in the Marketing Service. In effect, the new departmental organization placed each of the services more or less on an equal footing with respect to size. In the previous organization with nine branches, some were relatively small in comparison to the large and influential Experimental Farms and Stations. In forming Science Service and restructuring Experimental Farms into a Service, both of which had an agricultural research function, the principle said to have been followed was to separate the scientific aspects from the experimental aspects and place each under a different administration. A separation into basic and applied research is probably what was meant. Science Service staff was encouraged to delve deeply into the problems they studied. In doing so, they built a firm data base in many areas of biology associated with agriculture. Scientists today still use these basic data upon which to build their theories and develop technologies in production, protection, and utilization of crops and livestock.

Both services were restricted in their hiring practices with respect to the disciplines they could employ. Neither could recruit agricultural economists, these were the sole responsibility of Marketing Service. Experimental Farms could hire neither taxonomists nor chemists; these were the jurisdiction of Science Service. Finally, Science Service was not permitted to hire plant breeders, for they fell within the jurisdiction of Experimental Farms. Sister services were expected to provide professional help to the others and, in fact, Science Service provided bacteriologists and chemists to Experimental Farms Service, and that service reciprocated by providing plant breeders to Science Service. It was not until the 1960s that Economics Division seconded economists to other divisions or branches. Glen (15) aptly pointed out that the differences between Science Service and Experimental Farms Service lay more in the area of investigation than in the type of investigation (research versus experimentation) that each was said to be doing. Scientists in the Division of Botany and Plant Pathology, such as Dr. D.B.O. Savile, were said "to have been astounded to find [they] were in a different Service from the people in cereal, horticulture, and forage next to whom [they] would still be working." Similarly, scientists remaining in Experimental Farms Service ". . . were bitter because they seemed to have been made second-class citizens."

Science Service had five divisions. The largest was the Division of Entomology, which included staff and resources of the Entomological Branch except those dealing with foreign pest suppression. From Experimental Farms, Science Service took the divisions of Botany and Plant Pathology, including the Dominion Arboretum, under Dr. H.T. Güssow; Chemistry, headed by Mr. C.H. Robinson; and Bacteriology and Dairy Research, with Dr. A.G. Lochhead as its chief. To these, was added the Division of Animal Pathology, headed by Dr. E.A. Watson from the Health of Animals Branch. The Experimental Farms Service was left with 10 divisions and the 37 experimental farms and stations outside Ottawa (see Appendix II).

The five divisions of Science Service were reasonably autonomous. Each chief administered divisional funds, made decisions with respect to research projects of scientists within the division, assessed the quality of the research of the scientists, edited and authorized publication of scientific papers written by them, and selected staff for appointment by the Civil Service Commission. Generally there was little coordination in the research done by scientists of two or more divisions, even though frequently their laboratories were on the same experimental station or university campus. There were notable exceptions, of course. Plant pathologists and entomologists worked closely with forage crop breeders at the Dominion Forage Crops Laboratory, Saskatoon, Saskatchewan, and plant pathologists and plant breeders worked as a team at the Dominion Rust Research Laboratory at Winnipeg.

One of the first tasks confronting Swaine was to physically bring his new service together, because its elements were widely scattered. Entomologists and administrators of the service were on the sixth floor of the Confederation Building, adjacent to the West Block of the Parliament Buildings. The botanists and plant pathologists were in a building in the arboretum, and the chemists were in their 1899 building, both on the Central Experimental Farm. The dairy bacteriologists had a suite in an upper storey of an old building on Elgin Street, near the center of Ottawa. The veterinarians working on animal diseases had their laboratory in Hull, Quebec, and because of the infectious nature of their material, Swaine realized they had to remain isolated from other establishments with any livestock. Swaine wrote to Barton on 13 December 1937, requesting support for the construction of a Science Service Laboratory Building. He emphasized that by uniting the service in one building the principle underlying reorganization could best be served. At one point, Swaine suggested the department build a laboratory for Science Service at the National Research Council on Montreal Road, Ottawa, but the president of the Council, Major General A.G.L. McNaughton, explained that there was insufficient property available for such a large laboratory. How fortunate for agricultural research that McNaughton made this decision!

The new organization created some trauma for the staff of Experimental Farms Service. There is said to have been rivalries between faculties of agriculture at some universities and experimental farms. With the exception of Macdonald College at Sainte-Anne-de-Bellevue and the Ontario Agricultural College at Guelph, Experimental Farms predated all such faculties. Several of the deans and their staffs felt that an organization similar to that in the United States where Land Grant Colleges were financially supported directly from the Treasury of the U.S. Department of Agriculture under the Morrill Act (see Chapter 1), should prevail in Canada. Saunders, in his report of 1885, had recommended against such an organization and in doing so he prevented the balkanization of agricultural research in Canada, a process which was so evident in the United States. With the formation of Science Service, some universities are said to have commented to scientists at experimental farms that this was the end of the Experimental Farms Service. Such was not to be. The strength of experimental farms lay in the problem-solving capabilities of their staffs, the national unity of the organization, and the support received from each of the 37 communities and federal political ridings in which farms and stations were located.

We must now briefly examine the activities of each division of Science Service and meet their chiefs. Science Service had its greatest impact upon

Canadian agriculture during the second half of its life when, with a change of director, it truly became a unified service rather than a collection of divisions.

BACTERIOLOGY DIVISION

Studies on microbiology of soil and of food were the main thrusts of this division at the time it was moved from Experimental Farms to Science Service under Lochhead. One of the principal lines of research was to learn the manner in which plants resisted attack from soil-borne plant pathogens. Lochhead found that the differences in resistance to root rot diseases between plant varieties were related to various secretions from their roots. Dr. C.K. Johns, the food microbiologist, studied bacterial counts on frozen fruits and vegetables to determine methods that would minimize the number of bacteria in the finished product. During World War II, Johns did an heroic job inspecting powdered egg and milk plants to find and stop bacterial contamination, for which the consumers in the United Kingdom must have been thoroughly grateful. Today's high quality powered egg and milk stems largely from his work.

Dairy research, which was in the Dairy and Cold Storage Branch of the department, was moved from the center of Ottawa (28) to Science Service on the Central Experimental Farm and amalgamated with the Bacteriology Division in 1938. Dr. E.G. Hood's studies included methods of making accurate counts of bacteria found in milk and milk products, improving methods of pasteurizing milk, and causes of rancidity in cheese. He also provided an analytical service for the rest of the department. Even though housed with the Bacteriology Division, Swaine was concerned that dairy research was not really integrated with the division. Later, Drs. Neatby (1947), Archibald (1949), and Hopkins (1950) were of the same view. Some suggested the division be moved to the National Research Council, others that a union be effected with the Animal Husbandry Division. Neither proposal was acted upon.

In 1956 Lochhead retired as chief of the division, although he continued as a research scientist in soil microbiology. Dr. H. Katznelson replaced him as chief. Katznelson joined the department in 1940. He was a hardworking and meticulous scientist, preparing more than 120 scientific papers before his untimely death in 1965. By that time the Bacteriology Division was part of a food institute.

BOTANY AND PLANT PATHOLOGY DIVISION

When it was a division of Experimental Farms, Botany and Plant Pathology had three primary functions: development and management of an arboretum and herbarium; research on the biology and control of plant diseases; and inspection of plants for diseases. The arboretum initiated by Drs. Saunders and Macoun had become a truly national arboretum and a start had been made in bringing together specimens for a national agricultural herbarium. The division continued to perform all those functions after becoming part of Science Service. In 1940 alone, the division brought its collection of plant specimens up to 32 000 sheets, an increase of 6800 sheets or 27 percent in 1 year, an achievement brought about largely through the efforts of Dr. H.A. Senn.

World War II, which began 2 years after Science Service was formed, brought physical development largely to a halt but not the enthusiasm of its scientists. In addition to supplying its usual agricultural products, Canada sought to supply unusual items made unavailable by the disruptions of war. For instance, in 1942 the shortage of natural rubber was acute. The traditional source from southeast Asia was inaccessible, yet rubber was a vital commodity. The Botany Division, in cooperation with experimental farms and other scientific organizations across Canada, made a thorough survey of both native Canadian and other plants that could be grown in Canada in a search for a supplementary source of natural rubber. The most promising native plant was the common milkweed. The U.S. Department of Agriculture made a similar search and obtained seed of the Russian dandelion from the USSR. Canada received samples of the seed and grew them at 13 locations from coast to coast, mostly on experimental farms. Production of roots varied from 5000 to 9000 lb/acre (5600 to 6700 kg/ha). Even the lesser amount was more than that obtained in the USSR. The higher amount was 1000 lb (454 kg) better than that produced in the the United States. The rubber content of the roots, when analyzed by chemists from the National Research Council, varied from 1 percent to over 13 percent and the quality of the rubber was judged to be high. Meanwhile, chemists from the petrochemical industry developed artificial rubber in sufficient quantities to meet Allied needs.

Other activities directly associated with World War II included a search for a kapok substitute. Kapok, the floss from the kapok tree of the East Indies, Africa, and Latin America, was used in marine life preservers. Again, milkweed proved to be one of the most promising materials, this time the floss of the seed pods. With the cooperation of school children, 1200 lb (550 kg) of seed and 650 lb (300 kg) of floss were collected. The National Research Council and the Royal Canadian Navy found the floss to be a good substitute and so the following year school children collected over 120 000 bags of milkweed pods for the production of a kapok substitute. However, cultivation of milkweed proved difficult; it was more satisfactory to collect pods from natural sources than to grow the plant as a farm crop.

In the autumn of 1944, the first outbreak in Canada of Dutch elm disease was discovered in the Lake St. Pierre section of the St. Lawrence River about 50 miles (80 km) down river from Montreal. The discovery was made by forest pathologists of the Quebec Department of Lands and Forests. The disease, caused by a fungus, was first observed in Holland in 1921 and then in the United States in 1930. Because of its devastating effect on elm trees, watch had been kept by all forest pathologists in Canada for a possible outbreak. Immediately, steps were taken by the Plant Protection Division to determine the extent to which it had spread in Canada and to eradicate the disease. Unfortunately, in 1945 it was found to have spread over a large area extending from Lachine to Quebec City. The only known method of control was to fell diseased trees and strip their trunks and branches of bark. The procedure prevented the fungus-carrying bark beetle from breeding in the dead bark and thus infecting other trees. Pathologists felt there was a reasonable chance of keeping the disease under control in Canada, and they probably did slow its progress. However, real control was not accomplished, with the result that many thousands of our lovely umbrella-shaped trees have succumbed. The origin of the outbreak in Canada was traced to an importation from Great Britain of machinery crated in infected elm timber! Later, the disease moved into Canada from the United States.

Shortly after World War II, plant pathology laboratories in various parts of Canada cooperated with provincial departments of agriculture in developing programs to certify raspberry and strawberry plants as being free from disease. As expected, horticulturists found that healthy plants produced more and larger fruit than did diseased plants. Nova Scotia, Ontario, and British Columbia were the principal provinces concerned.

Craigie, the Dominion Botanist, retired in 1952. He was born in Nova Scotia in 1887, and graduated from Harvard and Minnesota before joining the Dominion Rust Research Laboratory, Winnipeg, when it opened in 1925. His major discovery was the sexual process in rust fungi. He became the principal plant pathologist at Winnipeg in 1928. When Güssow retired in 1944, Craigie moved to Ottawa as Dominion Botanist and Associate Director of Science Service. Retirement for Craigie did not mean leaving his laboratory. He continued his studies on the cytology of rusts until 1958. Craigie justifiably received many honors, among them being the Order of Canada and fellowships in both the Royal Society (of Great Britain) and the Royal Society of Canada.

Dr. W.F. Hanna followed in Craigie's footsteps as Dominion Plant Pathologist. He too was a Nova Scotian and joined the Rust Research Laboratory in 1928. Following a distinguished military career, he returned to Winnipeg in 1945 as officer in charge of the pathology laboratory. He also received many honors, including the Order of Canada.

CHEMISTRY DIVISION

As its name implies, the Division of Chemistry was expected to supply the technical expertise for Science Service and Experimental Farms Service in all matters related to chemistry. In theory, this may have been a good plan. In practice, it was insufficient because chemical analyses were needed at many more locations than there were chemists available. Hence, scientists with secondary training in chemistry applied that knowledge to solving problems in their primary research field such as soils, plant pathology, or animal science, often with the helpful advice from chemists in the division. For large national programs where hundreds of samples needed analyzing, the chemists were indeed an integral part of the program.

On 1 March 1944, Swaine suggested to Barton that a new building be constructed to house the Chemistry Division and part of the Bacteriology and Dairy Research Division. The estimated cost was $225 000. Barton responded on 3 March saying that he was in accord with the plan but was not clear if this was to be part of the original proposal for a Science Service building. He doubted, however, that the government would approve the construction of any building during the war. Six days later Swaine explained that the scheme was to build a wing on the Science Service Building. The need for chemurgic research in connection with agriculture was so acute that he felt the proposal should be placed before the federal government. Barton agreed to do so. However, nothing further came of this plan until 1956 when the Science Service Building (now the K.W. Neatby Building) was constructed as an addition to the records building of the Department of National Defence, built in 1937 on the Central Experimental Farm.

Mr. C.H. Robinson, who had replaced Shutt in 1933 as Dominion Agricultural Chemist, died suddenly on 11 April 1949 in his 61st year. Robinson had joined the division in 1910 following graduation from the University of Toronto. He became an extremely well-informed analytical chemist. Robinson built the division from one occupying a single laboratory to one housed in five buildings in Ottawa and supporting laboratories at Kentville, Nova Scotia, and at Summerland and Saanichton, British Columbia.

Dr. J.C. Woodward succeeded Robinson. Woodward was born on a mixed farm at Lennoxville, Quebec. He was educated at McGill and Cornell universities. Upon completion of his education in 1934, he joined the Chemistry Division when it was still with Experimental Farms. Woodward is credited with organizing the division into four sections—animal biochemistry, plant chemistry, soil chemistry, and analytical chemistry.

In 1955, Woodward was appointed Associate Director of Experimental Farms Service under Goulden. Dr. A.G.R. Emslie replaced Woodward. Emslie was born in China of Scottish parents. He came to Canada in 1923 and graduated from the Ontario Agricultural College in 1928, taking further degrees from the universities of Toronto and Aberdeen. He joined the Chemistry Division in 1935, a year after Woodward, and headed the animal chemistry unit in 1949.

Chemists of the division cooperated with scientists in other divisions of both Science Service and Experimental Farms Service, but by far the greater proportion of their work was with Field Husbandry and with Animal Husbandry divisions of Experimental Farms. One problem dealt with the fixation of phosphate in soil, making it unavailable to the plant. Chemists found that colloidal material in soil bound phosphorus. They were not able to change the colloids but did determine appropriate application rates of phosphorus for various soil conditions. Other work dealt with brown-heart disease in rutabagas, the solution for which was the application of boron to the soil.

In animal nutrition, one example of research done in cooperation with Experimental Farms involved the loss of iodine from iodized salt when salt licks were exposed to the weather, and another dealt with the feeding of vitamin D in poultry rations. By 1949, vitamin research was expanded to include vitamin A and various hormones.

ENTOMOLOGY DIVISION

One of the objectives of the new Science Service was to gather base line data on a number of subjects in order to better solve biological problems as they arose. In this connection, the Entomological Branch (now the Division of Entomology) paid particular attention to the study of life cycles of Canada's most important insect pests and to the improvement of the National Collection of Insects, for which it became responsible in 1917. In addition, of course, officers of the division were charged with providing control methods for both major and minor outbreaks of pests.

Grasshoppers continued to be a threat to crop production on the Great Plains and control with poison baits proved to be the only satisfactory method. Divisional scientists paid a good deal of attention to egg, nymph, and adult grasshopper surveys, and eventually were able to provide a reliable early warning system to provincial control officers.

However, many entomologists recognized that applying insecticides without due care also destroyed natural parasites of many target and other insects. Consequently, several entomologists, among them A.D. Pickett of Kentville, Nova Scotia, began thinking in terms of integrated pest management with the objective of finding ways to minimize the application of insecticides, thus maintaining the insect predators and parasites but still controlling undesirable populations. Another method of achieving the same goal was to artificially rear parasites and release them at an appropriate time to destroy target insects. It was here that the laboratory at Belleville with its secure rearing rooms came into play. In 1938, for instance, the laboratory reared 175 million parasites representing 14 species for release in eastern provinces against the European spruce budworm (an increase from 2.5 million raised in 1935!).

During the war, large shipments of grain and flour suffered damage from the hairy spider beetle, from the rusty grain beetle, and from various forms of mites. The Stored Products Insect Section gave assistance to the Plant Protection Division in inspecting storages and advising on control methods. The war made it necessary for the grain that was successfully shipped across the Atlantic to be delivered with as little infestation as possible.

Mr. H.G.M. Crawford, Dominion Entomologist from 1943, stepped down in February 1950. He had been with the department on a permanent basis since 1920. Dr. Robert Glen, whom we will meet again later, had been an adviser to the director on entomological matters, became Chief, Entomology Division. He retained this position until promoted to associate director, at which time Dr. B.N. Smallman replaced him. Smallman was born at Port Perry, Ontario. He was educated at the universities of Queen's, Western Ontario, and Edinburgh. In 1941, he joined the staff of the Board of Grain Commissioners as entomologist, transferring to the Insect Laboratory, Winnipeg, in 1945. Six years later Smallman was appointed Head, Entomology Section in the newly established Science Service Laboratory at London, Ontario. Smallman's major contributions were in the control of insects infesting stored products and the administration of research.

ANIMAL PATHOLOGY DIVISION

The transfer of animal pathologists from the Health of Animals Branch to Science Service did not immediately cause any great change in the veterinary research programs at laboratories in Hull, Quebec, Lethbridge, Alberta, and Saanichton and Milner, British Columbia. The main research programs dealt with equine encephalomyelitis (sleeping sickness) of horses, distemper of foxes, infectious abortion, and mastitis of cattle, as well as a number of other less prevalent diseases. The thrust in all the research was toward prevention rather than cure. In many instances, the division developed highly successful vaccines which, when used properly, gave good protection.

A serious outbreak of foot and mouth disease occurred in Saskatchewan in February 1952. The Health of Animals Division of Production Service had to destroy 1343 cattle, and a number of sheep, swine, and poultry. The disease was brought under control on 3 May, but 42 infected and contact farms remained under quarantine until mid-August 1952. The Animal Pathology Division, which had the responsibility for diagnosing contagious diseases, and the Health of Animals Division, which administered the regulations under the Animal Contagious Diseases Act, were in two separate services—Science and Production. It became apparent to the Minister, the Honourable James Gardiner, that only one person should be responsible for the administration of the diagnosis and control of such serious diseases. This decision followed a lengthy session (30 April to 16 June 1952 involving 465 pages of evidence) of the Standing Committee on Agriculture and Colonization of the House of Commons (22). As a result, the Animal Pathology Division was moved from Science Service to Production Service in 1952 and both came under the Veterinary Director General, Dr. T. Childs.

PLANT PROTECTION DIVISION

This division was moved to Science Service on 1 April 1942, having been with Production Service for only 4 years. At the time of the 1942 move its chief, L.S. McLaine, became chief of the Entomology Division, replacing A. Gibson, who retired. Mr. W.N. Keenan replaced McLaine as chief of the Plant Protection Division. The reason stated for the move was that the work of plant protection officers was more closely related to that of entomologists and plant pathologists in Science Service than to that of inspection officers in Production Service. In administering the regulations under the Destructive Insect and Pest Act, the division had plant inspectors at 12 major ports of entry from coast to coast and a plant fumigation station in Montreal. Supervision of the production of certified seed potatoes required that plant pathologists be located at Charlottetown, Kentville, Fredericton, Sainte-Anne-de-la-Pocatière, Guelph, Winnipeg, Edmonton, Vancouver, and Saanichton.

A CHANGE OF DIRECTORS

Like many others during the war years, Swaine remained in service well past the usual age of retirement, leaving after war's end in 1946, when he was in his 68th year. He had made an outstanding contribution to forest entomology in Canada. He was the force behind the establishment of Science Service in the department. Many in the service wondered who would replace him. Some of them took action.

The story is told of two keen entomologists, one from Saskatoon and the other from Lethbridge (Glen and Farstad), discussing Swaine's retirement as they rode a CPR transcontinental train westbound from meetings in Ottawa. By the time they neared Winnipeg they had decided that the right choice for the new director of Science Service was currently residing in that city. Their choice was the director of farm services for the North-West Line Elevators Association, Dr. Kenneth W. Neatby. Neatby was well known to scientists of both Science Service and Experimental Farms Service. Following graduation in plant breeding from the University of Saskatchewan, he had worked at the Dominion Rust Research Laboratory in Winnipeg for several years. As with many other breeders and pathologists in Canada, he had taken postgraduate studies under Drs. Hayes and Stakman at the University of Minnesota. For 5 years prior to joining Line Elevators, Neatby had been professor and head of the Field Crops Department at the University of Alberta. Accordingly, the two scientists detrained in the windy city to learn whether Neatby was aware of the vacancy and whether he might be interested in filling it himself. His response was cautious but tended to be positive. Neatby was interested. Events unfolded favorably and Science Service, the department, and Canadian agriculture were all the richer that Neatby was chosen in January 1946 as the second director of Science Service, under the Honourable James G. Gardiner, Minister of Agriculture, and Dr. G.S. Barton, Deputy Minister of Agriculture.

Neatby emigrated from London, England, with his parents, four brothers, and three sisters in 1906. The family lived at Earl Grey, Saskatchewan, for almost 2 years then, when Neatby was 8 years of age, homesteaded near Watrous. Their lives were not easy, for his father, a medical doctor by profession, learned to farm alongside his sons. It is said (32) that the Neatby childrens' early education came from their parents' love of learning and the 3000 books their father brought to the Canadian frontier. Neatby worked as a summer plotsman at the University of Saskatchewan while upgrading his academic standing. He was encouraged to pursue a career in genetics by one who was to become a lifelong friend and colleague, Cyril Goulden. Neatby was a gifted singer, golfer, and a genial, unforgettable host.

Following World War II, Canada entered a stimulating period of advancement and economic boom. The nation moved from a position of relative insignificance to that of a middle-level world power, replete with its own Nobel Peace Prize winner (Lester B. Pearson) and eventually its very own flag. Research reached new levels of public favor, having contributed so dramatically to wartime successes, that there was unprecedented postwar support for the enlargement of government research agencies. Neatby and his associates found themselves in the right place at the right time.

Many war veterans took their discharge credits in the form of education. Some completed their secondary schooling and went on to university, graduating in agriculture or biological sciences, and were subsequently hired by Science Service or Experimental Farms Service. In addition, the government of Canada formulated a generous educational leave program that encouraged employees who had their first degree to study at postgraduate schools in subjects related to their employment. Many young scientists took advantage of this program and used their research projects as theses material for advanced degrees.

REGIONAL LABORATORIES

Under Neatby's guidance Science Service grew and matured. After spending a few months familiarizing himself with the service he acted promptly and positively. On 22 August 1946, he sent a memorandum to all laboratories and offices asking for the viewpoints of their staffs on the future development and expansion of Science Service. He prompted them by asking questions such as:

- should we perpetuate the practice of individual laboratories of plant pathology, entomolgy, and chemistry, or should we aim at regional laboratories each with several disciplines?
- should Science Service laboratories be at or near universities or should they be close to Dominion Experimental Farms?
- would it be advantageous to have regional administration or is the present centralized administration satisfactory?

Early in 1947, Neatby brought Dr. W.E. van Steenburgh back to Science Service from the armed forces as his research adviser. Dr. Van, as he was fondly known, had joined the Entomological Branch in 1926 at Chatham, Ontario. In 1929, he moved to Belleville where he helped with the expansion of that entomology laboratory and also introduced parasites of a number of fruit insects. As a result of his Belleville contributions he was placed in charge of the fruit insect research in western Ontario, with headquarters at Harrow. He was an excellent administrator as well as being fully knowledgeable of how research operated. Van Steenburgh was appointed Associate Director, Science Service, in 1949 with the prime task of assuring that scientists in the service had adequate laboratory, library, and office accommodation.

Van Steenburgh's first task was to review and condense the answers received by Neatby from his 22 August questionnaire. This he did, and on 21 February 1947, Neatby sent a copy of van Steenburgh's paper to all officers in charge and to Deputy Minister Barton. Although there was some regional divergence of views, van Steenburgh said that generally everyone agreed on the following:

- there needed to be cooperation and integration of scientific effort within the service;
- a streamlining of administrative machinery would improve both the research and the administration of the service;
- well-equipped, centrally located regional laboratories were required on university campuses or on experimental farms if the latter were more advantageously situated;
- regional laboratories required land under their direct control for experimental work;
- close association was required between regional laboratories and field laboratories;
- all laboratories should be directly responsible to offices in Ottawa rather than have some report to regional laboratories, thus placing an extra administrative layer in the system; and
- it was essential that the closest working liaison and cooperation be maintained with experimental farms.

Drs. Neatby and van Steenburgh prepared a series of recommendations that became the policy for administering Science Service. The main thrust was the absolute necessity for close cooperation among all disciplines working on similar problems. On page 11 of his report, when referring to related research done by Experimental Farms and Science Services, van Steenburgh says that "the logical answer ... [is that] they should be working under unified direction, but ... this ... appear(s) impractical because of major reorganizational difficulties." Eleven years later the difficulties would be overcome.

Barton's reaction was favorable. Indeed, he said van Steenburgh had produced "a number of the most refreshing proposals" he had seen for some time. He thought that parts could be applied to other services.

The first action Neatby took was to bring divisions closer together in their research work. This was not easy, particularly when entomologists, pathologists, chemists, and bacteriologists were controlled by their respective divisional chiefs in Ottawa, and when they frequently occupied laboratories in different buildings, even though on the same experimental farm or university campus. Neatby decided that the best method of achieving this goal was to establish Science Service Laboratories, which would have scientists from various divisions, as needed, for the problems at hand. Regardless of discipline, and hence division, each scientist would be responsible to the laboratory director and not to the chief of the division as had been the case since 1910 when the first entomological laboratory was established at Bridgetown, Nova Scotia.

An opportunity presented itself when the Department of National Defence in Ottawa vacated its Records Building on the Central Experimental Farm. On 27 September 1947, two units of the Chemistry Division moved from their cramped and inadequate quarters to that site. Shortly after, the Entomology Division moved to the same building from the Confederation Building on Wellington Street. These two groups kept their divisional identity but Neatby now had most of the Ottawa Science Service staff on the same property.

Harrow was chosen as the test case for a combined Science Service Laboratory. At Harrow, there were only entomologists and plant pathologists, both were located on the Experimental Station, and both served the same flourishing vegetable and tree fruit industry. In 1948, Dr. L.W. Koch, the officer in charge of the plant pathology laboratory at Harrow, was given administrative responsibility for the first Science Service Laboratory. He worked closely with Dr. H.R. Boyce, who headed the entomological group. The laboratory occupied a renovated two-storey house made available to them on the Experimental Station. This proved to be a successful test, consequently Neatby was confident with his decision.

The next case was different. Dr. W.C. Broadfoot, a plant pathologist at the Edmonton laboratory, moved to Lethbridge to bring the science service laboratories together there. He had no place to accommodate Mr. C.W. Farstad and other crop entomologists who occupied offices and laboratories in the Lethbridge Post Office, nor Mr. R.H. Painter and other livestock entomologists who had accommodation at the Experimental Farm, and likewise the plant pathologists, some of whom would transfer from Edmonton. He needed a building and he needed it quickly. Instead of waiting for one to be built, he was able to move a surplus hospital from the Lethbridge airport. In March 1947, sod was turned on the Experimental Station and by December 1948 Dr. M.W. Cormack from Edmonton, head of the new Plant Pathology Section, moved into his laboratories. The other disciplines followed as the building progressed. Included among the imports were Mr. A.W. Platt and cereal breeders from the Swift Current Experimental Station who came to work with entomologists in the development of a wheat resistant to attack from wheat stem sawfly. This was the second instance of Science Service and Experimental Farms Service scientists being housed in the same building, the first being at Winnipeg.

The third case was different again. Neatby, after an initial proposal from Mr. W.A. Ross, who for several years was coordinator of the pesticide work for the Division of Entomology, decided that a special institution should be organized to study pesticides comprehensively. The campus of the University of Western Ontario at London was chosen as the site, because Neatby was convinced that the ties between scientists at universities and research workers in the Department of Agriculture should be strengthened. In London, Neatby started with a clean sheet. Van Steenburgh had the task of designing the general layout of the laboratories and offices. He recruited Dr. R. Glen from Ottawa, and Dr. L.W. Koch from Harrow. They conferred with Dr. W.H. Cook, director of the Division of Biological Sciences, NRC, and visited other research centers to glean ideas. The final design was novel, efficient, and economical. In subsequent years, the basic module became known as the "van Steenburgh unit" and was used for laboratories at many other locations. The London Science Service Laboratory was opened in 1951, with Dr. H. Martin from the Long Ashton Research Station near Bristol, England, as its director.

In September 1947, the same month that Science Service occupied the Records Building, Neatby wrote to Barton suggesting that a departmental

73

statistical advisory service be organized. Neatby said that both the Chemistry and
the Entomology divisions had asked to employ biometricians, but he felt that a
departmental service would be most effective in serving everyone. He noted that
Dr. C.H. Goulden, a plant breeder at Winnipeg, provided Experimental Farms
and other services with statistical advice. Goulden's textbook was the Canadian
authority on the use of statistical analysis of agricultural research data and he
taught the subject at the University of Manitoba. Barton agreed with Neatby,
asking him to confer with Archibald on how best to implement the proposal.
Dr. G.B. Oakland was appointed as the first chief of the Statistical Research
Service in 1950.

FOREST BIOLOGY DIVISION

Research on the protection of forest trees was divided between the Entomology Division and the Botany and Plant Pathology Division. J.J. de Gryse, a
forest entomologist, had moved from the laboratory at Indian Head to Ottawa in
1925 and in 1934 he succeeded Swaine as Chief, Forest Entomology, a division
of the Entomological Branch. After World War II, research in forestry increased in
both entomology and plant pathology.

In mid-December 1948, the Department of Mines and Resources proposed
that forest entomologists and forest plant pathologists be moved from the
Department of Agriculture to the Department of Mines and Resources. Drs.
Barton, Neatby, and Taggart met with Dr. H.L. Keenleyside, Deputy Minister of
Mines and Resources, and two of his advisers to discuss the proposal. At the
time, Taggart was an adviser to Deputy Minister Barton. Neatby reported that
neither Keenleyside nor his advisers understood what would be involved in such
a transfer. Nevertheless, they argued for the move because, they said, "the
Department of Agriculture was already big enough."

The debate continued and on 21 February 1950, Taggart, who by this time
was deputy minister, brought the Honourable J.G. Gardiner into the discussion.
Taggart had asked Glen to prepare a counterproposal suggesting that the
Forestry Branch, with the exception of research on forest products, be moved
from the Department of Mines and Resources to the Department of Agriculture.
Taggart proposed that before any decision was made, the whole matter be put
before the Privy Council Committee on Scientific and Industrial Research. In
June 1951, Taggart wrote to Major General H.A. Young, the new Deputy
Minister of the Department of Resources and Development, noting that the
subject had been dropped. The Department of Agriculture had again survived
inroads and was still intact.

Neatby then organized a Forest Biology Division by bringing the two groups
together under de Gryse in 1951. This action presented the forestry industry with
a single organization, and encouraged cooperative research between forest
entomologists and forest pathologists. Several modern laboratories were built in
support of its research, but the most unique one was the Forest Insect Pathology
Laboratory (1950) at Sault Ste. Marie, Ontario, which was built by the govern-

ment of Ontario and staffed by the government of Canada. Here, highly sophisticated facilities were installed to study viruses and other insect pathogens under controlled atmospheric conditions, with elaborate security provisions against the escape of such organisms. So novel was this development that it was discussed in detail with counterpart American authorities before implementation.

De Gryse retired in 1952 and Dr. M.L. Prebble became chief of the Division of Forest Biology. Prebble was born in Saint John, New Brunswick, received his education at the universities of New Brunswick and McGill, and then organized a regional forest entomology laboratory in Victoria, British Columbia. Five years later, in 1945, he transferred to Sault Ste. Marie as officer in charge. He was the fourth recipient of the Entomological Society of Canada Medal for outstanding achievement. Under Prebble's leadership and thorough knowledge of forests and forest insects, the division developed national programs of surveys, research, and advisory services relating to forest pathology and forest entomology.

ASSOCIATE COMMITTEES

Since December 1934, the Department of Agriculture and the National Research Council (NRC) had jointly sponsored Associate Committees on research dealing with cereal problems. In December 1950, Dr. G.A. Ledingham, Director of the NRC Prairie Regional Laboratory, Saskatoon, proposed to Neatby that Science Service and Experimental Farms Service join with NRC in support of a "conference of western animal science investigators." Animal research scientists at universities in Western Canada had been pressing NRC for such a conference because of the values they saw coming from similar meetings in various other disciplines. A series of letters was exchanged among Drs. Neatby, Taggart, Mackenzie (president of NRC), and Hopkins (newly appointed director of Experimental Farms Service). Taggart agreed that four departmental scientists could attend a conference. It came to light in some of the correspondence, however, that there was "a feeling in some quarters that the research activities of the department ought to be transferred to NRC or to a new organization . . . separately administered within the department." Taggart himself felt strongly that the department's leadership in agricultural research "and the publicity associated with it" should not go to any other organization. As a result, nothing materialized and agricultural research progressed as before.

PROGRESS AND COOPERATION

Before van Steenburgh left Science Service in 1956 to become Director General of Scientific Services in the Department of Mines and Technical Surveys, he had built 20 laboratory complexes at places such as St. Jean (1950), London (1951), Corner Brook (1952), Maple (1953), Guelph (1955), and Saskatoon (1957), and took pride in the fact that nearly a quarter of the Science Service professional staff were doing their research from joint laboratories. The main Science Service building on the Central Experimental Farm, facing Carling Avenue, was opened the year he left. Actually, it was an addition to the Department of National Defence records building. Dr. Van finished his career as Deputy Minister of the Department of Mines and Technical Surveys. Dr. Robert Glen, chief of the Division of Entomology, replaced van Steenburgh as Associate Director to Neatby.

Over the years, Glen had a remarkable influence on entomology, on agricultural science in Canada, and on science as a whole throughout the world. He moved to Canada from Scotland at an early age and received his higher education from the universities of Saskatchewan and Minnesota; at both institutions he was an outstanding scholar. His research at the Saskatoon laboratory, where he started as a student, dealt with the taxonomy, biology, and control of wireworms. He moved to Ottawa in 1945 as coordinator of research in entomology, where his administrative and diplomatic abilities became apparent. Glen holds honors from several universities and technical societies. He was the third recipient of the Entomological Society of Canada Medal for outstanding achievement.

In February 1958, Neatby wrote to all the Science Service Laboratory directors and reemphasized their duties and the conditions under which their scientific staff should operate. Before the development of joint laboratories in 1948, scientists reported through their officers in charge to the appropriate divisional chief—entomologists to Smallman, chemists to Emslie, bacteriologists to Katznelson, and plant pathologists to Hanna. In future, Neatby said, each scientist would report through the section head to the laboratory director, and each laboratory director would report to Neatby. An exception to this new ruling was made with the Forest Biology Division, where each laboratory retained a direct link to the chief of the division in Ottawa.

Divisional chiefs became technical coordinators on the staff of the Science Service director. They retained their control over single discipline laboratories, but were staff advisers for the regional laboratories. Neatby, the administrator, showed he understood that since plant and animal problems originated in the fields and forests, they could best be identified and solved by those who worked in the fields, the forests, the barns, and the greenhouses. The new scheme was nicely taking shape.

The joint regional laboratories encouraged a team approach. At Winnipeg's Rust Research Laboratory and at Saskatoon's Forage Research Laboratory, staffs of the Experimental Farms Service (plant breeders) and Science Service (plant pathologists) had worked closely together for years. They made excellent progress in their efforts to keep the agricultural industry provided with rust resistant wheats and forages suitable for the Canadian Great Plains. Neatby's creed was: Feed success and starve failure. The laboratory directors he appointed were excellent scientists who also had good administrative ability and he achieved a high rate of success.

The time was ripe for change. Before describing what happened in 1959, we must follow the development of the Experimental Farms Service through the dry thirties, the war, and its exciting postwar development.

Chapter 8
Activities of Experimental Farms
and Stations
1937–1958

The 40 experimental farms and stations, the nearly 200 illustration stations, and the Ottawa-based divisions continued, despite the 1937 separation, to serve a large and appreciative farming population. Deputy Minister Taggart maintained the open-door policy begun when he was superintendent of the Swift Current Experimental Station. The director saw the wisdom of having superintendents reside on their stations: it encouraged farmers to visit the stations to further their knowledge and to attend the stations' many field days. Sometimes on a warm Sunday afternoon, up to 2000 visitors would enjoy the picnic facilities. The Minister of Agriculture was a dominant figure in the government of the day with these well-established links to the grass roots of Canada.

During almost the whole of these two decades the Department of Agriculture had but one minister, the very well informed and active James Garfield Gardiner. Gardiner was born at Farquhar, Huron County, Ontario. Moving west, he attended secondary school in Clearwater, Manitoba, then graduated from the University of Manitoba. In 1904, he taught school in Lemberg, Saskatchewan, and started to farm in 1916. Saskatchewan elected him to its Legislature in 1914, where he became Minister of Highways in 1922 and Premier from 1926 to 1929 and again from 1934 to 1935. In 1935, he was elected to the House of Commons and appointed Minister of Agriculture, which portfolio he held until 1957.

Dr. J.G. Taggart was deputy minister to Gardiner during the postwar decade of 1949–1959. In Taggart, Gardiner had a person of broad knowledge and experience. He was born in Nova Scotia and trained at the Ontario Agricultural College, from which he graduated with a University of Toronto degree in 1911. He spent the next 10 years as an agricultural representative in Ontario, and then became principal of the Vermilion School of Agriculture in Alberta. Later, he worked with the tractor sales division of the Ford Motor Company in Regina, Saskatchewan. In 1921, Taggart was asked by Deputy Minister Grisdale to organize the Experimental Farm at Swift Current. After a job well done, he entered provincial politics in 1934 and was made Minister of Agriculture for Saskatchewan, a portfolio he held for 10 years. Starting in 1939, he held a number of appointments related to the war effort and federal agricultural boards until he was selected as deputy minister, replacing Dr. Barton in 1949.

The third person to play a key role in the activities of the Experimental Farms Service during this period was its director, Dr. E.S. Archibald. He was a hardworking and meticulous person, who had an instant recall memory and an encyclopedic knowledge of Canadian agriculture from coast to coast. With his vibrant and cheerful personality he was an excellent administrator and a strong leader. Archibald, therefore, had a prodigious impact upon the development of Experimental Farms during the depression, World War II, and the postwar era. He raised the educational standards of experimental farm staffs by hiring scientists with advanced degrees and by promoting educational leave for others. He was one of the principal architects in the formation of the Prairie Farm Rehabilitation Administration (PFRA), managing its operation until 1937. Under his direction 28 new experimental stations and laboratories were organized and staffed, including the Rust Research Laboratory, Winnipeg, the Forage Crops Laboratory, Saskatoon, and the Soils Research Laboratory, Swift Current. A magnificent snowcapped mountain of 8400 ft (2500 m) overlooking the experimental station at Mile 1019 on the Alaska Highway was named Mt. Archibald by the Geographic Board of Canada on 13 March 1947, commemorating a man of remarkable strengths. Archibald retired in 1951 after 39 years' service. Under the sponsorship of the Food and Agriculture Organization he went to Addis Ababa, Ethiopia, for 2 years where he advised their government on agricultural development.

HORTICULTURE

Mr. M.B. Davis, Dominion Horticulturist, was another force to be reckoned with. He too was a Maritimer, having been born in Yarmouth, Nova Scotia. He graduated from Macdonald College, with postgraduate work at the University of Minnesota and the University of Bristol, England. He managed a farm and was secretary of the United Fruit Company in Nova Scotia before joining the Horticulture Division under Mr. W.T. Macoun in 1914, whom he succeeded in 1933. Davis built one of the strongest divisions in Experimental Farms; a reason for this was his conviction that the growing of fruits and vegetables on all farms was both profitable and necessary for the well-being of the farm family. In this respect his philosophy echoed that of William Saunders.

Early in this period, Davis had a number of cold-storage rooms built in the Horticulture Building. These were to be used for experiments in low temperature storage of fruit and the study of methods of preserving fruit and vegetables by freezing. Because this was the first research done in Canada on frozen foods, the Honourable Eugene Whelan, Minister of Agriculture, placed a commemorative plaque on the Horticulture Building to that effect in 1982. In 1938, the division made extensive tests on many varieties of peas grown in the Gaspé district, the muck soils of Quebec, in the Port Arthur (Thunder Bay) district, and at Ottawa. The Telephone variety grown in the Gaspé District proved to be outstanding for freezing and was used as the standard of quality against which all other varieties were tested. The Horticulture Division also started research on gas or controlled atmosphere storage of fruits, which today provides Canadian consumers with high-quality apples throughout the year.

ILLUSTRATION STATIONS

Under Mr. J.C. Moynan, the Illustration Stations Division was at its peak just before the outbreak of World War II. Moynan was born in Sainte-Brigide, Quebec, in 1889, and, following graduation from McGill University, was involved with soldier reestablishment until joining the Illustration Stations Division in 1921. He was appointed chief of the division in 1928, occupying that position until retirement in 1953. Table 8.1 shows that in 1938 there were 192 illustration stations and 51 district experimental substations. All were under the direct control of illustration station supervisors, with headquarters at 17 experimental farms and stations. The district experimental substations were privately owned farms totaling over 25 000 acres (10 000 ha) in the drought areas of Manitoba, Saskatchewan, and Alberta. They were financed through PFRA but managed by Illustration Stations Division. The division organized a few additional substations to study special problems such as gray wooded soils in Alberta. The objectives at the substations were to test the results of experiments to control soil drifting under commercial farm conditions. Methods of strip farming and cultural practices to create a lumpy condition on the soil's surface were found to be effective.

Table 8.1   Illustration stations

Year	Illustration Stations	District experimental substations (PFRA financed)	Total
1915	21		21
1920	64		64
1925	142		142
1930	207		207
1935	184	39	223
1938	192	51	243
1940	171	46	217
1945	159	50	209
1950	162	54	216
1955	—	—	227
1958	211	26	237

Although office space and services were supplied by the superintendents of the 17 experimental farms, the illustration station supervisor was directly responsible to the divisional chief in Ottawa, not to the superintendent. The system resulted in some ludicrous personal relationships. At one experimental farm the superintendent and the illustration station supervisor chose to communicate with each other during working hours solely via written notes. On weekends, however, differences were set aside while they partnered a Saturday evening game of bridge.

The supervisors were generalists and had to know what research was being done in all disciplines on the experimental farms in their zone. They needed to know the results of the experimental work, and above all they had to translate the results into practical action. They then advised illustration station operators on how to test the new theories. Finally, they needed to be extension specialists because during July and August of each year, in cooperation with the provincial district agriculturist, a field day was held at each illustration station to show neighboring farmers the practical application of the year's work. Each year about 175 field days were held throughout Canada, attracting over 25 600 people.

Other divisions had their own systems of experimental substations. The Horticulture Division had officers of its Ottawa staff seconded to departments of horticulture in faculties of agriculture at five universities. It also seconded officers to the Ontario Horticultural Station at Vineland and the Experimental Station at Morden, Manitoba. In addition, the division was directly responsible for the total operation of the horticulture experimental stations at Smithfield, Ontario, and Sainte-Clothilde, Quebec. The Cereal Division had the Dominion Laboratory of Cereal Breeding, which was located on the campus of the University of Manitoba, under its direction. It was housed with the Dominion Laboratory of Plant Pathology, which was responsible to the Division of Botany and Plant Pathology. The two groups worked closely together and considered themselves to be integral parts of the Dominion Rust Research Laboratory. A similar situation existed between the Division of Forage Crops and the Division of Botany and Plant Pathology at the Dominion Forage Crops Laboratory, located on the campus of the University of Saskatchewan. The pilot flax mill at Portage la Prairie, Manitoba, was controlled by the Fibre Division. Although the Soils Research Laboratory, built and financed by PFRA, was located on the Experimental Station, Swift Current, it was responsible to the Field Husbandry Division. In addition, the soil survey units in each province were directly responsible to the Field Husbandry Division. The Bee Division retained a scientist at the Experimental Farm, Brandon, Manitoba. Finally, the Division of Animal Husbandry was responsible for the cooperative work with the Department of the Interior on cattalo (a cross between domestic cattle and buffalo, see Part II of this history) at the Buffalo National Park, Wainwright, Alberta, until 1950. The only divisions without any direct responsibilities for scientists or stations outside Ottawa were those of poultry and tobacco, although the Tobacco Division at one time (1912) had been responsible for its own outside stations.

ANIMAL HUSBANDRY

The appointment of a new chief for the Division of Animal Husbandry was made in 1951 in the person of Dr. H.K.C.A. Rasmussen. Born in Illinois, Rasmussen graduated from the universities of Manitoba, California, and Iowa State. He joined the Experimental Farms and Stations at Lethbridge in 1930 as head of the animal husbandry section. In 1949, he went to South Dakota State College as professor and head of its Animal Husbandry Department. Rasmussen was a tower of strength in the division. He continued the policy started by Muir of attracting well-trained scientists such as G.M. Carman, W.J. Pigden, and C.G. Hickman to his staff. Rasmussen restructured the divisions of Animal Husbandry and Poultry into one division in 1958, following the retirement of H.S. Gutteridge, Chief, Poultry Husbandry Division. Hopkins' recommendation to form a Division of Animal Industry was never acted upon. Nevertheless, Rasmussen did appoint Dr. J.A. Elliott, a dairy bacteriologist.

SOIL SURVEYS

The first people to formally survey Canadian soils were chemists from the Ontario Agricultural College (31) in 1914. Other provinces, through their provincial departments of agriculture or their universities, made reconnaissance surveys from time to time, but no attempt was made to put the system on a national footing until 1935, when PFRA money became available. The "cultural vote" of PFRA provided most of the funds for the soils departments of the universities of Alberta, Saskatchewan, and Manitoba to conduct surveys in their own provinces. E.S. Archibald must be given credit for instituting the National Soil Survey Committee. He was able to bring the provincial departments of agriculture, the soils departments of university faculties of agriculture, and the Experimental Farms Service together to form a unified national soil surveys program in the early 1940s. To begin with, Dr. E.S. Hopkins, Chief, Field Husbandry, Soils, and Agricultural Engineering Division, and later, in 1946, Dr. P.O. Ripley, supervised the program because the funds were administered by that division.

Hopkins was born in Lindsay, Ontario, in 1890, graduated from the Ontario Agricultural College, Guelph, and the University of Wisconsin. He was an agricultural representative in Peterborough, Ontario, for 3 years before teaching at the Alberta schools of agriculture at Vermilion and Olds. In 1920 he secured the position of Dominion Husbandman and Chief, Division of Field Husbandry, Central Experimental Farm. Archibald assigned him the added duties of Associated Director, Experimental Farms Service, in 1938. He relinquished his divisional chief's responsibilities in 1946. Hopkins followed Archibald as Director, Experimental Farms Service, in 1951.

Ripley was born at Port Perry, Ontario, and did his university studies at Guelph and at Michigan State College. He spent 3 years in the division at Ottawa before going to the Experimental Station, Lennoxville, Quebec. Five years later, in 1931, he returned to Ottawa in the Field Husbandry Division. Ripley followed Hopkins as Dominion Field Husbandman and Chief, Field Husbandry, Soils, and Agricultural Engineering Division.

The provincial headquarters for each survey team was usually situated at the provincial university. Teams consisted of equal numbers of provincial and Experimental Farm employees. The whole operated as a unit and the person in charge was either a provincial or federal soil surveyor, depending upon the leadership qualities of the people available. By 1944 the program was well in hand with about 125 million acres (50 million ha) having been mapped. The following year, with the cessation of World War II, the National Committee met for the first time. Dr. A. Leahey and Dr. P.C. Stobbe from the Field Husbandry Division were chairman and secretary, respectively. The reasons for bringing the provincial teams together were to decide upon a standard nomenclature for soils throughout Canada, and to follow the same chemical and physical analytical procedures when characterizing soil samples in the various provinces. The system has remained national in scope in all provinces with the exception of Quebec, which withdrew from the scheme almost at its inception in 1938. In 1975 Quebec reentered the program.

FOREST NURSERY STATIONS

The stations at Indian Head and Sutherland, Saskachewan, played an important part in the fight to control soil drifting as waged by PFRA in the late 1930s. These two stations, opened in 1903 and 1913, respectively, by the Department of the Interior, were turned over to the Experimental Farms Service in 1931. Their function was to grow and supply seedlings of many different species of evergreen and deciduous trees and shrubs to farmers on the Great Plains. The farmers, or planters, as the nursery station staff called them, were advised on where to plant, how to plant, and how to maintain their plantings. Trees were used as shelterbelts around farm homesteads, on the windward side of fields where soil tended to drift, and later on the windward side of dugouts to collect snow which, when melted, provided livestock with water. Table 8.2 summarizes the number of trees distributed and the number of planters receiving them between 1937 and 1958. The largest distribution was in 1937 after which the number of seedlings available was reduced because of drought and grasshopper damage. During World War II the number of planters was reduced, but by 1950 over 7000 farmers received about 6 000 000 plants per year.

Table 8.2   Distribution of trees from forest nursery stations to farmers

Year	Trees	To farmers
	(millions)	(thousands)
1937	8.1	7.0
1940	6.7	7.2
1945	4.0	5.8
1950	5.7	7.0
1958	6.3	3.7

Four field shelterbelt associations were organized in 1935 under the sponsorship of PFRA to reduce soil drifting and increase snow retention. At Conquest, for instance, 3 700 000 trees were planted by farmers over the 10-year period from 1937 to 1946. By the end of 1951, farmers had planted 533 miles (860 km) of hedges. The increases in yield as a result of the shelterbelts were calculated by Dunlop (9) in 1950 to be 16 percent for oats and 7 percent for wheat.

As the two stations matured, and particularly when Mr. John Walker became superintendent at Indian Head in 1958, activities increased to embrace more than the production of trees. Walker had received his education at the universities of Alberta and Minnesota. He was horticulturist at the Indian Head Experimental Farm for 5 years and then went to Winnipeg, where he became horticulturist for the Manitoba Department of Agriculture and, later, Professor of Horticulture at the University of Manitoba. In 1942 he returned to Indian Head as superintendent of the Forest Nursery Station. Scientists such as Drs. W.H. Cram and J. Wilner were attracted to the station by Walker's vision. Evergreen and deciduous species had already been tried from all the cold and dry climates of the world. Walker went further. Vegetables and ornamental plants could more readily be grown now that farms had shelterbelts and he recognized that breeding bushes and trees specifically for shelterbelts would be advantageous. A study of why some trees could withstand cold and drought, whereas others could not, would be useful. His expanded and well-trained staff responded to the challenges and were able to better the lot of the prairie farmer.

83

EXPERIMENTAL FARM LABORATORIES

The organization of the Dominion Rust Research Laboratory, Winnipeg, Manitoba, in 1925 was described in Chapter 5. From 1937 to 1958 its scientists developed rust resistant wheats that were widely grown and hence helped prairie farmers maintain their high productivity.

Dr. G.P. McRostie, Dominion Agrostologist and Chief, Forage Crops Division, Ottawa, accepted the position of professor of Agronomy, University of Manitoba in 1929. Archibald searched for the best replacement possible and settled on Dr. L.E. Kirk, professor of Field Husbandry, the University of Saskatchewan. Kirk, born in Bracebridge, Ontario, 5 days before the Experimental Farm Station Act was given royal assent in 1886, had attained international recognition as a grass and legume breeder. Kirk accepted Archibald's offer when it was agreed that the Forage Crops Division would establish a forage laboratory on the campus of the University of Saskatchewan, assume all the forage research, and teach the forage courses. The depression year of 1931 saw the Dominion Forage Crops Research Laboratory established, Saskatchewan students provided with a sound education at a reasonable cost to the university, and the laboratory provided with enthusiastic, competent, student help.

The Soil Research Laboratory located at the Swift Current Experimental Station in 1936, was an extension of the work started in 1922 by Mr. Sidney Barnes, a member of the Field Husbandry Division, Ottawa. It had been thought that capillarity was a factor in the loss of water from dryland soils. This misconception lead to the extensive use of dust mulch and the near destruction of soils within the Palliser Triangle. Barnes proved that the capillarity theory was erroneous. He was designated as officer in charge of the new Swift Current laboratory but unfortunately died in 1935. Dr. J.L. Doughty replaced Barnes. He and his staff made further important discoveries to impede the desertification of prairie soils. The Division of Field Husbandry retained responsibility for the laboratory until 1957 when it became the Soil Section of the Swift Current Experimental Station.

Fruit and vegetable processing laboratories were organized at Kentville, Nova Scotia, and at Morden, Manitoba, during this period. The laboratory at Summerland, British Columbia, had had its beginning in June 1929. At Lethbridge, Alberta, where irrigation made possible an intensive livestock industry, a wool laboratory (the only one in Canada) and an animal nutrition laboratory were set up. Each was staffed with well-trained specialists who paid particular attention to problems of the industry in their region and made themselves available to other parts of Canada as well.

WAR CONTRIBUTIONS

During World War II, divisions, experimental farms, and experimental stations were called upon in many ways to help with the war effort. Note has already been made of the search done by Science Service divisions to find kapok and rubber substitutes. Several farms and stations assisted in this endeavor by growing Russian dandelion for rubber and milkweed for kapok. In cooperation with the Agricultural Supply Board, they also grew large quantities of soybean, sunflower, and rapeseed to be used for oils, both industrial and edible. Horticulturists at many branch stations grew foundation vegetable seed for use in the production of commercial seed because the usual sources had been cut off.

When World War II began in 1939 and the Commonwealth air-training scheme commenced, the Forage Crops Division anticipated a need for information on establishing, enlarging, and maintaining airfields. They therefore collected all available information on these subjects in order to be ready for the questions they knew would arise. Within a year the questions came—and much more. By 1940, officers of the Forage Crops Division and forage specialists at Charlottetown, Scott, and Agassiz, were spending a great deal of their time advising both the Department of National Defence and the Department of Transport on how to prepare, seed, and maintain grass runways as well as grass around building sites. It developed that experimental farm staff were given the responsibility for all phases of turfing programs, which included preparing specifications, calling tenders, letting contracts, and directing field operations. This was no small task, as only four people handled all of Canada. At Ottawa one person spent nearly full time on the project, and the other three officers dealt with all airfields in their respective regions.

Another major contribution of experimental farms during the World War II period was investigational work on methods of dehydrating fruits and vegetables. First, the fruit and vegetable processing laboratories in Ottawa, Kentville, and Summerland, determined how fruits and vegetables might best be preserved. They then made this information available to commercial dehydrators and helped them apply the appropriate methods. In addition, Mr. J.A. Gilbey, a plant scientist from the Horticulture Division in Ottawa went to Labrador on behalf of the Department of National Defence to supervise the production of vegetables there. Although there was little or no soil and the temperatures were low, vegetables were successfully produced under completely artificial conditions, using hydroponics and artificial light, thus relieving an overburdened transportation system of supplying fresh vegetables to northern areas.

NORTHERN STATIONS

After World War II there appeared to be a need to increase food production in Canada. Many Canadians felt that unlimited agricultural land was available in the northern parts of the provinces and in the Yukon and Northwest Territories. It was believed that if Canadians could learn to manage this land it would provide a great new source of agricultural wealth for Canada.

Much was already known about the agricultural potential of the north. Missionaries had moved down the Mackenzie River valley, raising livestock and gardens to provide food. Saunders had crop adaptability tests started in 1905 in coooperation with the North-west Mounted Police and the first permanent northern station was established at Fort Vermilion, Alberta, in 1907. By the time W.D. Albright opened the Beaverlodge, Alberta, station in 1915, test plots were being grown at Forts Smith and Resolution in the Northwest Territories and at Grouard on Lesser Slave Lake, Alberta. Nonetheless, information on the potential of the Yukon and Northwest Territories was often fragmented or based on work done in the northern parts of the Prairie Provinces. With renewed interest in the production potential of the north following World War II, Archibald decided to open two permanent research locations in the territories.

One was established at Mile 1019 Alaska Highway in 1946 by J.W. Abbott. The U.S. Army had just completed construction of the highway and Abbott obtained their heavy equipment needed to start the station. Prior to 1946, Abbott farmed at Fort St. John, British Columbia, where he operated an illustration station for Beaverlodge. Following his retirement in 1956 there were successive officers in charge until Mile 1019 closed in 1970.

The second station was established by J.A. Gilbey in 1947 at Fort Simpson, Northwest Territories, on an island at the confluence of the Liard and Mackenzie rivers. It was located on a small farm that had been operated by the doctor associated with the Department of Indian Affairs. Gilbey was a graduate in horticulture at Macdonald College and had previously been with the Horticulture Division, Ottawa. Gilbey died suddenly in 1955 and the station was then operated by several officers in charge until it, too, was closed in 1970.

The stations at Mile 1019 and Fort Simpson grew test plots as far north as Inuvik and as far east as Yellowknife. When they were made substations of Beaverlodge in 1965 their scope was reduced to provide resources for research in the Slave River lowlands. The objective there was to determine the potential for grazing cattle on the one million acres (400 000 ha) of tall grass growing on a flood plain north of Fort Smith. During their 25 years of operation, these two stations determined the crop and animal production potential of the major northwestern arable areas. They developed basic production systems that could be used by those wishing to commence farming.

The stories have already been told of the openings in 1916 of the experimental stations at Kapuskasing, Ontario, and at La Ferme, Quebec. In these northern areas farm populations increased until 1930, but thereafter, many farms were abandoned. The interest in northern agriculture after World War II also resulted in the opening in 1956 of the last northern station at Fort Chimo, Quebec, near the shore of Ungava Bay. Ten years of research by agronomists clearly demonstrated that the seasons were too short for economic commercial farming. R.I. Hamilton and H. Gasser found that cool season garden vegetables could be grown but required plastic tunnels, mulches to capture and retain heat, and adequate fertilizer. The station was closed in 1965.

NEWFOUNDLAND

Canada expanded its borders in 1949 when Newfoundland was received as the 10th province. Since 1935, the government of Newfoundland had operated a demonstration farm and an agricultural training school. In 1942, the Experimental Farm, Nappan, cooperated with the Newfoundland demonstration farm in conducting plant variety trials and fertilizer tests. Shortly after Newfoundland became a province, Experimental Farms Service bought the demonstration farm near St. John's West and it became the most easterly experimental station in the system. Mr. I.J. Green remained as superintendent of the experimental station until 1956, when Mr. H.W.R. Chancey, head of the soil survey group in Newfoundland, became its superintendent; he later became its director. Chancey remained in that position until he retired in 1984, by which time he was known as "Mr. Agriculture" on the island. For a year he was on leave from the Research Branch to serve as provincial deputy minister of agriculture.

THE PACE ACCELERATES

The professional staffs in both the Experimental Farms Service and Science Service were expanding quickly with veterans who had taken their war gratuities in the form of education. From 1945 to 1955, the number of scientists increased from 370 to a peak of 1249 (see Appendix IV).

As both Experimental Farms and Science Service grew in the 1950s discussions took place between those who were of the view that "fundamental" research should be emphasized and those who favored the "practical" approach. In a letter dated 3 February 1951 to Taggart, Hopkins commented on a paper prepared by Glen, then chief of the Entomology Division, that listed the fundamental research planned for entomologists at the new Science Service Laboratory, London, Ontario. Hopkins agreed with the list but said that there seemed not to be much difference between fundamental and practical research projects. Some people in both Experimental Farms Service and in Science

Service were beginning to realize that all research in agriculture was done to solve problems and ranged, as needed, from fundamental to practical.

Hopkins retired in 1955 at the age of 65 and remained in Ottawa until his death early in 1960. Dr. Cyril H. Goulden, Dominion Cerealist, was chosen as his successor. Goulden, the quintessential scientist, had been transferred to Ottawa from Winnipeg to head the Cereal Division in 1948. He had been appointed as officer in charge and plant breeder at the Cereal Breeding Laboratory in Winnipeg when it opened in 1925. His contributions to cereal breeding and to the advancement of Canadian agriculture were numerous and outstanding, including an early influential textbook on statistical methods in research. Goulden was an obvious and happy choice for the position, and he continued his interest in the activities of the research bench and in the solution of statistical problems, while still administering the organization. Goulden immediately made an astute decision to appoint Dr. J.C. Woodward as his associate director. Woodward was chief of the Chemistry Division in Science Service when the move was made; however, he was familiar with Experimental Farms, for he had joined the Chemistry Division in 1934 before Science Service was formed. As a chemist, he had worked closely with both animal and plant scientists. The appointment gave Goulden an excellent additional line of communication with Science Service. Woodward was to make substantial contributions to future reorganizations.

Replacing Goulden as Dominion Cerealist was Dr. D.G. Hamilton. He was born on a farm near Fredericton, New Brunswick, and graduated from Macdonald College, Sainte-Anne-de-Bellevue, Quebec, and the University of Wisconsin. He joined the Cereal Division in 1938 and later served for 4 years in the Royal Canadian Artillery during World War II. Hamilton had successfully developed varieties of both oats and barley and proved to be a circumspect and thoughtful leader.

Also in 1955, Mr. M.B. Davis, Dominion Horticulturist, retired, having reached age 65. Dr. H. Hill, from the divisional staff replaced him. Hill, a native of Valleyfield, Quebec, was educated at McGill University, Montreal, and London University, England, in plant physiology. He was a potato seed inspector for 2 years before joining the Division of Horticulture in 1927. He was killed in 1959 in an automobile accident while driving between Smithers and Prince George, British Columbia, on business.

Goulden had the opportunity of working with a reasonably new group of divisional chiefs and superintendents of experimental farms and stations. In the 5 years prior to and including 1955 when he was appointed director, there were four new chiefs of divisions appointed in Ottawa and 12 new superintendents appointed in branch farms and stations. Over 35 percent of his senior staff were new appointments; most of them had taken advantage of the veterans' education allowance after the war. A new excitement was in the air, presaging changes to come.

The period from 1937 to 1958 in the history of Experimental Farms and Science Service saw Canadian agriculture change from an industry struggling with semiarid farming conditions and grasshopper infestations to a confident, optimistic, national enterprise. The increase in productivity per agricultural worker had been greater than any other industrial sector of Canada. Agricultural exports more than balanced Canadian food imports. Farmers actively sought the most recent plant introductions, the most up-to-date rations for cattle and poultry, and the newest disease and insect controls from their nearest experimental farm and science service laboratory. These were two decades of extraordinary progress with more to come.

88

Chapter 9
Two Services Become One—
Formation of the Research Branch

The viable seed of an idea to unite Experimental Farms and Science Service was present in 1947 when Neatby, who had been appointed only 2 years earlier, and van Steenburgh formulated a new Science Service policy. The seed may well have developed further in 1948 when Dr. Cyril Goulden moved to Ottawa from Winnipeg as chief of the Cereal Crops Division.

Drs. Goulden and Neatby had been scientists at the Cereal Rust Research Laboratory in Winnipeg from 1926 to 1935, Goulden as the wheat breeder and Neatby as his assistant, working on wheat resistant to rusts. They coauthored several papers, the first of which (16) was published in 1928. Both scientists had graduated in agriculture from the University of Saskatchewan, had taught at their alma mater, and each had received a doctorate in plant breeding and genetics from the St. Paul campus of the University of Minnesota. They were closely associated with the Agricultural Institute of Canada, both as national presidents and as Fellows. They were also Fellows of the Royal Society of Canada and were strongly dedicated to the well-being of Canadian agriculture and to Canada as a whole. The ground was being prepared.

In the fall of 1948, Neatby made a 6-week tour of agricultural research organizations in and around Washington, D.C., and eight northern states in the United States. He then prepared a voluminous report (58 pages) and a summary comparing the American organizations with the Dominion Department of Agriculture. He observed to Deputy Minister Barton that "consideration might well be given to radical changes in our administrative setup." He pointed out, as an example, that in Canada livestock research was distributed among the divisions of Animal Pathology and Chemistry in Science Service, Health of Animals in Production Service, and Animal Husbandry in Experimental Farms Service. He advocated for Canada something "comparable to the Bureau of Animal Industry." Without doubt Neatby was unhappy with the organization of agricultural research in the department, but because Deputy Minister Barton had agreed to the system devised by Swaine, Neatby probably was not going to change it until a new deputy minister was appointed, although he had planted the seed.

The first positive steps taken toward amalgamating Experimental Farms Service and Science Service were in the fall of 1956 when Drs. Goulden and Neatby invited Sir William Slater, the Secretary of the British Agricultural Research Council, to visit a number of Experimental Farms and Science Service Laboratories across Canada. Sir William arrived in Quebec City on 2 May 1957 and spent 6 weeks in the company of Neatby and Goulden, visiting agricultural research units en route to Vancouver. His report, received by Neatby in July, said in essence that excellent agricultural research was being done by the two services despite the dicotomous organization and strongly recommended they be brought together. The report was sent to the deputy minister, by now Dr. J.G. Taggart, who had a sympathetic understanding of agricultural research, after 13 years of experience as superintendent of the Swift Current Experimental Station.

By November 1957 Goulden, too, was firmly of the view that a union should take place. He wrote to Neatby and suggested a scheme calling for a "Bureau of Agricultural Research" that would contain two research services in addition to a separate service unit, which would provide photographic and computing capabilities. In the early years of electronic data processing, computers were both enormous and expensive and the idea of having one on each scientist's desk or even at every station was inconceivable. In Goulden's proposal one service would be for applied research and include such subjects as plant, soil, animal, forestry, and engineering research. The other, a specialized research service, would deal with basic research and include institutes such as entomology, plant pathology, botany, bacteriology, genetics, pesticides, herbicides, biological control, and biometrics. Goulden was concerned that without the specialized research service the basic or fundamental research would be pushed aside by the pressure from practical problems. Neatby reacted quickly and within the week thanked Goulden for his proposal and agreed with it except for the idea of setting up a separate administration for each of the two research services. Neatby was of the view that research in the department should be a mixture of basic and applied. He agreed with the perpetuation of research institutes that would investigate mission-oriented problems in depth. The strictly basic or fundamental research, in his view, should be left to universities, private foundations, and industry.

During the ensuing year the seed to unite the services began to germinate. Progress was made toward amalgamation. Science Service continued with its consolidation by bringing scientists from several divisions together under one director at both Saskatoon and Winnipeg. Consolidation at Lethbridge had been completed a decade earlier in 1947 and at London, Ontario, in 1951. By June 1958 at least two plans, A and B, were formulated for the establishment of a Canada Department of Agriculture Research Service. Plan A was accepted in principle and then was modified many times by a joint management committee of the two services. Lengthy discussions were held within each service and between the two services to define terms such as "institute" and "station." The main points, however, were fully agreed upon. These were that the two services should be united and have a common goal, any overlapping research should be brought together, and a direct reporting line should be established from the scientists to their directors and hence to the director general of the Research Service. The minister, the Honourable D.C. Harkness, and his deputy, Taggart, were fully supportive of Plan A. Indeed, Taggart chaired the various committee meetings of Drs. Neatby, Goulden, Woodward, and Glen.

The 28 June 1958 meeting (7) included Mr. G.G.E. Steele from Treasury Board, and Mr. M.A. Scobie from the Civil Service Commission. Both assured those present that there would be "no major disagreements (from the central agencies) on the basic principles of a Research Service as set out in Plan A." The forestry research interdepartmental relationships posed some difficulties because investigation of forest pathology and zoology problems was being done by the Forest Biology Division in Science Service, whereas research in silviculture and forest management was the responsibility of the Department of Northern Affairs and National Resources. The discussion dealt with the problem of research in forestry having a split administration. Those at the meeting agreed that all forest research except research dealing with forest products and economics should be transferred to the Department of Agriculture. Messrs. Steele and Scobie agreed to ask their principals to present the suggestion to the Committee of the Privy Council on Scientific and Industrial Research.

Glen was assigned the responsibility of preparing a formal submission to Treasury Board and the Civil Service Commission. Taggart recommended that the submission include the objectives of the new service (to be called the Research Branch), a description of the branch, the essential differences between the proposed organization and the organizations it would replace, and the reasons for the reorganization. In addition, Steele requested a second submission to Treasury Board for a change in the vote structure that required approval from Parliament. By 11 July 1958, Glen had prepared a four-page submission titled *Recommendation for the Establishment of a Research Branch*. Characteristically, it was a clear and succinct recommendation. Glen also pointed out to Taggart that the document "might well become (one) of some historical significance." The recommendation is reproduced in Appendix III.

On 12 September 1958, Drs. Goulden and Neatby sent a formal three-page letter to officers in charge of all Experimental Farms and Science Service units advising them that the Minister, the Honourable Douglas C. Harkness, had just released an announcement concerning the amalgamation of the two services, which would become effective 1 April 1959. Up to that point, all negotiations and discussions had been held at Ottawa, although rumors abounded elsewhere that a reorganization was in the works. The ministerial announcement envisaged a Research Branch to incorporate all units of Experimental Farms and Science Service into one efficient research organization. It went on to say that a team approach would be applied to the solution of problems, "something toward which departmental scientists have been trending more and more in recent years but which a unified administration and programming will make easier."

The following main points in the Goulden–Neatby letter expanding upon the ministerial announcement included the need for:
1. greater opportunity for research staff of different disciplines to work together in the solution of agricultural problems;
2. more decentralization of responsibility and authority to regional and institute officers in charge;
3. an opportunity for headquarters staff to devote more time to program planning and development;
4. a program directorate to develop and coordinate the whole research program on a problem rather than on a discipline basis, which was really the most significant change; and

5. research institutes to be formed from the research staffs of the various divisions.

The organizations dealt with in points 4 and 5 would share the two responsibilities previously held by divisions. The program directorate would have a purely staff advisory function, and the institutes would have line functions similar to those of research stations. The forest biology laboratories were least affected by the reorganization. The director of the forest biology program was assigned a line function for the management of the 10 regional forest biology research laboratories. In this regard the Forest Biology Division was unique.

The deputy minister selected the executives for the new organization. They were Neatby as Assistant Deputy Minister (Research), Goulden as Director General, Research Branch, Glen as Assistant Director General (Program), and Woodward as Assistant Director General (Administration). Goulden responded to the deputy and congratulated him upon his choice of Neatby as assistant deputy minister. However, he noted that Science Service staff had been awarded positions one and three, whereas Experimental Farms staff had been awarded positions two and four. Selections were also made for key positions such as program directors and directors of institutes.

Then tragedy struck on 27 October 1958. Dr. K.W. Neatby died of a brain tumor. Early in September he had found he was tiring easily and although he maintained his enthusiasm for the amalgamation scheme, he was not able to spend the time on its development that he felt he should. Needless to say, his death was a great blow to the fledgling organization. It followed that Goulden became the Assistant Deputy Minister (Research), Glen became Director General, Research Branch, Woodward became Assistant Director General (Program), and Mr. S.B. Williams, superintendent at the Nappan Experimental Farm, became Director of Administration.

Formal appeal notices for all the appointments were not issued until 4 December 1958. No appeals were received. The Program Directorate under Woodward was composed of Rasmussen (animals), Hamilton (crops), Smallman (entomology and plant pathology), Ripley (soils), and Prebble (forest biology). By early April 1959, appointments had been confirmed for all associate directors of programs and all directors of institutes and stations.

As with any reorganization there were aftershocks, and adjustments had to be made. The new organization and responsibilities within it were vastly different from those of the two parent services. Responsibility for deciding what research projects should be undertaken, within programs agreed to by the branch, now rested with each station director. Responsibility for planning, executing, and interpreting research now rested with individual scientists. This was in sharp contrast to both previous organizations where the responsibility for projects undertaken and the conduct of the research was vested with the discipline divisions headquartered in Ottawa. It took several months for scientists and headquarters staff to adjust to the change. At the new institutes a strong resistance developed among scientists. Throughout the new branch, old ties and long-established loyalties were shaken. Scientists now reported to a director whose discipline and experience might differ from their own. This was novel for many who had been in Science Service, and they sometimes expected the worst. A major impact was also felt by members of the Research Program Directorate who now had a staff advisory role rather than a line directional role. However, most people worked with a will and Woodward clarified many difficulties that arose during the first 6 months' operation with his memorandum dated 8 October 1959.

As the plans became reality, divisions disappeared and research institutes, mostly in Ottawa, were formed with their staffs. Appendix II summarizes the changes that occurred. Some divisions such as Chemistry from Science Service, and Illustration Stations from Experimental Farms Service were completely absorbed by different units. Other divisions such as Animal and Poultry Science of Experimental Farms Service, and Entomolgy of Science Service remained fairly well intact, forming research institutes with similar names and little change in staff.

At locations such as Nappan, Lennoxville, Vineland, and Agassiz where there were only Experimental Farms or Science Service staff, amalgamation seemed to make little difference except in name changes, many of which came after 1959. Experimental project farms was the new name given to illustration stations, and the responsibility for their management passed to the nearest director or superintendent. The number of experimental project farms was gradually reduced. At other locations such as Fredericton, Harrow, and Lethbridge, where both Experimental Stations and Science Service Laboratories were brought together under one director, amalgamation made possible the integration of scientific staffs and support staffs from the two original services into the new Research Branch. It resulted in the development of multidiscipline research projects. In many cases, this was just an extension of a process that had been growing during the preceding decade. Now it had the formal encouragement of the branch executive and, in most instances the endorsement of the scientists. New programs involving plant pathologists and plant breeders, plant breeders and animal nutritionists, animal nutritionists and veterinary entomologists, and entomologists and soil scientists became commonplace and quickly advanced the team approach to the accomplishment of research goals and the advancement of program objectives.

The well-nurtured seedling would become a fruitful tree.

93

Chapter 10
The First 10 Years of Research Branch
1959–1968

It soon became obvious that Drs. Neatby and Goulden's vision of "strength in union" was an inspired one. It is a scientist's duty to question, and as with any reorganization, a few doubts were raised—some scientists would have preferred to continue reporting directly to Ottawa. There were also some minor problems regarding the integration of both professional and support staffs. Generally, however, the amalgamation was viewed as a definite improvement in the administration of research.

The first structural change in the new branch came after 12 months when, on 1 April 1960, the administration of Forest Biology Program Directorate and its 10 Forest Biology Laboratories from Corner Brook, Newfoundland, to Victoria, British Columbia, were transferred to a newly created Department of Forestry. Previous discussions with Treasury Board and the Civil Service Commission involving forestry had centered around combining all forestry research and placing it within the Research Branch. Other plans prevailed.

Early in April 1962, Goulden, the first assistant deputy minister of the branch, retired. He had enjoyed a long and distinguished career with the department as officer in charge of the cereal breeding laboratory in Winnipeg, lecturer and author of text books on statistics as they applied to agricultural research, Dominion Cerealist in Ottawa, Director of the Experimental Farms Service, one of the two architects in the formation of the research branch, and finally, the first Assistant Deputy Minister (Research) (ADM(R)) of the new branch. Glen, the first director general, was the natural choice as Goulden's successor and the appointment was confirmed shortly after Goulden's retirement.

To fill the vacancy created by Glen's promotion, the position of Director General, Research Branch, was advertised and applications from the department throughout Canada were invited. On 27 November 1962, Mr. S.C. Barry, Deputy Minister of the Department of Agriculture, wrote a memorandum to the staff regarding the advertisement and advised that with the appointment of a new director general there would be a fundamental change in the organizational arrangement of the branch. As ADM(R), Goulden had been responsible for the operation of the Research Branch. In future, the Director General (Research Branch) would assume that responsibility. The ADM(R) would be a member of the departmental executive, and would also assume authority, on behalf of the deputy minister, for biological research within the department. Thus, in a sense, the situation of 1934 to 1937, when Swaine was appointed Chief Scientific Adviser to the department under Deputy Minister Barton, would return. In order to perform his duties as ADM(R), Glen moved from the Central Experimental

Farm to the departmental headquarters located in the Confederation Building, Wellington Street, Ottawa.

The Experimental Farm Station Act of 1886 and each of its subsequent amendments specifically stated that all research done by the department would be the responsibility of the Experimental Farms. However, from time to time, ministers decided that some research would become the responsibility of the Health of Animals Branch, the Economics Branch, the Board of Grain Commissioners of Canada, or the Prairie Farm Rehabilitation Administration. Thus, Glen coordinated the total departmental research effort and liaised with other federal departments on research matters. He was the principal spokesman for the department on research when dealing with provincial departments, universities, industry, and farm organizations, and gained the confidence of the deans of agriculture who invited him to attend their annual meetings. He also concerned himself with developing an appropriate public image of the department and contributed to the development of departmental policies, objectives, and procedures. In 1967, Canada's Centennial year, Glen moved again, this time to the recently opened Sir John Carling Building on the campus of the Central Experimental Farm, to be with the rest of the departmental headquarters staff.

Dr. J. Ansel Anderson, Director, Winnipeg Research Station, won the competition for the position of Director General, Research Branch and moved to Ottawa in February 1963. He had been employed by the National Research Council of Canada from 1931 to 1939, and then by the Canadian Wheat Board as Chief Chemist. He had held the position of Director, Winnipeg Research Station, for almost a year. Anderson was no stranger to the scientific staff of the branch nor to the way in which its scientists conducted their research. He had chaired the Associate Committee on Grain Quality of the National Research Council from 1946 to 1951. The committee was the watchdog for the quality of new cereal varieties to be licensed in Canada and it maintained extremely high standards. Every cereal breeder in the Research Branch knew Anderson. They respected him for his knowledge of agriculture, his writing skill, his power of critical analysis, and his high standard of scientific accomplishment. He held an LL.D. from each of the three prairie universities. He was a scientist's scientist as well as a friend and promoter of Canadian agriculture. Much was expected of the new director general.

Anderson spent the first 6 months learning about the branch by visiting its institutions. During the next 5 years he knew all of them from Newfoundland to Vancouver Island, as far north as Mile 1019 on the Alaska Highway and Fort Simpson at the confluence of the Liard and Mackenzie rivers, and as far south as Manyberries and Harrow; indeed, he visited most of the larger institutions two or three times.

On 3 July 1963, he appointed Woodward as his Associate Director General, and gave the four directors, Drs. Hamilton, Rasmussen, Ripley, and Smallman, specific line functions in addition to their staff responsibilities.

Further alterations resulted from the 1960 Royal Commission on Government Organization headed by J.G. Glassco. The commission, appointed by

Order-in-Council on 16 September 1960, submitted its reports (13) to the government in February 1963, just as Anderson was moving from Winnipeg to Ottawa to assume his new duties. The report commended the Department of Agriculture for the high quality of research and the record of "important and tangible contributions to the agricultural community as a result of its research." The commission went on to agree that the formation of the Research Branch with its centralization of administrative and common services and its freedom for research units to plan and execute their own projects, were major steps forward. However, the commission noted that "the Department appears to have accepted the initial reforms as final." It criticized the organization of the Research Branch for having 54 units reporting to a single office—that of the Assistant Deputy Minister (Research). (The report was written while Goulden was the incumbent of that position.) The final comment of the commission was that many small units were carrying on research in several different areas, lacked sufficient scientific staff, had inadequate equipment, and could afford only limited library facilities. The Glassco Commission strongly recommended that "regional laboratories be formed by consolidating present research units."

Anderson was well aware of the comments made in the commission's report and realized that because of the newly assigned responsibility attached to the position of director general the initiative to make improvements must be his. Thus, on 10 March 1964, he verbally proposed to Deputy Minister Barry, that the span of control of the director general be reduced from 54 to 5 units. To accomplish his goal he planned to retain the central administration and the service and program coordination functions as they were but to divide the remaining portion of the branch into three. Accordingly, the institutes, the eastern section, and the western section would each report to an assistant director general. Deputy Minister Barry agreed to the proposal and on 1 April 1964, Anderson notified heads of all establishments that the revision would become effective on 15 April 1964. He noted that although an assistant director general had been placed between himself and the station directors, and his correspondence with them would be reduced, other means such as regional meetings and visits to research stations on his part would provide ample opportunity for individual consultation. Although not stated in the announcement, it was clear that the three assistant directors general would have their offices at the Ottawa headquarters and represent the geographic areas at branch executive meetings. Rasmussen, an excellent livestock scientist, was selected for the west; Hamilton, a highly experienced plant researcher with a good knowledge of French, was selected for the east; and 5 months later Migicovsky, a world-renowned biochemist, was selected for the institutes.

Dr. B.B. Migicovsky was a native of Winnipeg, and a graduate of the universities of Manitoba and Minnesota. He joined the vitamin assay laboratory at the Central Experimental Farm in 1940, served in the military for 3 years overseas, and returned to the Chemistry Division of Science Service in 1945. Migicovsky's crucible yielded glowing results. His most notable scientific contributions won him international recognition. They were the development of a method for removing strontium 90 and caesium 137 from milk, and improvement in our knowledge of calcium and vitamin D utilization in livestock.

The change also affected the program directors. Ripley retained his function of coordination, but his title was changed to Assistant Director General (Coordination). The title of associate program director was changed to research coordinator. This more clearly indicated their staff rather than their line function.

The response from station and institute directors was immediate and positive. They saw that they would have a strong regional voice at the branch executive level, and they generally welcomed the fact that the person to whom they reported would be responsible for no more than 21 other officers in charge. The breakdown was 22 western stations, 19 eastern stations, and nine institutes.

Two additional appointments should be noted. In the fall of 1959, Deputy Minister Barry was so impressed with S.B. Williams' administrative capabilities and encyclopedic knowledge of Canadian agriculture, that he persuaded Williams to join his staff at departmental headquarters. Williams eventually succeeded Barry as deputy minister. In 1959 Dr. D.S. Laughland won the competition for the position of Director of Administration vacated by Williams. Laughland was born in Collingwood, Ontario, where his father was a district agriculturist. His family moved to Guelph at the time Laughland entered school. He graduated from the Ontario Agricultural College, Guelph, served in the Royal Canadian Navy, and returned to the Chemistry Division, Science Service, Ottawa, as a biochemist in the new vitamin unit in 1945. In 1963, when Laughland transferred to the Civil Service Commission, Anderson persuaded Dr. R.A. Ludwig, Director, Plant Research Institute, to apply for the position. Ludwig did, and was appointed as Assistant Director General (Administration). He moved to a suite of offices adjacent to Anderson, thus facilitating close collaboration on major administrative problems.

Ludwig, a native of Calgary, received degrees in plant pathology from the University of Alberta and from Macdonald College. He joined the faculty of Macdonald College in 1940, and in 1951 went to the Science Service Laboratory, London, Ontario, under Dr. H. Martin. Upon amalgamation of Experimental Farms and Science Service in 1959, Ludwig was appointed Director, Kentville Research Station. He moved to Ottawa when the directorship of the Plant Research Institute became available in 1961. Anderson is alleged to have said that Ludwig was an intelligent, hardworking scientist, with a strong personality, and a fertile imagination.

Most ministers took a keen interest in the activities of Research Branch as they had in those of its predecessors. In 1963, the Honourable Harry Hays, dairy farmer, auctioneer, and ex-mayor of Calgary, succeeded the Honourable Alvin Hamilton as minister. Hays wanted visitors, particularly international visitors, to see samples of the best Canadian cattle; therefore in 1965 he arranged to have dairy and beef breeds assembled as showcase herds. Experimental Farms released two barns and the necessary pasture to house and feed the stock. Production and Marketing Branch provided the management and funds. The minister often visited the barns early in the morning before the herdsmen

arrived. He also visited research stations whenever possible—particularly
Lacombe and Lethbridge, the two nearest Calgary.

Anderson next concerned himself with the final comment of the Glassco
Commission Report on agriculture research, which dealt with "many small units
carrying on research . . . and lacking suitable facilities." He consolidated stations
by closing some and transferring their staff and function to larger and more
favorably located institutions. He made the Experimental Station at Scott a
substation of the Saskatoon Research Station, the Experimental Farm at Indian
Head a substation of the Regina Research Station, and Fort Vermilion, Alberta,
Fort Simpson, Northwest Territories, Mile 1019, Yukon, and Prince George,
British Columbia all substations of Beaverlodge, which became the Northern
Research Group. Because of local pressures beyond his control, however, he was
unable to close either Scott or Indian Head. He was able to terminate the
Experimental Fur Farm at Summerside, Prince Edward Island, the experimental
stations at St. Charles de Caplan and Fort Chimo, Quebec, the entomology
laboratories in Guelph and Chatham, Ontario, and the plant pathology laboratory in Edmonton. Operations were concluded at the experimental stations at
Wabowden, Manitoba, and Smithers, British Columbia, and finally the Forest
Nursery Station at Sutherland, Saskatchewan, was closed. The consolidations
provided needed scientific and support staff for the larger, more productive, and
better equipped stations.

During Anderson's tenure, and as a further result of the Glassco Commission, Treasury Board staff asked government research organizations to measure
their expenditures of human and financial resources for each of their research
goals. Although information on finances was readily available, it seemed to be a
poor base upon which to measure and classify research effort. A better base
seemed to be the number of scientists engaged in each discipline such as soils,
animal science, or crops, which could then be further divided into subgroups, as
needed. This was the start of Management by Objectives in the branch.

In April 1963, Glen met with Anderson, S.B. Williams, Director of Administration for the department, A.H. Turner, Director, Economics Branch, and Dr. B.
Kristjanson, from the Dominion Veterinarian's office to discuss future policies for
agricultural research and how to satisfy the request from Treasury Board. The
National Productivity Council, the Economics Research Council, the National
Research Council, and the Glassco Commission were each inquiring into the
efficacy of federal spending for national research programs. Glen expected the
federal government to initiate an economic planning authority and he wanted
Research Branch to show how its research benefited farmers and consumers
alike.

Research Branch management agreed to devise and test a system of
measuring each research project in terms of the cost, time, and number of
scientists involved. Often a scientist divided his time among several projects.
Rasmussen and Dr. A.S. Johnson devised a scheme and tested it on the Plant
Research Institute, Ottawa, the Pesticide Research Institute, London, and the
research stations at Winnipeg and Lethbridge. The four directors found little

difficulty in supplying the needed information and each made some suggestions for minor modifications. These modifications were incorporated into the scheme and, in December 1963, Anderson requested all directors to supply a breakdown of the cost and time each scientist spent on each research project.

After the returns were received in January 1964, Glen asked deans of faculties of agriculture and schools of veterinary medicine to provide similar data. He made the same request of provincial deputy ministers of agriculture for their programs. By July 1964, all the data were tabulated and copies sent to contributors. The information was used not only by the branch and the department but also by the Science Council to judge the relative effort expended on each broad subject. Treasury Board asked that estimates of expenditures be classified according to the same subject breakdown. Here the matter rested until 1968 when management by goals and objectives came into vogue. In October 1968, Drs. J.W. Morrison and C.J. Bishop wrestled wih the problem of integrating the project system with the research and development survey and the ways in which directors could utilize Management by Objectives (MBO). By the end of 1968, directors prepared goals and objectives for their programs and Morrison became the branch and department leader in the art of coordinating the system.

In 1968, Glen accepted the opportunity to become Secretary of the Commonwealth Scientific Committee and Scientific Adviser to the Commonwealth Secretary General, with headquarters in London, England. He was admirably suited to this position because of his sound research background, his knowledge of the administration of scientific research, and his strong empathy with people. He served the Commonwealth with distinction for 5 years. He then retired to Victoria, British Columbia.

At this time, Anderson was approaching his 65th birthday and would also soon retire. He performed the duties of both Assistant Deputy Minister (Research) and Director General, Research Branch, during his last few months with the branch.

The first 10 years of Research Branch brought research scientists together from all disciplines with the exception of veterinary medicine and economics. Even there, close cooperation was achieved in a few carefully selected programs. A unified interdisciplinary agricultural research organization proved more productive than several single discipline groups.

Chapter 11
Research Branch in Today's Society 1969–1985

In recent times the Research Branch has experienced numerous organizational changes. Some have resulted from shifts in government policy, others have come from changes in the organization of the department itself. With each change, the expectation has been to increase the response of research to the needs of all segments of the Canadian food production industry. Central agencies of government have sought ways of predicting the benefits of research before it was started and of measuring its effectiveness before it was finished. Research Branch has responded positively, and often showed the way in which a new system or organization could function or be made more effective. Changes have sometimes resulted in higher administrative costs per scientist, a decentralization of research programs, or an unfortunate duplication of effort.

101

The grapevine telegraph worked overtime during the summer of 1968, following the retirements of Glen and Anderson. Would new appointments be from within the branch, from within the department, or from outside? By August 1968 Dr. J.C. Woodward, one of the four principals in building the Research Branch in 1959, had been appointed Assistant Deputy Minister (Research). He was born on a beef farm near Lennoxville, Quebec, and graduated from Macdonald College and from Cornell University in animal nutrition. He was an assistant agronome for the Quebec Department of Agriculture before joining the Division of Chemistry in 1934. Following military service, where he was awarded the Military Cross and Bar, he went to Portage la Prairie, Manitoba, to set up the fiber flax station for Experimental Farms. Returning to Ottawa, he was appointed Dominion Chemist, Science Service, in 1949, and Associate Director of Experimental Farms Service in 1955.

Dr. B.B. Migicovsky moved from Assistant Director General (Institutes) to Director General, Research Branch, replacing Anderson. Dr. E.J. LeRoux, a coordinator of entomology, replaced Migicovsky at the Institute desk. He was born in Ottawa, served with the Royal Canadian Navy, and graduated from the universities of Carleton in Ottawa and McGill in Montreal. LeRoux was part of a northern biting fly survey in 1949 with the Division of Entomology and then joined the Science Service Laboratory, Harrow, in 1950 as an entomologist. He transferred to the Saint-Jean, Quebec, Science Service Laboratory in 1953. At Saint-Jean he earned an international reputation for his study of insect population dynamics and the effect of insects on the ecology. In July 1962, he left Research Branch to teach entomology at Macdonald College, returning to Ottawa as entomology coordinator in December 1965, succeeding Dr. A.P. Arnason.

Dr. T.H. Anstey, director of the Research Station, Lethbridge, replaced Rasmussen, who was promoted to the position of senior Assistant Director General. In that capacity, Rasmussen was responsible for program coordination and a backup to the director general. In July 1968, Anstey accepted a British Nuffield Fellowship to study the organization and management of agricultural research in Great Britain. He reported to Ottawa in April 1969. Anstey had joined the Experimental Farm, Agassiz, British Columbia, in 1941 as a summer student, returning in 1946. Subsequently he was appointed superintendent, Experimental Station, Summerland, moving to Lethbridge in 1959.

Woodward moved his office from the K.W. Neatby Building to the Sir John Carling Building in order to be close to departmental headquarters. Indeed, he followed Glen in this regard, although Anderson, when he became Acting Assistant Deputy Minister had kept his office in the K.W. Neatby Building.

Migicovsky, recognizing the value of Management by Objectives (MBO) as started by Anderson, had Morrison and his team refine the system. By June 1969 Morrison was able to list all scientists, their objectives, and the portion of their time devoted to each research project. Managers now had firm information, rather than estimates, of the amount of research being done in Canada in each agricultural discipline. Judgment was still needed as to the appropriateness of the ratios among disciplines.

Having reached this point, Migicovsky called a special directors' conference for the fall of 1969 to explain how the final step in MBO would be taken. The outcome was that each station director was to receive commitments from his scientists to meet specific goals and objectives, to establish a precise system of accountability, and to integrate management by objectives with estimates of monetary requirements and personnel assessments.

Migicovsky set out to further develop an esprit de corps and a cohesion in the branch. He wondered why it was that each research station had a good image in its own community, but as a branch the image was not the sum of its parts. As a consequence he launched a branch newspaper, *Tableau*.

The first issue of *Tableau* appeared in October 1970, with Mr. D.B. Waddell as editor. The editorial board was composed of Dr. D.G. Hamilton, Mr. D.G. Peterson, and Dr. C.E. Chaplin, Chief, Scientific Information Service as chairman. The first issue featured an editorial *From the D.G.'s Desk* in which Migicovsky discussed science policy in the government of Canada. He commented upon the various reports that had been or were about to be issued, such as those by Smallman (37) and by Shebeski (36), and emphasized that change was the order of the day. That first issue of *Tableau* also contained an article by Dr. K. Rasmussen, who was soon to retire: he wrote about opportunities ahead; there was a front-page report on the opening of a new electron microscope center at the Cell Biology Research Institute; the Ottawa Research Station was featured; and there were items on newsworthy people. Although *Tableau* was not an official organ of the branch, it was, and still is, a form of communication to help further understanding and raise morale among the people of the branch.

When Rasmussen retired in 1970, Dr. A.E. Hannah replaced him. Born at Rouleau, Saskatchewan, Hannah graduated from the universities of Manitoba and Minnesota in plant genetics, following active service with the Royal Canadian Air Force. He is a member of the international Silk Worm Club, having saved his own life by using a parachute. He was the oilseed breeder at the Dominion Cereal Breeding Laboratory, Winnipeg, from 1949 to 1955 and then moved to the Cereal Division, Ottawa, as the barley breeder. When Research Branch was formed he became coordinator for cereal breeding. Following Anderson's move from Winnipeg to Ottawa, Hannah was appointed director of the Research Station, Winnipeg, from which position he moved to Assistant Director General (Coordination).

In late 1971, Deputy Minister Williams selected Hannah to chair a task force to develop a food system approach to planning and coordinating part of the activities of the department. To begin with, a seven-person team dealt with meats, oilseeds, and feed grains. Dr. K.W. Downey, rapeseed breeder from Saskatoon, and Dr. D.P. Heaney, from the Animal Research Institute, Ottawa, were the other two members from the branch. Later, Dr. A.S. Johnson from the coordinators' group replaced Heaney. The team developed a workable plan, and in July 1972 the minister, the Honourable H.A. Olson, organized a Food Systems Branch, with Hannah as its assistant deputy minister. The objectives of the branch were to review, evaluate, and monitor federal government food programs as they related to the production and marketing of agricultural products. Hannah collected a small staff from different disciplines and needed the full cooperation of all assistant deputy ministers of operating branches to achieve the goals. In 1977 Mr. G. Lussier, a new deputy minister (Mr. L.D. Hudon had replaced Mr. Williams in 1975), convinced the minister, the Honourable Eugene Whelan, that, in fact, each branch in the department was part of the food system, and he felt that the formation of a Regional Development Branch would be more appropriate. Lussier therefore amalgamated the Food Systems Branch with other parts of the department and organized a Regional Development Branch. In 1977, Hannah transferred to Revenue Canada as an assistant deputy minister.

During the early 1970s the department developed five new programs in an attempt to reduce a large volume of surplus grain. Included was a grassland incentive program under which Research Branch sought ways and means of encouraging grain farmers to seed their poorer land to permanent grass. In addition, scientists in the branch made a concentrated effort to find new uses for surplus cereals and identify innovative crops to be sold on both Canadian and world markets. The thrust developed by the Minister, the Honourable H.A. Olson, was toward having all branches act in unison for the betterment of Canadian agriculture.

Growth within government had been restricted from 1970, Research Branch not excepted. Migicovsky was hard pressed to stretch his limited resources and continued to seek ways and means of reducing overhead in order to retain essential programs. One way of accomplishing this goal was to reduce the number of research stations, and combine their staff and facilities with others.

103

In 1971, he made the difficult decision to close the Institute for Biological Control at Belleville. Most of the scientists and staff were moved to the Research Station, Winnipeg, where a new section was organized to investigate integrated pest management in cereal crops. Those scientists at Belleville who were investigating the use of insects to control weeds were transferred to the Research Station, Regina, which several years before had been designated as the western weed control station. The few remaining staff either moved to Ottawa or left the branch in favor of university careers.

By the early 1970s MBO was entrenched as the operative system in Research Branch and was being tried by other branches with some success. Initially, some scientists were not entirely happy with the new system. They expressed the view that nothing was really new, that over time research had already developed its goals and objectives in a natural way. Was not the new system of classification oversimplification? Scientists worried because it tended to submerge details of research and the identities of individuals and disciplines. Despite their doubts, the scientific staff recognized that some method was needed to classify what was being done and to identify what benefit research was having upon food production in Canada. Therefore everyone yielded a little to make MBO work.

In 1978 Dr. J.W. Morrison, who at that time was Director General of Coordination, received a Public Service Commission of Canada Merit Award in recognition of his initiative in establishing management by objectives in the Research Branch. Morrison was born in Hanna, Alberta, and graduated from the universities of Alberta and London, England. He was a cytogeneticist by training and worked in the Cereal Division, Ottawa. After spending a few months at the Experimental Station, Mile 1019, as acting superintendent, he was appointed superintendent of the Experimental Station, Morden. He left Morden in 1966 to become coordinator of cereals at branch headquarters.

Dr. R.M. Hochster, director, Chemistry and Biology Research Institute died on 16 September 1971, following a brief illness. Hochster had been with the Research Branch for only 6 years, having come from the National Research Council in 1965. Early in the following year Dr. G. Fleischmann, a plant pathologist working on crown rust at the Research Station, Winnipeg, was selected as Hochster's replacement. Fleischmann was an imaginative, able, and busy scientist who remained in the directorship for only 2 years before moving to the Department of the Environment. We will learn more of Fleischmann later.

During the summer of 1972 the government announced a "make or buy" policy with respect to its research activities. Each time a new program was planned for mission-oriented research, the department proposing the research had to determine if it could be done satisfactorily under contract by the Canadian private sector or by universities. The objectives were to stimulate research in the private sector, to assure that the results were used quickly and effectively in the economy, and to increase employment among Canadians in various fields of technology. Research Branch immediately identified agricultural mechanization as one subject upon which research was needed but for which the branch did not have the staff or adequate facilities for large-scale development. Dr. J.R. Aitken, a coordinator, was assigned the responsibility of managing the total program. In 1981 the program was turned over to the Atlantic, Quebec, Ontario, and Western regions. The number of subjects upon which research was needed increased to 21. The amount spent from branch funds increased from $90 000 to $9 202 000, and the number of firms and universities involved with the program steadily increased, as summarized in Table 11.1.

Table 11.1 Contract research

Year	Expen- diture	Percentage spent by: Univer- sity	Other	Number of contracts	New subjects undertaken
	($000)				
1973	90	4	96	3	Mechanization
1974	365	90	10	34	
1975	921	69	31	69	Livestock
1976	1 685	68	32	112	Insects, Land Evaluation
1977	2 520	66	34	148	Crops, Feed
1978	3 501	46	54	204	Energy, Processing
1979	4 259	48	52	286	Cooperation with Industry
1980	4 914	50	50	308	
1981	6 460	47	53	387	Nutrition, Vertebrate Pests
1982	10 420	43	57	348	
1983	10 572	40	60	357	
1984	12 207*	—	—	357	Toxic chemicals

* Estimate

Contracts were awarded for short-term projects on specific problems such as the development of a new piece of equipment or a new product, utilization of a by-product, or development of a system for land evaluation or crop information. Under the mechanization program, a unique ditch digger, a lowbush blueberry harvester, a cauliflower harvester, and a zero tillage seeder were developed. Other accomplishments included progress on controlling wild oats, a system to produce methane from manure, and a change in pig diets that increased the size of swine litters. New sources of energy and new methods of conserving energy for farm use were identified.

LeRoux, who had been Assistant Director General of Institutes since 1968, replaced Hannah in 1972 as Assistant Director General (Coordination). In June 1973 he organized the coordinators into four groups, increasing their number from 12 to 15. The groups and their leaders were as follows: crops, Dr. J.W. Morrison; animals, Dr. W.J. Pigden; production, Dr. W.S. Ferguson; and protection, Dr. H. Hurtig. In addition, four special advisers were appointed on a part-time basis. They were the directors of the Engineering Research Service,

Mr. C.G.E. Downing, the Ornamentals Research Service, Dr. Alan Chan, the Statistical Research Service, Mr. L.P. Lefkovitch, and an economist from the Economics Branch, Dr. D. Ware. The following year Dr. Julius Frank, director of the Animal Pathology Division of the Health of Animals Branch was included as the fifth special adviser.

When the Research Branch was formed, both plant taxonomy and the herbarium were part of the Plant Research Institute. Taxonomy on insects, and the insect collections were retained in the Entomology Research Institute. Dr. W.B. Mountain vacated the directorship of the Entomology Research Institute in 1973 to replace LeRoux as Assistant Director General of Institutes. Mountain had been concerned with the dicotomy of the taxonomic function and decided that it would be best to amalgamate the activities of the two institutes. Thus he created the Biosystematics Research Institute, with Dr. D.F. Hardwick, an entomologist, as its director. Staff in both institutes not involved with either plant or insect taxonomy moved to the Ottawa Research Station.

Although Mountain was born in Kamsack, Saskatchewan, he was raised in Stratford, Ontario, where he received his elementary schooling. He served with the Royal Canadian Air Force from 1942 to the end of World War II, then graduated from the universities of Western Ontario and Toronto in plant pathology and mycology. After working as a summer student at the Science Service Laboratory, Harrow, he joined its staff in 1951, becoming head of the nematology section. In 1964 he was appointed Director, Vineland Research Station, and 5 years later he was appointed Director, Entomology Research Institute, Ottawa.

There were 2 years of relative stability. Woodward retired late in 1974 and Deputy Minister Williams retired in March 1975, at the age of 63 after 44 years service. Mr. L.D. Hudon, deputy secretary to the Privy Council, replaced Williams. Hudon had had experience at the management level in both the Department of External Affairs and the Department of Finance before going to the Privy Council Office. In July, Migicovsky was appointed Assistant Deputy Minister (Research) but retained the duties of the Director General. He kept his offices in the K.W. Neatby Building rather than in the Sir John Carling Building. In order to accommodate the change, Migicovsky organized the branch into an Operations Directorate and a Planning and Evaluation Directorate, each with its own director general. LeRoux was assigned the responsibility for Operations and had with him three assistant directors general: Dr. J.J. Cartier for the Quebec and Atlantic Region, Dr. W.B. Mountain for the Central Region, including the six institutes, and Dr. T.H. Anstey for the Western Region. Dr. D.G. Hamilton who previously had been responsible for the Eastern Region, became Director General of Planning and Evaluation. Dr. R.A. Ludwig retained his responsibility for administration. Directors of research stations were now three steps away from their chief executive officer, the Assistant Deputy Minister, rather than just two.

Cartier came to Ottawa in 1969 from the Saint-Jean Research Station, where he studied host plant resistance to aphids, particularly on pea vines where the peas were grown specifically for canning purposes. He was born in Beauharnois, Quebec, and received degrees from the universities of Montreal and Kansas State in entomology. Cartier replaced LeRoux as coordinator for entomology when he moved to Ottawa and then replaced Hamilton in the 1975 reorganization.

Early in January 1977 Dr. R.A. Ludwig died suddenly. He had experienced a serious fall the previous year from which it was hoped he would recover. Ludwig was Assistant Director General for Administration for 12 years. In a sense, Ludwig was a second van Steenburgh of Science Service days, with the operation of what was known as "the Ludwig Construction Company," because he organized the building of so many laboratories. Mr. J.E. Ryan, Ludwig's assistant, became director of the Finance and Administration Division in the branch. Ryan was born in Pendleton, Ontario. He worked in Research Branch, then for the Finance and Administration Branch, returning to Research Branch to succeed Mr. J.P. McCrea in 1970 as chief of administrative services.

Mr. Hudon left the department in April 1977 to become Secretary of the Ministry of State for Science and Technology. He was succeeded by Mr. Gaétan Lussier, a graduate in agriculture from the colleges of Oka and Macdonald. He taught at the Institute of Agricultural Technology, Saint-Hyacinthe, and eventually became deputy minister of the Quebec Department of Agriculture, which position he held for 6 years prior to joining Agriculture Canada. This was the start of many more changes in the department and in the branch. Before the end of the year, Dr. G. Fleischmann returned to a newly created position as Assistant Deputy Minister, Planning and Evaluation. Fleischmann had been extremely active since leaving the Chemistry and Biology Research Institute 2 years earlier. From Environment Canada, where he was Director General for Policy and Program Development, he moved to the Program Branch of Treasury Board in 1976. The following year he was appointed Assistant Secretary of Treasury Board but remained in that post for less than a year before returning to Agriculture Canada.

At the end of 1977 Hamilton retired and Morrison became Director General of Planning and Evaluation. On 30 December 1977, Migicovsky also retired. He had been with the department for 37 years, the last 10 of which were as director general and assistant deputy minister. Fleischmann was appointed acting assistant deputy minister on 1 January 1978. It was not until March that Migicovsky's successor was chosen in the person of Dr. E.J. LeRoux. In the meantime, however, many other changes were made in the branch. The notion of regionalization became government policy. Fleischmann and his Planning and Evaluation Group judged that decision making was too centralized. He therefore arranged for the directors general of the Eastern and Western regions to have their offices in Quebec City and Saskatoon, respectively. Cartier moved to Quebec and Dr. A.A. Guitard, Director of the Research Station, Swift Current, was appointed Director General of the Western Region. Mountain remained as Director General for the Central Region and Institutes, and Morrison remained as Director General for Planning and Evaluation. Anstey became

107

senior adviser on international research and development because of his considerable experience with various Canadian International Development Agency aid programs.

Guitard was born at Carstairs, Alberta, and attended the universities of Alberta and Nebraska. He joined the Experimental Station, Beaverlodge, Alberta, in 1947 as cerealist, and became its director in 1962 when Beaverlodge was made a research station. In 1969 he moved to Swift Current as director, following Andrews. While there, he was appointed a member of the International Garrison Diversion Study Board. The new organization resulted in the establishment of three administrations in addition to the branch administration: one in Quebec City, one in Saskatoon, and one in Ottawa to handle the Central Region. LeRoux changed his executive by dispensing with the operations directorate and having the director general of each region and of Planning and Evaluation reporting directly to him.

In late 1977 the department, responding to government regional development policies, decided that there needed to be a greater degree of cooperation in planning programs and policies among the federal Department of Agriculture, provincial departments of agriculture, and the agriculture and food industries. In a move to accomplish this goal, Minister Eugene Whelan appointed a senior person in each province as chief liaison officer between the federal department and the various organizations in each province. Of the 10 officers appointed, the following eight were directors of research stations: Mr. H.W.R. Chancey, St. John's West; Dr. L.B. MacLeod, Charlottetown; Dr. J.R. Wright, Kentville; Dr. G.M. Weaver, Fredericton; Mr. J.J. Jasmin, Saint-Jean; Dr. A.J. McGinnis, Vineland; Dr. W.C. McDonald, Winnipeg; and Dr. M. Weintraub, Vancouver. These directors continued to manage their research stations and in addition reported to Director General Brouillard in the Policy, Planning, and Evaluation Directorate. Within 2 years this directorate was elevated to the status of a branch with Brouillard as its assistant deputy minister. All but two of the Research Branch directors had shed their liaison duties and others were appointed in their stead.

A new research institute combining the functions of the Engineering Research and the Statistical Research services was organized in late 1977. Although both organizations did cooperative research with other parts of the branch and the department in their respective fields of expertise, their scientists also did considerable independent research as well. To bring the name more in line with their actual function, the united services were renamed the Engineering and Statistical Research Institute. Mr. P. Voisey was appointed its director effective 1 January 1978.

Voisey was born and and raised in England, where he worked as an aeronautical engineer until moving to Canada in 1955. He developed engine and flying controls for Canadair until 1960, at which time he joined the Engineering Research Service under Mr. W. Kalbfleisch.

In November 1976, because of severe criticism from the Auditor General, the government became concerned about the growth of the Civil Service and the "unprecedented demands upon the structure, organization and process of administrative management and control" within the service. As a result, Privy Council appointed a Royal Commission on Financial Management and Accountability, chaired by A.T. Lambert, a retired banker, who had considerable business experience. Lambert published his Commission report (26) in March 1979. It resulted in the appointment of a Controller General for the government who required each department to prepare an action plan of control and financial management. A committee of the department prepared the plan for Agriculture Canada and Dr. W.B. Mountain, Director General, Central Region, was asked by Deputy Minister Lussier to head an implementation team. Therefore in November 1979, Mountain moved to departmental headquarters on the understanding he would return to Research Branch when he had completed the task. In October 1980, however, Deputy Minister Lussier made a further reorganization in the department, including Research Branch, which left Mountain without his position. He therefore transferred to Environment Canada as Assistant Deputy Minister of Environmental Conservation Service.

The branch was further decentralized in 1980 when the Eastern Region was divided into two, forming the Atlantic and the Quebec regions. Dr. E.E. Lister moved to Halifax from Ottawa as Director General of the Atlantic Region. He was born at Harvey Station, New Brunswick, and graduated from the universities of McGill and Cornell in animal nutrition. He worked as an animal nutritionist for Ogilvie Flour Mills, Montreal, for 5 years before joining the Animal Research Institute in 1965.

Mr. J.J. Jasmin was appointed Director General of the Quebec Region and moved its headquarters from Quebec City to Montreal. Jasmin was born in Paris, France, of Canadian parents, and came to Canada at the age of 6 years. He graduated from Macdonald College and Michigan State University. Jasmin's experience within the government and in the private sector was both wide and varied. He joined the staff of the Horticulture Division in 1947 and in 1956 became officer in charge of the Experimental Station, Sainte-Clothilde, Quebec. In 1963 he was hired by the government of Quebec, then moved to private industry in 1964. He returned to Research Branch from the presidency of CAGRIC Inc., a Quebec agricultural consulting firm, in 1972 as Director, Research Station, Saint-Jean.

The Central Region, which had included institutes as well as the research stations in Ontario, was also divided. The institutes were handled by Morrison in addition to his Planning and Evaluation responsibilities. In 1982, however, he spent a year's sabbatical at the University of Manitoba and Dr. R.L. Halstead took the Planning and Coordination job.

Born in Oyen, Alberta, Halstead received his schooling in Gleichen, near Calgary. He graduated from the universities of Manitoba and Wisconsin in soil science, then joined the Chemistry Division of Science Service in 1954, moving to the Soils Research Institute in 1959. Halstead became associated with the coordination group in 1971 as acting soils coordinator, remained with the group in various capacities, and became its director general in 1982.

109

Ontario became a region in 1980 to which the Pesticide Research Institute in London and the Animal Research Institute were added and renamed as Research Station, London, and Animal Research Centre, Ottawa, respectively. Cartier returned to Ottawa from Quebec City as Ontario's new director general.

In 1981 LeRoux was appointed a member of National Research Council for a 3-year term. The Council consisted of scientists from private industry and universities; however, appointments from government departments were occasionally made. LeRoux was only the third appointee from the Canada Department of Agriculture over the life of the Council, the other two being Deputy Ministers Grisdale from 1923 to 1932 and Barton from 1933 to 1936.

Guitard, in Saskatoon, talked about retiring early. LeRoux convinced him to spend 6 months as his special assistant to revise the postgraduate training strategy of the branch. At this point in 1981 Dr. J.E. Andrews, director of the Research Station, Lethbridge, replaced Guitard. Andrews was born in Selkirk, Manitoba, and graduated from the universities of Manitoba and Minnesota in plant breeding and genetics. He joined the Cereal Breeding Laboratory, Winnipeg, in 1945 and then went to Lethbridge as a winter wheat breeder. It was here that he gained his scientific reputation for developing and introducing Winalta winter wheat. This variety quickly became the standard, for it was as hardy as any other variety and had almost the milling and baking characteristics of a hard red spring wheat. No other winter wheat had all these qualities. In 1960 Andrews was appointed Director, Research Station, Brandon, and then moved in the same capacity to Swift Current in 1966 and to Lethbridge in 1969. While at Lethbridge, he made an outstanding contribution to international aid by managing Canadian International Development Agency (CIDA) projects in India and Sri Lanka.

Andrews operated the region by commuting between Lethbridge and Saskatoon for 2 years, when he too stepped down to become the senior adviser to LeRoux on special projects. As senior adviser he devoted most of his time to managing international aid programs, such as the India Dryland Project, and negotiating agricultural technical exchange agreements with other countries. Dr. W.L. Pelton, Director, Swift Current Research Station, replaced Andrews in August 1983 as director general. Also in 1983, the Western Region was divided into two portions: Pelton became responsible for the Prairie Region and Dr. S.C. Thompson for the British Columbia Region, with headquarters in Vancouver. These moves caused shifts in directorships at a number of research stations, which are recorded in Appendix II.

Pelton was educated as an agricultural engineer and soil physicist. He was born in Simcoe, Ontario, and attended the Ontario Agricultural College and the universities of Toronto and Wisconsin. He joined the Swift Current staff in 1958 to do research in agrometeorology and was consulted widely on micrometeorology and its relation to agricultural production. He spent 3 years at Hyderabad, India, as the Canadian Joint Coordinator of the Indo-Canadian Dryland Research Project. He returned to Swift Current, but in 1975 he was named Assistant Director, Lethbridge Research Station. In 1978, he once again went to the research station at Swift Current, this time as its director.

Thompson was appointed from the position of Deputy Director, Animal Research Centre, Ottawa, to Director General, British Columbia Region. He was born in England and obtained his agricultural degrees from the University of Reading. Prior to moving to Vancouver he attended a year's course in Canadian domestic and foreign policy designed for senior officers and civil servants at the National Defence College, Fort Frontenac, Kingston, Ontario. During this period Dr. W. Baier, the coordinator for international programs, was the acting Director General, British Columbia Region.

Toward the end of 1984 the newly elected federal government agreed to proceed with the construction of office–laboratory buildings at two locations that had been in the planning stages for many years. At Summerland, British Columbia, construction started in 1984 on a complex which, in a few years, will bring the scientific staff together from five buildings. They will have the added advantage of provincial extension agrologists being in the same complex, a situation similar to those already existing at Kentville, Nova Scotia, and Lethbridge, Alberta. In Ottawa, sod was turned in 1984 for the construction of an office–laboratory building for the Animal Research Centre. Ever since the mid-1960s when research livestock was moved from the Central Experimental Farm to property 15 km south of Ottawa (in the Greenbelt), scientists have been hampered by the separation of their laboratories and offices from their experimental animals. The new building will remedy this situation by bringing all the Animal Research Centre staff together close to the animals.

Early in 1985 the last organizational change to be recorded in this history was precipitated by the retirement of Dr. J.W. Morrison, Director General, Institutes, and Dr. T. Rajhathy, Director, Ottawa Research Station. The Ottawa Research Station and the Chemistry and Biology Research Institute were united under the directorship of Dr. I.A. de la Roche. The Plant Research Centre, as it is called, together with the Animal Research Centre and the other research instiues in Ottawa were placed under the new Director General, Institutes, Dr. R.L. Halstead, who now has the responsibility for the total Central Experimental Farm. Plans are under way to move the headquarters of the Ontario Region (Dr. J.J. Cartier and staff) to Toronto. Dr. W. Baier replaced Halstead as Director General, Program Coordination.

Baier was born in Liegnitz, Germany, and received his undergraduate and doctorate degrees from the University of Stuttgart-Hohenheim. In South Africa he received a further master of science degree in agronomy from the University of Pretoria. Baier came to Canada in 1964 where he was employed by the Plant Research Institute in the Agrometeorology Section, becoming its section head in 1969. In 1978 he was appointed Assistant Director, Land Resource Research Institute, then Research Coordinator in 1981.

Economic, sociologic, and political concerns were evident in the closing 15 years of the Research Branch's first century. Well-intentioned micromanagement regulations were introduced that tended to overload directors with reports and details. The Collegiate Theory, advanced by the Harvard School of Business Administration, was adopted throughout the government of Canada. It hypothesized that a person trained in management could manage anything, even technically and scientifically based organizations.[1]

It is a tribute to the resourcefulness of our scientists that despite administrative buildup, most of the research undertaken in the branch has been both well targeted and exciting. Some of the problems faced, the solutions arrived at, and the benefits derived are sketched in the following pages.

[1]Thomas J. Peters and Robert H. Waterman, Jr., *In Search of Excellence*, Harper & Row, New York, 1982, deals with the theory and offers good reading on modern methods of managing technically based organizations.

112

References, Part 1

1. Anonymous. 1789. Plan of a society for promoting agriculture in the province of Nova Scotia. The Royal Gazette and the Nova-Scotia Advertiser. I(23):3.
2. Anonymous. 1982. Photos ... breeds of cattle horses swine. Agric. Can. Publ. 1749B.
3. Beattie, Gladys Mackey. 1982. The "Canadian horse" is our very own. Can. Geogr. 105(5):34–39.
4. Bethune, C.J.S. 1908. Dr. James Fletcher. Can. Entomol. 40:432–436.
5. Broadfoot, Barry. 1976. The pioneer years, 1895–1914. Doubleday Canada Ltd., Toronto. 403 p.
6. Canada Agriculture. 1967. The first hundred years. Can. Dep. Agric., Hist. Ser. 1. 82 p.
7. Canada Department of Agriculture. 1958. Minutes of meetings chaired by Dr. Taggart. File 170.1/03, Vol. 3. Dep. Agric., Ottawa.
8. Clark, A.H. 1959. Three centuries and the island. Univ. Toronto Press., Toronto. 287 p.
9. Dunlop, R.H. 1950. The origin and development of field shelterbelt projects. Seminar, Forest Nursery Station, Indian Head, Saskatchewan. 22 February.
10. Estey, R.H. 1983. James Fletcher (1852–1908) and the genesis of plant pathology in Canada. Can. J. Plant Pathol. 5:120–124.
11. Fredeen, Howard T. 1984. Lacombe Research Station 1907–1982. Agric. Can., Hist. Ser. 18. 71 p.
12. Gibson, Arthur. 1951. Entomology in Canada, with special reference to developments within the Dominion Department of Agriculture. Manuscript located at Agric. Can. Library, Ottawa.
13. Glassco, J. Grant. 1963. Report of the royal commission on government organization. 5 vols. Ottawa.
14. Glen, Robert, compiler. 1965. Entomology in Canada up to 1956. Can. Entomol. 88:290–371.
15. Glen, Robert. 1962. Organization of the research branch of the Canada Department of Agriculture: An historical review. Agric. Inst. Rev. 17(3):18–46.
16. Goulden, C.H.; Neatby, K.W.; Welsh, J.N. 1928. The inheritance of resistance to *Puccinia graminus tritici* in a cross between two varieties of *Triticum vulgare*. Phytopathology. 18:631–658.
17. Gray, James H. 1978. Men against the desert. Western Producer Prairie Books, Saskatoon. 270 p.
18. Gridgeman, N.T. 1979. Biological sciences at the National Research Council of Canada: The early years to 1952. Wilfred Laurier University Press, Waterloo. 153 p.
19. Hardwick, D.F. 1976. The history and objectives of the Biosystematics Research Institute. Bull. Entomol. Soc. Can. 8(2):15–21.
20. Hewitt, C.Gordon. 1914. The Canadian entomological service. Can. Entomol. XLVI:214–216.
21. Hopkins, E.S.; Barnes, S.; Palmer, A.E.; Chepil, W.S. 1935. Soil drifting in the prairie provinces. Can. Dep. Agric. Bull. 179 n.s.
22. House of Commons. 1952. Standing committee on agriculture and colonization. 6th Session, 21st. Parliament. Ottawa.
23. Johnson, T. 1967. The Dominion Rust Research Laboratory. Manuscript located at Research Station Library, Winnipeg.

24. Johnston, Alex. 1977. To serve agriculture. The Lethbridge Research Station 1906–1976. Can. Dep. Agric., Hist. Ser. 9. 59 p.
25. Kilbourn, William. 1973. The making of the nation. The Canadian Centennial Publishing Co. Ltd., Toronto. 127 p.
26. Lambert, A.T. 1979. Royal commission on financial management accountability. Final report. Queen's printer. Ottawa. 586 p.
27. L'Arrivée, J.C.M. 1984. Bee research by Agriculture Canada—early years 1891–1914. Can. Beekeep. 11(3):63–65.
28. Lochhead, A.G. 1968. A history of the Bacteriology Division of the Canada Department of Agriculture 1923/1955. Can. Dep. Agric., Hist. Ser. 5. 56 p.
29. MacEwan, Grant. Undated. The sodbusters. Thomas Nelson and Sons, Toronto. 240 p.
30. MacEwan, Grant. 1983. Charles Noble, guardian of the soil. Western Producer Prairie Books, Saskatoon. 208 p.
31. McKeague, J.A.; Stobbe, P.C. 1978. History of soil survey in Canada 1914–1975. Can. Dep. Agric., Hist. Ser. 11. 30 p.
32. Neatby, L.H. 1979. Chronicle of a pioneer prairie family. Western Producer Prairie Books, Saskatoon. 93 p.
33. Pomeroy, Elsie M. 1965. William Saunders and his five sons. The Ryerson Press, Toronto. 192 p.
34. Reaman, G. Elmore. 1970. A history of agriculture in Ontario. Vols. I & II. Saunders, Toronto. 180 p. & 264 p.
35. Saunders, William. 1886. A report on agricultural colleges and experimental stations, with suggestions relating to experimental agriculture in Canada. Report of the Minister of Agriculture for the Dominion of Canada for calendar year 1885. Appdx. 54. Ottawa. 221 p.
36. Shebeski, L.K., et al. 1971. Two blades of grass: The challenge facing agriculture. Report No. 12, Science Council of Canada. Ottawa. 61 p.
37. Smallman, B.N., et al. 1970. Agricultural science in Canada. Special study No. 10, Science Council of Canada. Ottawa. 148 p.
38. Spence, George. 1967. Survivial of a vision. Queen's Printer, Ottawa. 167 p.
39. Strange, H.G.L. 1954. A short history of prairie agriculture. Searle Grain Co., Winnipeg. 104 p.
40. Thistle, Mel. 1966. The inner ring. Univ. Toronto Press, Toronto. 435 p.

PART II
SCIENCE AT WORK

Research Branch
Agriculture Canada
1886–1986

Chapter 12
Meeting the Challenge

As Canada's Experimental Farm System has developed through the efforts of its ministers, its administrators, its scientists, and its technicians, its research has had a profound effect upon the food production system. Many Canadians are of the view that agricultural research is performed on behalf of our farmers and they are correct, but its value extends well beyond the producer through a chain of beneficiaries including the consumer.

Because the farming population in Canada has declined from 32 percent in 1931 to its current level of 4 percent, there are some who surmise that farming must be of less importance today than it once was. In relation to the country's population the proportion of Canadians now living and working on farms has declined, but those Canadians who make up this smaller proportion provide more food than did their parents or grandparents. The conclusion we must draw from this is that today's farmers are exceedingly productive. Not only are there fewer farmers, but there is less high-quality land available! Nonetheless, the largest single activity in Canada involves agriculture and food. Reliable and conservative estimates tell us that producing, transporting, processing, servicing, retailing, and researching within the food chain are the vocations of at least one-quarter of the Canadian population. Hamilton (1) estimates that this 25 percent of our population accounts for about 40 percent of Canada's economic activity. One also must conclude, therefore, that farmers are growing appropriate crops and raising the right livestock, provincial departments of agriculture are providing the advisory and regulatory support where needed, provincial schools of agriculture and universities are offering suitable educational programs, and the Canada Department of Agriculture is fulfilling its role in providing solutions to problems through research, in addition to its responsibilities for supporting world trade of agricultural products. In short, throughout the past century, or more, many Canadians have been making correct decisions. The perpetual production of nutritious, wholesome, and affordable food for ourselves and for many others in less fortunate parts of the world has been the hallmark of Canadian agriculture for nearly a century.

Thomas Robert Malthus, in his 1798 writing of *The Principle of Population*, hypothesized that population, when unchecked, increases in a geometric progression, whereas sustenance increases only in an arithmetic progression.[1] Malthus was correct in his prediction of world population, because it increased geometrically and is now eight times what it was when he wrote those words. And what of food production? It has more than kept pace. The only way

117

---

[1]In geometric progressions each term, but the first, is greater than its predecessor by a constant ratio such as 2, 4, 8, 16. In arithmetic progressions, each term, but the first, is greater than its predecessor by a constant quantity such as 2, 4, 6, 8.

Malthus believed that food production could be doubled was to double both the number of farmers and the area farmed. As a direct consequence of agricultural research, other means have been found to double and triple food production. The second part of this 100-year history details some of the ways in which scientists have met the challenge.

The 10-year mean-yields of spring wheat in Saskatchewan are an example of how agricultural research has contributed. In the period 1910–1919 the provincial range varied from 0.57 to 1.69 t/ha with a mean of 1.11 t/ha (see Table 12.1). During the 1930s and 1940s both the range and the mean fell as a result of drought and stem rust. When research brought these two problems under control in the early 1950s both indicators increased in each subsequent 10-year period.

Table 12.1 Range and mean-yield of spring wheat in Saskatchewan

Year	Range	Mean-yield
	(t/ha)	(t/ha)
1910–1919	0.57–1.69	1.11
1920–1929	0.68–1.57	1.11
1930–1939	0.18–1.18	0.68
1940–1949	0.75–1.65	1.08
1950–1959	0.65–1.78	1.27
1960–1969	0.57–1.87	1.41
1970–1979	1.41–2.09	1.59

Statistics Canada (2) provides three general measures of the importance of agriculture to the total economy. The first measure is the domestic product, which shows that the agricultural proportion of the gross national product has fallen from 17 percent in 1926 to 7 percent in 1955 and to 3 percent in 1976, even though the agricultural gross domestic product rose from $884 million to $5905 million during the same period. Although the proportion of the gross domestic product supplied by agriculture declined, it was still the second most important primary industry in 1975. (Mining was slightly greater in financial return.) The second measure shows that exports of agricultural products have increased in relative importance. In the 1965–1974 period they accounted for approximately 5 percent of world trade. However, Canada's share of world trade in flaxseed, rapeseed, barley, and wheat during the same period was 80, 60, 20, and 20 percent, respectively. In the 1977–1982 period the agricultural trade surplus, defined as agricultural exports minus agricultural imports, accounted for more than 50 percent of the total Canadian trade surplus. The third measure indicates that expenditures for food in the home expressed as a percentage of disposable income have fallen from 21.4 percent in 1947 to 15.4 percent in 1968 and to 12.0 percent in 1982.

Part I of this book examined the organizational metamorphosis of agricultural research, starting with the Experimental Farm System developed by

Dr. William Saunders and culminating with the modern Research Branch under Dr. E.J. LeRoux. Occasionally, it highlighted scientific accomplishments to show how the system functioned.

Part II deals almost entirely with science at work and its powerful influence upon food production. Brief reference is made when necessary to the organization of the Experimental Farm System, the Entomological Branch, Science Service, or Research Branch.

During the first century, the Canadian treasury provided $3254 million in support of agricultural research (see Appendix IV). Has our science justified the expenditure? If it has, will it continue to do so? What would our farmland have produced without scientific intervention? No attempt is made here to balance returns against costs, but readers will be provided with a sufficient number of examples to enable them to form their opinions.

References

1. Hamilton, D.G. 1980. Evaluation of research and development in agriculture and food in Canada. Can. Agric. Res. Counc., Univ. Guelph. 185 p.
2. Leacy, F.H., editor. 1983. Historical statistics of Canada. 2nd Ed. Stat. Can., Ottawa.

Chapter 13
Soils Research

In 1984 the importance of soils and soil conservation was brought forcefully to the attention of all Canadians through the study conducted by Senator Herbert O. Sparrow (93). He and his committee concluded that soil degradation is "a problem which is already costing Canadian farmers more than $1 billion per year in [lost] farm income," that research in all phases of soil conservation should be expanded, and that unless immediate action is taken, Canada risks losing a large portion of its agricultural capability. This chapter reviews a few of the ways in which Research Branch contributed to our knowledge of soils and to our understanding of the need to conserve this resource for the use of future generations. Although progress has been made in learning how best to manage the soils of Canada, much remains to be achieved.

SURVEY

The survey of Canada's soil resources is one of the most harmonious and successful cooperative programs between the federal and provincial governments. In each province, both parties have contributed staff and financial resources, and both have conducted surveys, using identical methods and procedures. In fact, soil surveyors have national rather than 10 provincial soil survey plans. McKeague and Stobbe (73) recorded the history of soil surveying in Canada from 1914 to 1975—its first 60 years. The following is a brief summary of that record, updated to include the past decade.

The objectives of soil surveys are to characterize and to understand the extent and the qualities of the soils being examined. Qualities of soils are judged for their food and timber production potential. Soils thought to be most suitable for food production and that were also reasonably close to urban centers were surveyed first. Some surveys described soils on a numerical basis, assigning a sequence of digits for the way in which soils were deposited, their mineralogical composition, their drainage class, and their texture. Other surveys used names for the various kinds of soils within each major classification level. Many of the names were the same as those used by Russian surveyors, because the science of soil surveying was pioneered by Russian scientists in about 1870. The soils of southwestern Ontario were the first in Canada to be surveyed. Surveys were started in 1914 by the Department of Chemistry of the Ontario Agricultural College, Guelph, with other provinces following from time to time.

Following the first International Congress of Soil Science held in Washington, D.C., in 1927, several of the world's leading soil scientists visited British Columbia, Alberta, and Saskatchewan where they stimulated a great deal of interest in classifying and mapping soils. At that time each province conducted its own survey, each had its own system, and each published its own maps on a scale of its choosing. E.S.

Archibald brought the soil surveyors of the provinces and experimental farms together in 1935. The need was great because of the drought and dust storms in the Prairie Provinces. In 1934, with Prairie Farm Rehabilitation Administration (PFRA) funding, Experimental Farms entered the picture and cooperated with the Prairie Provinces in surveying their soils in order to have factual information for the solution of the dust bowl problems. PFRA funding lasted until 1940 when funds in the Field Husbandry Division under E.S. Hopkins and, later, under P.O. Ripley were assigned to soil surveys. The survey then became national in scope with the formation of the National Soil Survey Committee.

The National Soil Survey Committee met in Ottawa for the first time in May 1945. Again, it was Archibald who was the prime force in bringing together provinces and experimental farms. McKeague and Stobbe remarked on the "large measure of agreement" reached by surveyors from all provinces and the experimental farms on technical matters relating to soil classification. They attributed this to the skill and understanding of A. Leahey, Chairman of the Committee, who had been with the Alberta soil survey until 1936 when he joined the Field Husbandry Division to begin the correlation of soil surveys across Canada. His objectives were to standardize the nomenclature of soil names, mapping units, and laboratory analytical procedures. P.C. Stobbe joined him in 1939, after gaining survey experience in New Brunswick, Quebec, and Manitoba and he became Secretary of the National Committee. Leahey was chairman until 1965; Stobbe took the chair until 1969 when he retired. Both Leahey and Stobbe earned credit for the development of a nationally accepted soil classification system. As secretary, Stobbe played the key role in achieving accord among surveyors from all provinces.

The survey entered a new and responsive phase to meet the needs of all Canadians. In 1963 the Canada Land Inventory (CLI) began under the terms of the Agricultural and Rural Development Act (ARDA). The objective of this program was to determine the suitability of lands for agriculture, forestry, recreation, and wildlife. Adequate information was available to compile such inventories based on recently surveyed areas, but additional field information was required for the early surveys. During its first 12 years, CLI was moved from the Department of Agriculture to the Department of Regional Economic Expansion, and finally to the Department of the Environment. Although soil surveyors were well aware that Canada had limited amounts of soil suitable for growing high-value crops such as fruits and vegetables, Canadians assumed that land used for expansion of cities and highways could readily be replaced with land farther from urban centers. They failed to recognize major Canadian cities had been located where they were so that their citizens could be fed from the highly fertile surrounding land.

W.A. Ehrlich, coordinator for soils at Research Branch headquarters, replaced Stobbe as chairman of the National (by this time, Canada) Soil Survey Committee in 1969. The Canada Committee broadened its discussions beyond soil taxonomy to encompass crop yield assessments and remote sensing. Membership expanded to include foresters, environmental specialists, and urban and recreational planners. The committee decided that Research Branch was the appropriate agency to retain control. In 1973 J.S. Clark, Director, Land Resource Research Institute, replaced Ehrlich as chairman and immediately started to develop a computerized soil information system, code-named CanSIS. This system has dramatically changed the method of producing maps for special purposes. Now, once the basic data are in storage, maps of all types are produced and drawn by computer.

By 1975 more than 300 million hectares had been surveyed, covering 35 percent of Canada's land mass, of which 60 percent had been rated under the Canada Land Inventory. Over 200 maps had been published since 1920—between 1960 and 1974 they were published at the rate of six a year. In the early years maps were drafted and prepared for printing by the Department of the Interior. When that department was abolished in 1931, cartographic services were supplied by the Department of Mines and Technical Surveys. In 1945 Archibald provided space for the Cartographic Section at the Central Experimental Farm, where it remains today. The technology of map production has advanced dramatically. In the early days, base maps were not always available and frequently had to be prepared by soil surveyors. Following World War II the Topographical Survey of the Department of Mines and Resources embarked on a program of aerial photographic mapping of Canada. The use of such photographs by soil surveyors greatly accelerated their work. They were able to predict the type of soil and hence reduce the number of digs, a time-consuming activity. Soil surveyors quickly adapted to the computer age and maps are now prepared with the aid of this powerful tool. Transportation modes used by surveyors have advanced during the years from packhorses, canoes, railway handcars, automobiles, trucks, and four-wheel-drive vehicles to aircraft, including helicopters. Helicopters were used first by J.D. Lindsay for surveys in northern Alberta in 1955.

Recently Canada's surveyed area has expanded to include federal lands such as Indian reserves, national parks, and the Yukon and Northwest territories. These areas have been surveyed for the purpose of establishing power transmitters and pipeline routes. Survey activities have expanded in southern Canada in response to the demand for detailed information relating to irrigation and urban projects. The increased scope calls for more correlation and quality control, which requires three regional correlators plus the national correlator. Increased effort has been devoted toward assessing soil degradation, its causes, severity, extent, and risk. Special surveys have mapped surface soil acidity, salinity, land use, and degradation. Such information provides guidance on the use and maintenance of Canada's land resource. Surveying, classifying, and mapping soils remain the prime functions of the Canada Soil Survey Committee.

THE DUST BOWL

When the Great Plains were first cultivated, toward the end of the nineteenth century and the early part of the current century, dust mulches were recommended (27) to reduce moisture loss from summerfallowed wheat fields. Dust mulches, developed by repeatedly discing and harrowing, reduced evaporation; however, in windy areas, valuable topsoil containing precious organic matter was lost.

The reasons the dust bowl developed in Manitoba, Saskatchewan, and Alberta between 1930 and 1940 are explained by Anderson (2), Palmer (79), and Gray (27). They describe how the use of plows, the overuse of summerfallows, and the cropping of large contiguous areas, together with low precipitation, high temperatures, strong winds, excessive evaporization, and topsoil removal exacerbated the condition. The resulting loss of crop is shown by the data in Table 13.1.

Table 13.1 Yields of wheat in Saskatchewan[1]

Years	Yearly range	10-year average
	(t/ha)	(t/ha)
1910–1919	0.57–1.69	1.11
1920–1929	0.68–1.57	1.11
1930–1939	0.18–1.18	0.68
1940–1949	0.75–1.65	1.08
1950–1959	0.65–1.78	1.27
1960–1969	0.57–1.87	1.41
1970–1978	1.41–2.09	1.59

[1]Source: Statistics Canada.

In the midst of the dust bowl of the 1930s, the Experimental Farm at Indian Head and stations at Swift Current, Scott, Rosthern, and Lethbridge became the main combatants in the fight against the blowing soil. E.S. Archibald took positive steps by encouraging superintendents and staff of experimental farms to continue their efforts in finding solutions, and by passing these solutions to farmers promptly. He was a man of action who sorted out the administrative niceties after a job was done.

Blowing soil had been experienced before on the prairies but only in isolated areas for short periods. The first experimental farm to report drifting soil was Indian Head, in 1887. By 1918 Monarch, between Calgary and Lethbridge, experienced blowing soil regularly.

One of the first research programs at Swift Current started in 1922 and dealt with moisture conservation, control of wind erosion, and weed eradication. J.L. Doughty and his colleagues (19) showed soil drifting to be a complex problem varying according to the conditions of the air, the ground surface, and the soil. Air turbulence increases the speed of wind near the soil surface and therefore is the primary cause of dust storms. The degree of turbulence is governed by the friction of air striking surface obstacles and by convectional eddies caused by temperature differences between soil and air. Convectional eddies increase the wind's velocity and occur over bare fallow, the soil situation most prone to drifting. The presence of clods, plant stubble, or other obstructions lessens the speed of air movement near the surface and thus reduces its erosive effect.

The composition of soil was shown by Swift Current scientists to have a direct bearing upon its erodability. Freshly broken sod containing much root fiber rarely drifts. As the fiber decomposes into humus it brings about the aggregation of soil particles, but for most prairie soils the aggregates are too small to resist the force of the wind. Therefore, to preserve the soil, it is imperative that fibrous organic matter be continually added to the soil. Calcium carbonate (lime), which is present in most prairie soils, was shown to cause soil aggregates to break into small granules and then to be lifted by wind.

W.H. Fairfield and A.E. Palmer at the Lethbridge Experimental Station, cooperated with farmers in the Monarch district as early as 1920 to combat soil drifting. Some farmers noticed that the first place on their fallowed fields to start drifting was the east side, the west side being the last to drift. They reasoned that since the prevailing winds were from the west, wind needed to blow across a portion of fallowed land before it started to lift soil particles. The solution was strip farming, that is, fallowing narrow strips running north and south and cropping the alternate strips. Palmer found the most effective width of strips depended upon the texture of the soil—light soils required narrow strips (5 rods or approximately 25 m), whereas heavy soils could sustain strips 10 times as wide. The disadvantages encountered with strip farming were soil buildup on the eastern sides of strips, encroachment of weeds from fallowed strips to seeded areas, and increased infestation of wheat by wheat stem sawfly. These disadvantages were overcome with research that led to new soil management techniques and new crop varieties.

Farmers controlled drifting within strips by leaving as much stubble standing in the field as possible. Before this control method was used, farmers plowed stubble into the soils immediately following harvest, as had been the custom in Eastern Canada and in Europe, where soils are heavier and rainfall is plentiful. Because of the soil conditions peculiar to Western Canada, it was necessary to devise a way of controlling weed growth. C.S. Noble (54), an innovative farmer near Monarch, decided to use a California blade, which was originally designed to lift sugarbeets, to kill standing weeds. He adjusted the blade to slice the soil just under the surface, cutting roots but leaving plants firmly in the soil. Noble modified his blade many times and now manufactures it for sale by the thousands throughout the world. The damaging plow was retired, stubble-mulch farming, or trash farming, had its start and today, with further modifications, it is recognized as the only way to handle fallow on prairie soils. For their pioneer work on soil conservation, Noble and Fairfield were each awarded the Member of the British Empire (M.B.E) in 1934.

In 1931 a substation of Swift Current opened in Regina, with W.S. Chepil in charge. The primary purpose of the Regina substation was to study weed control and soil drifting under the heavy clay conditions of the Regina plains, an area of about 600 thousand hectares. Shortly after the Prairie Farm Rehabilitation Administration (PFRA) was established within the Canada Department of

125

Agriculture in 1935, a Soil Research Laboratory was built on the Swift Current Experimental Station, using PFRA funding. Unfortunately Barnes, who was the director designate at Swift Current, died when the laboratory was in its planning stage. J.L. Doughty, who had extensive experience in soil research, soil survey, and teaching in Canada and the United States, was appointed director. Chepil moved from the Regina substation to the new laboratory; several other well-qualified soil scientists were appointed to the staff.

By 1935 scientists at Swift Current, Lethbridge, and Ottawa prepared a definitive manual (36) that informed farmers of ways to manage their soils. During the next 11 years this publication was modified four times and reprinted four times. In addition to describing strip farming and trash farming, the manual contained advice on the use of cover crops (a late-summer seeding of spring grain), the benefits of establishing shelterbelts, weed control on fallow land, and appropriate emergency measures should fallow land start to drift.

A manual was not enough to encourage all farmers throughout the Great Plains to use cultivating methods to protect their soils against the ravages of wind and drought. Many farmers in the most seriously affected areas simply abandoned their farms and moved north. In 1921, Noble in Alberta successfully pressured the provincial government to appoint well-trained practical people as district agriculturists, a service already available in Ontario. Saskatchewan farmers did not get this service until after 1934 when Taggart, superintendent at Swift Current, urged their provincial government to place an agricultural representative in each municipality. The research staff at experimental stations, particularly at Swift Current, who provided this service, spent much of their time talking to farmer groups and showing them how to adjust and operate machinery.

When PFRA was established in 1935 to assist prairie farmers, Archibald's appointment as its director gave him the joint responsibility for two large organizations. The arrangement provided for experimental farms on the Great Plains to build laboratories, to obtain equipment, and to provide help to fledgling provincial extension services. G.L. Spence was appointed director of PFRA in 1937, but the financial relationship binding PFRA and Experimental Farms remained until 1943. During that 8-year period more than 50 district experimental substations were organized and operated by practical farmers, each under the supervision of an experimental farm; 17 reclamation projects in the most severely drifted areas were established to study methods of handling drifted land; nearly 600 regrassing projects allowed scientists to determine methods of establishing satisfactory stands of grass under drought conditions; four large tree-planting programs and thousands of smaller ones demonstrated the value of trees for shelter and the control of soil drifting (see Table 8.2); and production of crops under irrigation was developed.

To fully acquaint farmers in the Prairie Provinces with the new methods of soil management, PFRA organized 228 Agricultural Improvement Associations, in which 35 800 people held membership. It was imperative that every farmer adopt the new stubble-mulch farming (trash farming) and strip farming techniques to avoid damaging neighbors' fields. PFRA was the catalyst upon which those farmers who stayed in the dust bowl depended. Cooperation between PFRA, Experimental Farm, and Science Service officers continued even after PFRA was moved from the Department of Agriculture. The story of its achievements then is told by others (4, 5, 94).

The perils of the prairie dust bowl were conquered just in time, because in 1940 many young scientists from experimental farms and universities joined the Armed Forces to fight in World War II. Those who stayed, continued to improve upon the techniques for preventing soil loss and, in 1945, they were ready with many projects for the veterans returning from military service. D.T. Anderson, at Lethbridge, pioneered studies of the conservation of surface trash. He found that the wide-blade (Noble) cultivator destroyed less stubble than any other machine used to control weeds (3). K.E. Bowren at Melfort, P.J. Jantzen at Swift Current, and D.A. Brown at Brandon as well as Anderson, determined which machines best maintained mulches while killing water-sucking weeds. Soil scientists were concerned about the proportion of land left as fallow and the rapid decline of organic matter in prairie soils. They sought ways to keep weeds under control yet not damage the standing stubble.

New chemical herbicides held the answer. In 1966 and 1967, C.H. Anderson at Swift Current, used herbicides alone to control weeds in fallow following winter wheat. No tillage implements were employed from the preceding crop harvest until the seeding of winter wheat, about a year later. The chemical fallow or zero-tilled fields produced over 30 percent more wheat than the mechanically fallowed fields. The increase was attributed to better moisture reserves resulting from snow trapped by the stubble on the zero-tilled plots. Further research revealed that a combination of chemical and mechanical fallow was required to control all kinds of weeds, including grass.

Lindwall and Anderson (50) started similar investigations at Lethbridge in 1967, and after 4 years of tests on a clay loam soil in southern Alberta concluded that chemical- or zero-tillage would be uneconomical unless the price of herbicides decreased significantly. They continued to study zero-tillage, particularly ways of seeding through a heavy accumulation of trash.

The potential of zero-tillage for saving energy, reducing soil erosion, conserving moisture, and maintaining yields is being tested by producers. Organic matter has been increasing in zero-tilled fields for 15 years now, whereas previously it had been declining for 75 years. The final results of this research will be beneficial in reducing soil drifting, improving soil structure and fertility, and maintaining crop production. Today, therefore, we are on the fourth major phase of tillage practices on the prairies (51); the first three were plowing, plowless farming, and trash farming.

ROTATIONS

For generations agriculturists have recognized the need to rotate crops, that is, change the kind of crop grown from year to year on a given field. Growing the same crop each year removes the same plant nutrients and is conducive to an increase of diseases and insect pests peculiar to that crop. In a rotation, particularly if legumes such as peas, beans, or alfalfa can be included, some soil nutrients can be maintained or perhaps increased. Determining the best sequence of crops for each soil type and set of weather conditions is a long-term task. Several cycles in a rotation are necessary before reliable conclusions can be reached. Each full cycle may require from 5 to 10 years. However, for tree fruits, rotation in the short-term is not possible, because an apple orchard, for example, may yield profitably over a 50-year period.

The Broadbalk fields at Rothamsted, England, are the oldest continuously recorded rotations. They were initiated in 1843 by Sir John Lawes. Canada has some noteworthy long-term rotations. The first ones were located at the Central Experimental Farm, Ottawa, in 1888 (84). They were laid out in single plots and, thus, not amenable to critical mathematical analyses; however, after 55 years and confirmation of their results by more sophisticated short-term experiments, conclusions were drawn with reasonable confidence.

The oldest irrigated rotation in North America, perhaps in the world, is Rotation U at Lethbridge, Alberta (21, 22). In 1911 W.H. Fairfield laid out a 10-year rotation consisting of 6 years of alfalfa followed by 1 year each of potatoes, wheat, oats, and barley. Until 1933, the only added nutrients were 26 t/ha of manure applied to plots growing 5th-year alfalfa. Varieties of all crops and cultivating methods advanced over time to support farmers' changing requirements. Today, yields of alfalfa exceed the original by approximately 10 percent. Soil analyses for organic matter and plant nutrients following each 10-year cycle show that the soil characteristics remain relatively constant, and with phosphorus, have increased on fertilized portions of each plot. Properly managed, southern Alberta soils under irrigation promise to remain productive indefinitely.

Dryland rotations under prairie conditions are not as complicated as irrigated rotations because of the limited crops that can be grown. Again at Lethbridge, Fairfield set up three wheat rotations: continuous wheat (rotation A); wheat, fallow (rotation B); and wheat, wheat, fallow (rotation C). Rotations A through C had one, two, and three fields, respectively; thus on A, wheat was always grown on stubble, on B wheat was always grown on fallow, and on C, one-half of wheat was always grown on stubble and the other half was always grown on fallow. Freyman and coworkers (26) report that yields on all rotations decreased rapidly during the first few years and then varied considerably, depending upon annual precipitation. Yields on fertilized plots have exceeded those on nonfertilized plots. The largest increase, starting in the mid-1960s, resulted from the use of herbicides for the control of broad-leaved weeds and wild oats. It then was possible to seed shallowly into a moist seedbed immediately following one, rather than several, spring cultivations. Improved cultivars of spring wheat also have increased yields, but advances in weed control have been the most effective means of netting higher yields.

At Swift Current, Campbell and coworkers (18) found that continuous wheat treated with a nitrogen–phosphorus fertilizer outproduced wheat grown under fallow in a 2-year rotation by 50 percent. The application of nitrogen increased yields by 5 percent. More importantly, because wheat often is sold based on its protein content, nitrogen improved protein levels from 13.5 percent to nearly 16 percent. Over the 12-year period, yields were directly related to the amount of rainfall during the growing season. This confirmed Ripley's (84) conclusions that fallow serves as a hedge against possible crop failure. Biederbeck, Campbell, and Zentner (13) showed that the amount and quality of soil organic matter can be improved with the use of annual cropping to cereals, proper fertilization, and stubble-mulch tillage. Zentner and coworkers (108), analyzing the same Swift Current data from an economic perspective, concluded that cropping in Saskatchewan should be more intensive than was traditional cropping. They recommended a 3-year fallow–wheat–wheat system. Continuous wheat netted a fluctuating annual income.

The prime lesson learned from rotations is that soil moisture is the most important factor in obtaining a satisfactory dryland wheat crop. Lethbridge scientists, therefore, have developed an "if" rotation: if soil moisture extends down at least 600 mm at seeding time, then a crop may be sown on fallow or on stubble. Many dryland farmers follow this practice, which requires careful decisions at seeding time. When properly managed, with all plant refuse being returned to the soil, levels of nitrogen, and hence organic matter, can be maintained.

In Eastern Canada, where moisture is rarely a limiting factor, rotation and crop-sequence experiments were started at Ottawa, Kapuskasing, and Sainte-Anne-de-la-Pocatière (84) during the period 1934–1941. They were terminated in 9–12 years when it became apparent that legumes such as alfalfa, sweetclover, and red clover were by far superior preceding crops to cereals or grasses. Ripley (84) concluded that a regular cropping sequence was less significant than arranging crop sequences to suit each individual farming system.

FERTILIZERS

Soils differ in their structural nature and in their chemical content, so their capacity to provide nutrients for plants is different. The addition of organic material such as animal manures and crop residues improves the structure of soil and adds various nutrients for plant use, but there is not sufficient organic material readily available from these sources to satisfy the demands of the world's ever-increasing population. As a consequence, many soils are losing their productive capacity, since their nutrients are sold in grain and other produce and thus are removed from the farm. It is obvious that other sources of plant nutrients must be found and used to maintain the soil's fertility.

The response of plants to one element is not independent of their response to other elements. There is instead, a close interrelationship that renders the study of plant nutrient needs more difficult but also more interesting. Zentner and Read (107) at Swift Current, for instance, showed that under most situations if optimum returns are to be realized adjustments need to be made in the quantity of applied fertilizer according to the soil moisture available.

The champions of the profitable use of fertilizers were the supervisors of the illustration stations. These were the dedicated scientists who, for 44 years during the life of the Illustration Stations Division, interpreted and demonstrated to the farmers—operators of these stations and their neighbors the results of the sophisticated plant nutrition experiments of their erudite colleagues at experimental farms. Their field demonstrations were designed to be of immediate value to the farmer and to provide him with greater dollar returns on his investment.

Research with fertilizers can be divided into four sections:
- primary elements such as nitrogen, phosphorus, and potassium;
- secondary elements such as magnesium and sulfur;
- minor elements such as boron, manganese, copper, iron, lead, and zinc; and
- soil amendments such as lime, manures, and mulches.

Fertilizer experiments were started on Experimental Farms by F.T. Shutt, the first chief of the Chemistry Division, in 1893. His direct influence lasted for 40 years, culminating with his final publication (88) in 1933.

Primary elements

Nitrogen

Until German scientists learned how to combine nitrogen with oxygen and hydrogen at the beginning of the twentieth century, all nitrogen added to soils came from natural sources such as electrical storms, legume-associated nitrogen-fixing bacteria, manures, and nitrate of soda mined in Chile. Phosphorus and potash were mined and converted into fertilizers in forms such as superphosphate and potassium salts.

As early as 1905 Shutt and Charron (87) of the Division of Chemistry showed that successive cropping and poor farming methods caused the loss of nitrogen and organic matter from soils. Angus MacKay collected samples of soil at Indian Head that had been cropped for 22 years. These soil samples showed a loss of more than 30 percent of the soil nitrogen content, of which only one-third could be attributed to removal by crops. Shutt and Charron concluded that the remainder had been lost through cultural practices such as summerfallow. The value of leguminous green-manure crops in maintaining and increasing soil fertility was already known, but in order to demonstrate the phenomenon under Canadian conditions, Shutt and Charron conducted a 12-year series of experiments on each of the five experimental farms. These experiments conclusively showed that plowing in clover increased yields by 30 to 50 percent, and these effects lasted for at least 3 years. Data obtained 20 years later by Shutt (85) showed that after 38 years the loss of nitrogen was 40 percent, with the same proportion as before being lost by poor cultural practices.

At Swift Current, Barnes (10) in 1934, Doughty, et al. (19) in 1943, and Doughty, Cook, and Warder (20) in 1954 showed that nitrogen was lost through summerfallowing, but the rate described by Shutt reduced with continued cultivation. They found that nitrogen in the nitrate form, the form available to plants, was greater in soils following summerfallow than it was in soils with stubble until the end of June. Available nitrogen, which in native prairie soils was high, increased following harvest when soil moisture increased. They made extensive studies of the effects of straw and other organic material worked into soil during the summerfallow operation and concluded that if the amount of straw and stubble is extremely high, the following crop would benefit from the added material. By 1960, farmers on the dry cereal-growing areas of Canada found it profitable to apply nitrogen fertilizer in ever increasing amounts to compensate for its loss during the decomposition of organic matter. Ferguson and Gorby (23) found that under some conditions in Canada, contrary to other studies, the incorporation of straw into the soil did not generally induce a nitrogen deficiency. They reasoned that the low soil-moisture content and the long period of winter freezing reduced saprophytic microorganism activity that is associated with nitrogen utilization during decomposition of straw.

At Summerland, British Columbia, Mason and Miltimore (69) applied 12 rates of ammonium nitrate to a natural range consisting of beardless wheatgrass, *Agropyron inermis*, and sagebrush, *Artemisia tridentata*. Nitrogen was applied at rates varying from 25 to 450 lb/acre (28 to 500 kg/ha). The yields of wheatgrass more than doubled. They repeated the experiment at four other test sites and obtained about the same results. In addition to increased yields of grass, the nitrogen content of the grass, and hence its nutritive value, increased by a factor of three. From a practical viewpoint, applications of more than 50 lb/acre (55 kg/ ha) were uneconomic. Mason (68) applied five rates of nitrogen varying from 0.15 kg to 0.90 kg per tree in a McIntosh apple orchard under irrigation and found no effect on yield but there was a slight reduction in the firmness and grade of fruit. He concluded that nitrogenous fertilizers may not always be desirable for McIntosh apple trees.

On the west coast, Maas, Webster, Gardner, and Turley (53) found that frequent applications of nitrogen gave more uniform, but not greater, yields of grass than a single application of an equal amount. Mack (56) at the Soil Research Institute found that low winter temperatures may benefit the production of nitrogen from soil organic matter. Soil subjected to an extremely low temperature ($-196°C$) for just 10 minutes produced 50 percent more nitrate nitrogen (an indication of ruptured bacterial cells) than when the temperature was held close to freezing. This led Mack to further research on soil temperatures during the growing season when the effect was reversed, as measured by plant growth. The phenomenon was confirmed by Biederbeck and Campbell (12) at Swift Current both in the laboratory and in the field.

Soil scientists at research stations on the dry Great Plains have found that moisture is a limiting factor to plant growth when fertilizers are used. Lehane and Staple (49) showed that each millimetre of moisture in the soil at seeding time increased wheat yields by 5–8 kg/ha, depending upon the soil texture and variety. Bole and Pittman (16) took the matter a step further with barley and determined efficiencies of added water in the form of both soil moisture and precipitation. They related these to the use and cost of nitrogen fertilizer showing, as an example, that

131

nitrogen fertilizer should be increased by about 0.3 kg/ha for each additional millimetre of spring soil moisture. This relationship was refined by Zentner and Read (107) who showed that water supplied during the growing season was three times more effective than stored soil moisture, but because of the unpredictability of adequate soil moisture, adding nitrogen fertilizer to optimize its use would be an economic risk. Bole and Pittman have made similar predictions for spring and winter wheat, based on 15 years' of fertilizer trials. Recommendations to farmers from provincial soil-testing laboratories are based largely upon this and similar work from experimental farms and research stations.

Phosphorus and potassium

The desert soils of British Columbia's Okanagan Valley were among the first in Canada to be carefully studied for their nutritional content. J.C. Wilcox of Summerland started his studies in 1931 and by 1947 (101) had found that although nitrogen fertilizer was needed by some Okanagan orchards, phosphorus and potassium fertilizers generally were not needed to improve tree vigor and yield.

The levels of sulfur, phosphorus, and potassium in alfalfa were studied on 210 sites throughout the Okanagan Valley by Mason, van Ryswyk, and Miltimore (70). Their results showed that there was little difference in potassium content among alfalfa plants grown on the four major soil types, but that there were marked differences in the content of sulfur and phosphorus. They concluded that Podsolic soils would more likely benefit from applications of sulfur and that Brown soils would probably benefit from applications of phosphorus. However, alfalfa grown on Dark Brown or on Black soils would likely not respond to either of these elements.

In other parts of Canada the situation was different. MacLean, Bishop, and Lutwick (60) from the Field Husbandry and Chemistry divisions, showed in 1953 that oats and alfalfa responded to applied phosphorus on about 90 percent of the soils near Ottawa. They also showed that oats responded to applied potassium (61) on only 20 percent of the soils, but that alfalfa responded to applied potassium on 80 percent of the soils. Three years later, in 1956, Halstead teamed with MacLean and Lutwick (32) and learned that the phosphorus content of alfalfa could be increased threefold by the application of a phosphatic fertilizer to the soil. This resulted in nearly triple yields on three different soil types. In 1955, Heeney, Bishop, and Hill (34) developed a curvilinear relationship (logarithmic) between percentage yield of canning tomatoes grown in Prince Edward County, Ontario, and soil tests for phosphorus. Ward (99), during a greenhouse experiment at the Chemistry Division of Science Service, found that the yield of potato tubers was directly proportional to the amount of potassium applied. He also studied the effects of different levels of applied potassium on the amount of other elements in the tissue of the potato plant, thus starting extensive investigation at Ottawa, and later at Harrow, for which he gained considerable recognition.

On the Great Plains, phosphorus studies at Swift Current by Beaton, Read, and Hinman (11), Guitard (28) at Beaverlodge, Anderson (1) at Fort Vermilion, and Spratt and McCurdy (95) at Indian Head, showed that yields of alfalfa, wheat, and barley were increased with appropriate applications of various forms of phosphatic fertilizers. In 1967, at Brandon, Ferguson and Gorby (24) found that the response of wheat to applied phosphorus varied considerably, depending upon the type of soil used. On some soils, responses were similar to those observed by Heeney, Bishop, and Hill (34), whereas on others the responses were different, one being an unusual sigmoid curve, which they explained was the result of a complex interaction of microclimatological factors. In the late 1960s, Read (82) at Swift Current related the response of forage grasses and legumes to the amount of available phosphorus in the soil. Those crops grown on soils with less than 10 parts per million (ppm) of available phosphorus produced about 150 percent more forage when phosphorus was applied to the soil than did crops grown on soils with 20 ppm, or more, of available phosphorus. The results of laboratory analyses of soils were continually being made available to farmers to help them decide on the kind and amount of fertilizer to apply to particular fields for each crop being grown.

A 20-year study, started in 1947 by Lawrence and Heinrichs (48) at Swift Current, showed that both the amount of applied ammonium phosphate and the spacing of Russian wild ryegrass rows affected yields. Spacings of 0.6 m produced better forage yields than both narrower or wider spacings. For optimum seed production slightly wider spacings were needed. Both forage and seed production increased by factors of two and three as fertilizer application was increased. Toward the end of that study, a six-western station cooperative experiment in 1965, led by M.R. Kilcher of Swift Current (47), showed for the first time that on alkali soils both nitrogen and phosphorus had a 3- or 4-year residual effect. This meant that instead of applying fertilizers each year, the operation could be done once every 4 years with relatively larger amounts applied at those times. This observation was confirmed by Read and coworkers at Swift Current (83), and Bailey and coworkers (8) at Brandon in 1977, following 8 years of continuous cropping.

In Quebec and the Maritime Provinces, soil scientists at Saint-Jean, Nappan, Kentville, and Charlottetown made their contributions to the understanding of the intricacies of soil phosphorus and potassium in plant nutrition. MacKay and Munro (59) reviewed the fertilizer situation for the Atlantic region in 1964.

Jasmin, Heeney, and Tourchot (41), working on organic soils in Quebec, found that levels of both phosphorus and potassium in the soil and in potato plants increased as these two nutrients were added to the soil. Strangely, the yield of potatoes did not increase, although when these nutrient levels were in excess, the result was a reduction of the boiling quality of the tubers. Jasmin and coworkers knew that phosphorus in organic soil was present in several complex compounds, learning later that it would continually become available without further additions.

At Kentville, MacKay, MacEachern, and Bishop (58), and later Bishop and colleagues (14) studied the response of potatoes to applications of potassium and phosphorus fertilizers. They observed a yield increase from fertilizer in all instances on mineral soils. Predictions with potassium were reliable from soil analyses, provided a polynomial curve was used. Although already known, the dry matter content of the tubers, and hence their cooking quality, reduced on a straight line as potassium fertilizer, in the chloride form, was increased. It therefore became vital that a soil test should accurately indicate the optimum amount of fertilizer for both yield and quality. Predictions with phosphorus were not as accurate as those for potassium, but it was a step in the right direction to ensure adequate nutrition for optimum production.

L.B. MacLeod's first work (66) dealt with the uptake of potassium by alfalfa and orchardgrass under different soil acidities. At Nappan he made a detailed study (65) of the relationships between nitrogen and potassium fertilization of alfalfa–grass mixtures, finding that alfalfa contributed nitrogen at a rate of 45–90 kg/ha each year to the grass portion of the mixture. When only nitrogen fertilizer was applied, the growth of alfalfa was depressed because of competition from the grasses due to increased tillering. Such growth gave grasses the capacity to compete with alfalfa for the available phosphorus. On Prince Edward Island, working with R.B. Carson of the Analytical Chemistry Research Service in Ottawa, MacLeod (67) found that although the percentage of protein content of barley plants was reduced by applying phosphorus and potassium, the total amount of protein metabolized per plant was increased because of the increased production of plant material. Potassium had a greater effect on the nitrogen metabolism of barley than did phosphorus.

Secondary elements

Magnesium

Tobacco grown on sandy soils in the United States was the first crop in which plant physiologists recognized a magnesium deficiency. The symptom of this diseased condition, known as chlorosis, is a loss of green color in the leaves. It was first reported in Canada at Fredericton in 1934 by MacLeod and Howatt (64) in potatoes, and later confirmed by E.M. Taylor and J.L. Howatt, also of Fredericton, in both potatoes and cereals. After 4 years of careful investigation Taylor and Howatt (96) reported in 1937 that magnesium sulfate sprays corrected the condition in 10 days. The following year Horton (37) of Delhi, Ontario, found tobacco suffering from sand drown, a magnesium deficiency disease characterized by chlorosis, which was first reported from the United States. Hill and Johnston (35) of the Division of Horticulture identified magnesium deficiency in 1940 in fruit trees. They showed that magnesium sulfate sprays, not soil applications, were needed to correct this deficiency in trees.

In the years that followed, much more research was done on magnesium deficiency by McEvoy (72) of the Tobacco Division, Woodbridge (102), and

Ashby and Stewart (7) of Summerland, on apple and cherry, Halstead, Mac-Lean, and Nielsen (33) of Ottawa, on alfalfa, and Ward and Miller (100) of Harrow, on greenhouse tomatoes. The solution to the problem for each one of these crops was to adjust the ratio of magnesium to other soil nutrients or to apply appropriate sprays to the foliage.

Sulfur

Although sulfur is a secondary element, it has more influence on other nutrient elements than it has itself on plant growth. Three examples from the many available will illustrate: McEvoy (72) noted that a magnesium deficiency in tobacco plants caused a yellowing of the leaves. The addition of large amounts of ammonia and sulfates made magnesium even less available, and thus increased the chlorosis condition of the plants and decreased their growth. Reference has already been made to the conclusions of Mason, et al. (70) that sulfur might be of benefit to alfalfa on Podsolic soils because these soils had developed where annual precipitation was 50 cm, sufficient to leach out the sulfates. Nielsen, Halstead, and MacLean (77) found that sulfur had a synergistic relationship with potassium, and an antagonistic relationship with phosphorus, judging by the increase of the former and the decrease of the latter in oat grain and its straw. Sulfur ions in the soil had no effect on the yield of oat grain or its straw, even though sulfur itself was deposited into the grain.

Minor elements

Boron

In 1928, brown-heart disease in turnips was causing serious loss to turnip growers throughout the Maritime Provinces (17). H.T. Güssow, Dominion Botanist, assigned R.R. Hurst of the Plant Pathology Laboratory, Charlottetown, the task of designing experiments to combat this disease. In cooperation with 10 Prince Edward Island illustration stations and R.C. Parent, from the Experimental Station, Hurst conducted a number of manure and fertilizer experiments. Some of the treatments included borax and these proved to be successful; turnips from all plots receiving borax were free from the disease. At the same time D.J. MacLeod and J.L. Howatt of the Plant Pathology Laboratory, Fredericton, found a similar solution to the problem (63). Boron in the form of borax or boric acid has provided a simple solution for brown-heart disease in the turnip fields of the Maritime Provinces, Quebec, Ontario, and British Columbia ever since.

In 1935 McLarty (74) and McLarty, Wilcox, and Woodbridge (75) at Summerland found that the solution to the problems of drought spot disorder and corky core disease that were affecting the fruit of apple trees was to inject the trees with boric acid. As a result of this finding, the production of saleable fruit more than tripled. This one discovery, the result of 10 years' work, saved the apple industry of the Okanagan Valley. With continued use of boron these problems have not recurred. Boron deficiency causes damage to apricots (25),

135

alfalfa (76, 78), and many other crops (57) which, in all instances, is correctable with applications of small amounts of the element. Too much boron can be damaging or even fatal to all crops. In 1969 Ashby (6) solved this problem of oversupply by applying boron in perforated polyethylene capsules that allowed for its slow release, a system used also for the release of nitrogen. U.C. Gupta found that only a small portion of the total boron in soil was actually available to plants. He (29) prepared an excellent monograph on boron in 1979, summarizing our current knowledge on the subject.

Manganese

Gray speck disorder of oats occurs on all continents and is caused by a deficiency of manganese. Its first occurrence in Canada was reported by Prof. J.D. MacLachlan of the Ontario Agricultural College, Guelph, in 1941. He showed how a single spray of manganese sulfate eliminated all evidence of the disorder. Similar symptoms had been observed on oat plots at the Central Experimental Farm as early as 1923, when it was found that sterilizing the affected soil at 100°C for 2.5 hours, or permitting it to dry at room temperature before seeding, readily controlled this disorder. A severe outbreak of gray speck occurred in 1940, but it affected only some varieties. M.I. Timonin of the Bacteriology Division researched the problem and concluded that soil microflora were involved in most manganese deficiencies. By careful experimentation he found (97) that susceptible varieties of oats harbored a dense population of manganese-oxidizing bacteria, rendering soil manganese unavailable to oat plants. Resistant varieties, however, were relatively free from these bacteria and therefore grew normally. Soil sterilization (and soil fumigants) reduced or completely eradicated the oxidizing bacteria. The final solution to the problem was to breed oat varieties that were unable to harbor the damaging bacteria. Five years later, Woodbridge and McLarty (103) at Summerland found peach and apple trees suffering from manganese deficiencies. They were able to correct this condition with manganese sulfate sprays to the trees but, as Hill and Johnston (35) had found with magnesium, when they adjusted the ratio of manganese to other soil nutrients they were not able to correct the condition with soil applications.

Other minor elements

Textbooks have been written on effects minor element deficiencies and excesses have on plants and their biological value to animals whose diets include such plants. Among the elements studied by Research Branch scientists from coast to coast are copper, iron, molybdenum, selenium, and zinc. In all situations where the causal factor has been identified, satisfactory remedies have been found.

Minor elements that are not essential to plant growth can cause damage, even though they are in trace amounts. M.K. John at Agassiz, British Columbia, studied the retention of metallic soil pollutants from industries in the lower Fraser

Valley. He found (44), for instance, that mercury was concentrated in the roots rather than in the leaves and seeds of plants. Lead reacted in a different way; it tended to concentrate in the leaves of lettuce and spinach and in the tubers of radishes and carrots. The edible portions of broccoli and cauliflower, however, remained reasonably free from lead when grown under the same conditions (45).

Soil amendments

Lime

Lime, used to reduce soil acidity, is one of the most common of soil amendments. Much has been learned about lime since 1900, and soil scientists continue to learn more. Men like F.T. Shutt, working on the chemistry of soils, learned much about soil acidity and its causes and were able to provide information on how to amend such soils for improved crop production. Many wet, low-lying, and poorly drained soils contain large amounts of organic matter, as do muck and peat soils. These soils contain no calcium carbonate causing them to be sour or acidic. Conversely, many light upland soils may be acidic because the carbonates have been leached and, if cropped, the lime present has been used by the plants. Shutt prepared a number of bulletins on the subject, the first in 1914 as part of C.E. Saunders' summary of results with cereals. One of his last bulletins (86), published in 1928, provides farmers with comparative values of different sources of lime and gives examples of improved yields (fourfold increases with barley at Cap Rouge, Quebec) following liming. Subsequent departmental publications on the same subject were written by L.E. Wright and H.S. Hammond (No. 585 in 1937), C.D.T. Cameron and L.S. Hamilton (No. 1086 in 1960), H.J. Atkinson (No. 869 in 1961), P.B. Hoyt, M. Nyborg, and D.C. Penney (No. 1521 in 1974), and K. Bruce MacDonald (No. 1731 in 1981).

In 1914, one of the first experiments laid out by superintendent Blair at Kentville dealt with the application of lime and fertilizers to a 3-year rotation. Blair and Leefe (15) reported that, after 24 years, the application of limestone was the greatest single factor influencing the yields of all crops. They also noted that potato scab increased as the the soil was neutralized with lime. A similar experiment (106), started at Sainte-Anne-de-la-Pocatière in 1932, produced only slightly increased yields with lime, whereas manure and lime or any combination including phosphatic fertilizers increased the yield of turnips three to five times over that without added amendments.

At Nappan in 1957, on dikeland soils, MacLeod (62) found increases of 10 percent or more in grain yields, and an astonishing increase in clover yields of four to eight times on limed soils in comparison with that of unlimed soils. The amount of lime applied in order to obtain optimum results depended upon both soil and crop. Similar results were obtained by Jasmin and Heeney (40) on organic soils in Quebec. However, they took their 1957 research a step further and found that although liming increased the available soil phosphorus, it materially reduced the amount of nitrogen, phosphorus, and potassium in the plant's tissue, even though, as might be expected, calcium levels were increased. They concluded that unless adequate amounts of lime were used, high applications of fertilizer could in fact be detrimental to plant growth.

One of the questions to answer now was how did the increase in available soil phosphorus occur? Other scientists had shown that organic forms of soil phosphorus underwent mineralization, its extent depended upon various soil conditions. The nature of the release of phosphorus from its organic form following the application of lime, however, was limited. Halstead, Lapensee, and Ivarson (31) of the Soil Research Institute resolved the problem in 1963 when they learned that there was only a slight mineralization of organic phosphorus in unlimed soil over a 9-month period, whereas in limed soil there was a marked increase in microbial activity, which was associated with a reduction in the amount of extractable organic phosphorus and an increase in available soil phosphorus.

Professor F.A. Wyatt, at the University of Alberta, had learned by 1945, that liming wooded soils in the central part of his province doubled the yield of alfalfa hay. Not until Hoyt, Hennig, and Dobb (38) studied 28 wooded and parkland soils of the Peace River region was it realized that liming affected yields of both alfalfa and barley under the climatic conditions existing in these northern areas. The two crops were affected in different ways, however. Unless phosphorus fertilizer as well as lime was added to barley, crop yields were actually depressed. Alfalfa benefited from lime whether phosphorus was added or not. Hoyt, Hennig, and Dobb also found that the amount of soluble aluminum in the soils related closely to the yields of alfalfa and barley. This confirmed an observation made in 1947 by V.A. Chernov, a Russian soil scientist. In the coastal region of British Columbia, John, Case, and Van Laerhoven (46) provided further evidence on the ways in which lime reduces the toxicity of aluminum and manganese, hence increasing yields of alfalfa.

For some time, soil scientists have recognized that most acid soils are saturated with aluminum rather than hydrogen ions. The acidity of the soil is therefore a result of hydrolysis of aluminum. Canadian soil scientists, particularly at the Land Resource Research Institute, have studied this phenomenon in relation to the liming of soils. Turner and Clark (98) gained international reputations in 1966 for introducing the concept of "corrected lime potential" to define the degree of base saturation in soils. This became the basis for procedures now used in soil-testing laboratories to determine the "lime requirement" of soils. More recently Singh (89) investigated the chemistry of aluminum in the presence of sulfur and found that a substantial amount of aluminum is in a form that can move through the soil into water courses, causing aluminum toxicity to aquatic plants and animals.

In Prince Edward Island, where soils are known to be acidic and crops respond to applications of lime, U.C. Gupta suspected from his earlier research and also from some done in other countries that the reaction of plants to molybdenum might be affected by the acidity of the soil, and hence to liming.

His judgment was correct, for when he put his theory to the test in 1969 (30) he showed that both cauliflower and alfalfa crops failed on two of the three soils tested unless both molybdenum and lime were applied. On the third soil, as found by some other workers, lime alone gave a yield response, indicating that the soil contained sufficient molybdenum.

Climatic conditions are major factors in the development of soil acidity, as are a number of agricultural practices such as fertilization, irrigation, and improved drainage. One current threat to further acidification of soils is acid rain. Singh and Coote (90) have recently concluded that up to 1985 the impact of acid rain on agricultural soils, and hence crops, is modest when compared to natural processes of acidification. They warn, however, that a continued watch needs to be kept on the rates and amounts of acid-forming pollutants affecting agricultural soils.

Manure 139

Farmyard manure and sewage effluent (see Chapter 14) have been recognized for generations as valued by-products of the farm. Well before the turn of this century William Saunders, F.T. Shutt, J.W. Robertson, and the superintendents of each of the four other experimental farms examined the ways farmyard manures should be stored (kept under cover) and added to soils to improve their nutrient status and tilth. Green-manure crops, those which are plowed into the soil, are usually legumes or mixtures of legumes and grasses, which, like farmyard manure, add both nutrients and organic matter to the soil, improving its structure and aiding the growth of plants. All forms of organic matter such as seaweed, vegetable composts, and sawdust, can be used as manures but their value as a soil amendment varies, depending upon their location and composition.

Shutt published Dominion Experimental Farms Bulletin No. 31 on barnyard manure in 1898. It was reprinted several times and rewritten in 1937 by Wright and Hammond (105). It resembles a textbook and discusses, in terms understandable to all familiar with soil, the best husbandry practices of the day. As far as manures are concerned these practices still remain.

Experimental farms, stations, and illustration stations conducted hundreds of experiments, including a 10-year experiment by Woods (104), to learn more about the action of farmyard manure on soils and crops. Woods compared cover crops (green-manure), farmyard manure, and fertilizer by measuring the production of raspberries in the lower Fraser Valley of British Columbia. Winter injury to canes was reduced when cover crops were plowed under each year, which resulted in increased yields compared with crops treated with farmyard manure or fertilizer. Farmyard manure increased the organic content of the soil more than any of the other treatments but had no effect upon the yield of raspberries because, as with fertilizer applications, cane growth increased by 8 and 24 percent, respectively, but resulted in winter injury and a reduction of crop.

Woods concluded that climatic conditions at Agassiz influenced the yield of raspberries as much as cultural conditions.

Mulches

Grass, hay, straw, and dust have universally been used by gardeners and farmers as mulches to reduce evaporation from the surface of soils and as a blanket to inhibit weed growth. Grass, hay, and straw are used as soil additives to increase organic matter and hence improve the tilth and fertility of soil. The mulch known to most of us is the straw placed between the rows of strawberry plants. This is done to keep the soil away from the berries while they ripen. Straw, however, often contains many weed seeds and therefore, rather than acting as an inhibitor of weed growth, it frequently adds to the weed population.

J.J. Woods at the Saanichton Experimental Station was one of the first scientists in Canada to investigate the use of different kinds of mulches on horticultural crops. Because sawdust was in plentiful supply on the west coast, he compared sawdust with straw and hay in 1948. He and his colleagues found that hay and straw markedly improved the growth of pear and cherry trees and in some years doubled the yield of pears when compared with trees grown under clean cultivation. He found sawdust to be as good as or better than hay or straw with loganberries and gooseberries, and of considerable benefit to cool-season vegetable crops such as radish, spinach, and lettuce. Because the decomposition of sawdust used nitrogen from the soil, extra nitrogenous fertilizer had to be added. His results with strawberries were not as good, but H.F. Fletcher and J.A. Freeman at Agassiz, as well as others, found sawdust to be of particular value on strawberries.

Twenty years after this work was done greenhouse vegetable growers on Vancouver Island had trouble with soil-borne diseases in their tomato crops. There were two known methods of solving the problem: sterilize the soil either with a fumigant or with steam, or grow the plants in sand and feed them with an aqueous solution of plant nutrients. Both methods required handling heavy soil or sand. Was there a better method? E.F. Maas and R.M. Adamson of Saanichton thought so. Sawdust had been used as a mulch. Could it be used instead of soil? It was light, disease free, and inexpensive, and it adsorbed moisture reasonably well. They prepared a standard plant nutrient solution, then half filled a number of plastic garbage bags with sawdust and planted a tomato plant in a hole cut in the side of each bag. Each bag was laid on its side on the floor of the greenhouse. Nutrients were automatically fed through plastic tubing to each plant. Maas and Adamson were amazed with their success! Since then, the method has been refined several times. Some growers use either peat or mixtures of peat and sawdust. Other crops such as cucumbers and chrysanthemums also grew well when tested in this way. Today, it is the method (52) used by nearly all greenhouse growers in many parts of Canada and the United States.

SALINITY

The excess accumulation of calcium sulfate (gypsum), magnesium sulfate (Epsom salts), sodium sulfate (Glauber's salt), potassium sulfate, and other salts in soil solutions causes a reduction of plant growth, often to the point of preventing all growth. In the Prairie Provinces 80 000 ha of irrigated land and 2 million hectares of nonirrrigated land are saline (92).

In irrigated districts, the main causes of salinization are seepage from canals, poor irrigation practices (usually over-irrigation), and inadequate drainage. On nonirrigated land, the main causes are weather patterns that cycle excess water to prairie root zones already aggravated by the introduction of cereal crops in place of native range, summerfallowing of cereal cropland, and man-made obstacles that impound water and so impede drainage. Excess water dissolves, transports, and concentrates salts in the soil, usually in low areas. Water in excess of crop usage raises the water table. When the water comes to within about 100 mm of the soil surface it moves to the surface by capillary action, carrying salts with it. At the surface, water evaporates and the salts remain in the soil. In the root zone, water is used by the plants; thus more water and salts are drawn from below the root zone and concentrated in this critical area. Salinization on both irrigated and nonirrigated lands is serious, but since 4 percent of the arable land in Alberta, for instance, is irrigated and it produces 20 percent of the province's agricultural products, salinity of irrigated soils is the most economically damaging.

When irrigation was started in southern Alberta in 1900, soil scientists from the United States warned of the possible dangers of salinization. Because of long winters, however, Canadian soil scientists were of the view that salinization would not become a problem. Except for a few locations where irrigation canals leaked and flowed down to nondrained basins, they were correct—for about 50 years. Then, in the early 1960s, saline areas began to build on both irrigated and dryland. Today, salinization is a serious problem.

Scientists at Swift Current and Lethbridge have studied ways of combatting salinization to both prevent and correct it. It is a problem experienced wherever surface water is applied to irrigate soils. Such locations include arid and semiarid areas in the United States, Mexico, Chile, Australia, Iran, Pakistan, India, the USSR, and China. Sommerfeldt and MacKay (91) conclude that by discontinuing summerfallow, recropping and growing perennial forages, removing retainers of excess water and snow, and using drainage water for irrigation would, in most instances, arrest and even reduce the salinity problem in dryland areas. For irrigated areas, Sommerfeldt and Rapp (92) emphasize the urgent need for good drainage, followed by leaching with successive floodings, and seeding salt-tolerant crops such as barley, wheatgrasses, and sweetclover. McElgunn and Lawrence (71) tested several forage species for their tolerance to salt and found that tall wheatgrass, *Agropyron elongatum*, and Alti wild ryegrass, *Elymus angustus*, are more tolerant than a number of other species.

Other studies done by Jamie, et al. (39) at Swift Current indicated that if drainage is inadequate when using saline sewage effluent for irrigating, salts are

141

not removed by leaching. However, when drainage is adequate, steady state salinity profiles develop in most soils and allow for good plant growth.

The problem of soil salinization is still not well defined for Canadian conditions. It is under concentrated study by soil scientists at universities, provincial departments of agriculture, and Research Branch.

DIKELANDS

The dikelands of New Brunswick and Nova Scotia have been formed over the centuries by each receding tide of the Bay of Fundy, leaving a rich deposit of silt. Baird (9) reports that as early as 1604 Acadian colonists reclaimed some lands from the sea by building mud dikes and draining the resulting flats. Today, 32 000 ha are diked and an additional 4000–6000 ha could be reclaimed. Beef was produced on the dikelands from the time they were first formed until 1892, when the British placed an embargo on the import of Canadian cattle. Although a domestic market developed for beef, the demand for hay became greater. This remained the prime commodity until the price of hay dropped drastically in 1933. As a consequence, many of the dikes were neglected by farmers and swept into the sea. Drainage ditches became filled with mud and the area quickly lost its capacity to produce.

As a result of appeals from the Maritime Beef Cattle Committee and the New Brunswick and Nova Scotia Departments of Agriculture for assistance in reclaiming the dikelands, E.S. Archibald, director of Experimental Farms, met in 1943 with representatives of the two groups to decide what might be done to rejuvenate the agricultural potential of the area. Archibald formed the Maritime Dikeland Rehabilitation Committee with representatives from property owners, provincial departments of agriculture of New Brunswick and Nova Scotia, and the Canada Department of Agriculture. W.W. Baird, Superintendent, Experimental Farm, Nappan, was selected by the committee as its chairman. The committee obtained the advice of B. Russell and L.B. Thomson, both of whom had helped to establish PFRA and had had reclamation experience in the western provinces. The committee obtained $10 000 from Archibald to survey the area and by that fall more than 6000 ha had been examined. During the next 4 years Baird directed 130 projects to reconstruct dikes in both New Brunswick and Nova Scotia. The work was financed jointly by owners, the two provinces, and the government of Canada.

The Experimental Farm, Nappan, had 28 ha of dikeland included in its original property purchased in 1887. It had been laid out in narrow dales, with drainage ditches between the dales. In 1917 it was reconstructed with wider dales, thereby reducing the number of ditches and increasing the cultivated area. Some of their first experiments showed that dikeland soils were higher in nitrogen, phosphorus, magnesium, potash, and manganese than upland soils. In consequence, hay from the dikelands was higher in protein and ash (minerals) than that grown on uplands. Many other experiments involving the application of manures and fertilizers, cultural methods, and kinds and varieties of crops, helped the committee advise farmers on appropriate practices for their rehabilitated lands.

The Baird committee, probably with full support from Archibald, persuaded the government of Canada to pass the Maritime Marshland Rehabilitation Act (MMRA) in 1948. Under this Act, the federal government assumed responsibility for the construction of the main protective dike works, whereas the provinces and landowners assumed the cost of land drainage and utilization. Between 1948 and 1968 MMRA built protection dikes in all major areas along the Bay of Fundy. However, land use did not follow as expected (55), largely because modern farm machinery could not be manoeuvered on land with open drainage systems. This problem created renewed interest in soil fertility and drainage on the part of scientists.

The first experiments in soil fertility and drainage were started at Nappan in 1922. At that time, narrow straight-sided ditches were dug by hand about one chain (22 m) apart. Thirty years later, some dales were constructed by crowning the land between the ditches, thus eliminating the need for straight-sided ditches. The shallow depressions between dales carried off the water and permitted machines to operate in any direction across the land. This was a simple, effective, although costly way of solving the problem of moving machinery about on dikeland. Shortly after the reconstruction program was started under the new MMRA, scientists at the Experimental Farm, Nappan, conducted experiments by laying tile in the bottom of drainage ditches, covering them with soil, and again forming land with depressions on the tile-drain line. The system worked well for a few years until the tiles became plugged with silt and failed to function. Further experiments showed that if land forming was adequate, flats up to 50 m wide without tile drainage were satisfactory. This work, together with the rehabilitation of dikes and main drainage ditches, has now rejuvenated the thousands of productive hectares of dikelands surrounding the Bay of Fundy. It supplies hay, pasture, and some coarse grains for the Maritime beef and dairy industries.

PEAT AND BOG SOILS

Canada has 130 million hectares (14 percent of its land area) of undifferentiated peat land, some of which is currently used. Such land is found in all 10 provinces and the territories. Peat lands in Quebec and Ontario are the best developed because of their proximity to centers of population and because the climates under which they are found are more conducive to plant growth than those in many other areas of Canada.

F.T. Shutt, the first Chief, Chemistry Division, studied peat and mud deposits shortly after his appointment in 1886. He realized the value of such material for improving soils when incorporated, and in 1933, with L.E. Wright, one of his colleagues, he published his final bulletin on the subject (88). They had learned much about the variability of peat and related soils and how essential analysis was before such soils could be extensively used for agricultural production.

In developing these organic soils Canadian scientists have built on Shutt's work and have borrowed and adapted from other parts of the world. In 1937, scientists at the Illustration Stations Division completed experiments on several peat bogs. They found that the common practice of burning the top 100 to 200 mm of peat following clearing reduced the yields of corn and potatoes by about 50 percent. Yields of most crops responded well to applications of potassium and in some instances to boron. In 1939, the Sainte-Clothilde Experimental Substation was established as a consequence of these findings at the Illustration Stations Division. Research at Sainte-Clothilde covers soil subsidence, drainage, compaction, pesticides, variety selection, and soil fertility. Jasmin and Heeney (40) showed that for carrots and onions lime is needed at the rate of 10 t/ha but that for potatoes lime is only needed at the rate of 3.4 t/ha. In 1977, Jasmin, et al. (42) summarized 40 years' of research at Sainte-Clothilde.

At Saint-Jean, Quebec, Jasmin and Hamilton (43) showed that the availability of nutrients to plants was different on organic soils than on mineral soils. The practice of applying lime on mineral soils to reduce acidity was often more important on peat soils to supply calcium to plants than to reduce acidity. Applications of other mineral elements such as phosphorus and potassium, plus minor elements, were needed for proper plant growth. Nitrogen, however, was a different matter. These soils generally contained more than 80 percent organic matter that decomposes through microbial activity to produce nitrogen. Soil temperature, acidity, and moisture levels all affect microbial activity and nitrogen availability. When microbial activity is too great, nitrogen available to plants may be reduced to the point where extra nitrogen must be added in the early spring.

Research on peat soil at the St. John's West Research Station, Newfoundland, has dealt with the production of both forage and vegetable crops and included work on drainage, cultivar evaluation, fertility, and weed control. Rayment and Campbell (80) concluded that covered rather than open ditches were needed for vegetable culture because of the great numbers of ditches required by Newfoundland's high precipitation. They found that land ridging provided local drainage and aeration for improved vegetable crop production. Studies by O.A. Olsen conducted between 1958 and 1960 showed that vegetable production on Newfoundland peat soil was equal or superior to that on mineral soil, and that application of minor elements was essential.

Forage production presents the most promising potential for the extensive use of Newfoundland peat soils; research has delineated the drainage and fertilizer requirements (81) compatible with economic production of pasture and winter feeds.

The development of Bradford Marsh, north of Toronto, Ontario, and the solution to problems associated with vegetable production there was accomplished through collaboration with the University of Guelph and the Ontario Department of Food and Agriculture.

British Columbia, Manitoba, New Brunswick, and Nova Scotia, too, have organic soils currently in production. Peats and bogs, although underutilized in Canada, are valuable natural resources that require careful development if they are to remain useful for future generations.

References

1. Anderson, C.H. 1963. The precision of rod-row tests with barley grown at three topographical locations as influenced by alleviating a phosphorus deficiency. Can. J. Plant Sci. 43:519–521.
2. Anderson, C.H. 1975. A history of soil erosion by wind in the Palliser triangle of western Canada. Can. Dep. Agric. Hist. Ser. No. 8. 25 p.
3. Anderson, D.T. 1961. Surface trash conservation with tillage machines. Can. J. Soil Sci. 41:99–114.
4. Anonymous. 1961. PFRA, the story of conservation on the prairies. Queen's Printer, Ottawa. 71 p.
5. Anonymous. 1969. Prairie resources and PFRA. Queen's Printer, Ottawa. 78 p.
6. Ashby, D.L. 1969. Controlled release of magnesium, zinc, boron and nitrogen from polyethylene capsules. Can. J. Plant Sci. 49:555–566.
7. Ashby, D.L.; Stewart, J.A. 1969. Magnesium deficiency in tree fruits as related to leaf concentration of magnesium, potassium and calcium. J. Am. Soc. Hortic. Sci. 94:310–313.
8. Bailey, L.D.; Spratt, E.D.; Read, D.W.L.; Warder, F.G.; Ferguson, W.S. 1977. Residual effects of phosphorus fertilizer. II. For wheat and flax grown on chernozemic soils in Manitoba. Can. J. Soil Sci. 57:263–270.
9. Baird, W.W. 1954. Report on dykeland reclamation 1913 to 1952. Experimental Farm, Nappan, N.S. 63 p.
10. Barnes, S. 1934. Legumes and the nitrogen problem in the dry farming areas of western Canada. Sci. Agric. XIV:459–465.
11. Beaton, J.D.; Read, D.W.L.; Hinman, W.C. 1962. Phosphorus uptake by alfalfa as influenced by phosphate source and moisture. 1962. Can. J. Soil Sci. 42:254–265.
12. Biederbeck, V.O.; Campbell, C.A. 1973. Soil microbial activity as influenced by temperature trends and fluctuations. Can. J. Soil Sci. 53:luenced by temperature trends and fluctuations. Can. J. Soil Sci. 53:363–376.
13. Biederbeck, V.O.; Campbell, C.A.; Zentner, R.P. 1984. Effect of crop rotation and fertilization on some biological properties of a loam in southwestern Saskatchewan. Can. J. Soil Sci. 64:355–367.
14. Bishop, R.F.; MacEachern, C.A.; MacKay, D.C. 1967. The relation of soil test values to fertilizer response by the potato. IV. Available phosphorus and phosphatic fertilizer requirements. Can. J. Soil Sci. 47:175–185.
15. Blair, W.S.; Leefe, J.S. 1939. Influence of lime in crop rotation with a note on the occurrence of boron deficiency in mangels. Sci. Agric. XIX:330–341.
16. Bole, J.B.; Pittman, U.J. 1980. Spring soil water, precipitation, and nitrogen fertilizer. Effect on barley yield. Can. J. Soil Sci. 60:461–469.
17. Bourdon, Mary B. 1984. Charlottetown Research Station 1909–1984. Agric. Can. Hist. Ser. No. 19. 83 p.
18. Campbell, C.A.; Read, D.W.L.; Zentner, R.P.; Leyshon, A.J.; Ferguson, W.S. 1983. First 12 years of a long-term crop rotation study in southwestern Saskatchewan—yields and quality of grain. Can. J. Plant Sci. 63:91–108.
19. Doughty, J.L., et al. 1943. Report of investigations, soil research laboratory, Swift Current, Sask. Dep. Agric. 63 p.
20. Doughty, J.L.; Cook, F.D.; Warder, F.G. 1954. Effect of cultivation on the organic matter and nitrogen of brown soils. Can. J. Agric. Sci. 34:406–411.

21. Dubetz, S. 1983. Ten-year irrigated rotation U. 1911–1980. Agric. Can. Contrib. No. 1983-21. 17 p.
22. Dubetz, S.; Oosterveld, M. 1979. Sixty-six-year trends in irrigated yields—barley, wheat, and oats. Can. J. Plant Sci. 59:685–689.
23. Ferguson, W.S.; Gorby, B.J. 1964. Effect of straw on availabilaity of nitrogen to cereal crops. can. J. Soil Sci. 44:286–291.
24. Ferguson, W.S.; Gorby, B.J. 1967. The effect of soil catena and phosphorus on the yield of wheat in western Canada. Can. J. Soil Sci. 47:39–48.
25. Fitzpatrick, Randal E.; Woodbridge, C.G. 1941. Boron deficiency in apricots. Sci. Agric. 22:271–273.
26. Freyman, S.; Palmer, C.J.; Hobbs, E.H.; Dormaar, J.F.; Schaalje, G.B.; Moyer, J.R. 1981. Yields trends on long-term dryland wheat rotations at Lethbridge. Can. J. Plant Sci. 61:609–619.
27. Gray, J.H. 1978. Men against the desert. Western Producer Prairie Books, Saskatoon. 270 p.
28. Guitard, A.A. 1960. The precision of rod-row trials with wheat as influenced by alleviating a phosphorus deficiency. Can. J. Plant Sci. 40:539–541.
29. Gupta, Umesh C. 1979. Boron nutrition of crops. Adv. Agron. 31:273–307.
30. Gupta, Umesh C. 1969. Effect and interaction of molybdenum and limestone on growth and molybdenum content of cauliflower, alfalfa, and bromegrass on acid soils. Soil Sci. Soc. Am. 33:929–932.
31. Halstead, R.L.; Lapensee, J.M.; Ivarson, K.C. 1963. Mineralization of soil organic phosphorus with particular reference to the effect of lime. Can. J. Soil Sci. 43:97–106.
32. Halstead, R.L.; MacLean, A.J.; Lutwick, L.E. 1956. Availability of phosphorus in soils pre-treated with superphosphate. Can. J. Agric. Sci. 36:471–482.
33. Halstead, R.L.; MacLean, A.J.; Neilsen, K.F. 1958. Ca:Mg ratios in soil and the yield and composition of alfalfa. Can. J. Agric. Sci. 38:85–93.
34. Heeney, H.B.; Bishop, R.F.; Hill, H. 1955. The relationship of soil phosphorus tests to crop yield and to the requirement of the tomato for phosphatic fertilizer. Can. J. Agric. Sci. 35:11–18.
35. Hill, H.; Johnston, F.B. 1940. Magnesium deficiency of apple trees in sand culture and in commercial orchards. Sci. Agric. XX:516–525.
36. Hopkins, E.S.; Palmer, E.A.; Chepil, W.S. 1964. Soil drifting control in the prairie provinces. Dom. Can. Dep. Agric., Publ. 568 rev. 4. 59 p.
37. Horton, H.A. 1938. Magnesium relations in new belt tobacco soils. The Lighter 8(2):11–15.
38. Hoyt, P.B.; Hennig, A.M.; Dobb, J.L. 1967. Response of barley and alfalfa to liming of solonetzic, podzolic and gleysolic soils of the Peace River region. Can. J. Soil Sci. 47:15–21.
39. Jamie, Y,W,; Biederbeck, V.O.; Nicholaichuk, W.; Korven, H.C. 1964. Salinity and alfalfa yield under effluent irrigation in southwestern Saskatchewan. Can. J. Soil Sci. 64:323–332.
40. Jasmin, J.J.; Heeney, H.B. 1962. The effect of lime on the status of nitrogen, phosphorus, potassium, calcium, and magnesium in a few vegetables grown on acid peat soils. Can. J. Plant Sci. 42:445–451.
41. Jasmin, J.J.; Heeney, H.B.; Tourchot, R. 1964. Potassium and phosphorus accumulation in organig soil and their influence on potato crops. Can. J. Soil Sci. 44:280–285.

146

42. Jasmin, J.J.; Hamilton, H.A.; Millette, J.; Hogue, E.J.; Bernier, R. 1977. Mise en production des sols organiques. Agric. Can. tech. Bull. 11. 38 p.

43. Jasmin, J.J.; Hamilton, H.A. 1980. Vegetable production on organic soils in Canada. Pages 51–58 *in* The diversity of peat. Newfoundland and Labrador Peat Assoc.

44. John, Matt K. 1972. Mercury uptake from soil by various plant species. Environ. Contam. Toxicol. 8(2):77–80.

45. John, Matt K. 1972. Lead distribution in plants grown on a contaminated soil. Environ. Lett. 3(2):111–116.

46. John, Matt K.; Case, V.W.; Van Laerhoven, C. 1972. Liming of alfalfa (*Medicago sativa* L.) I. Effect on plant growth and soil properties. Plant Soil 37:353–361.

47. Kilcher, M.R.; Smoliak, S.; Hubbard, W.A.; Johnston, A.; Gross, A.H.; McCurdy, E.V. 1965. Effects of inorganic nitrogen and phosphorus fertilizers on selected sites of native grassland in western Canada. Can. J. Plant Sci. 45:229–237.

48. Lawrence, T.; Heinrichs, D.H. 1968. Long-term effects of row spacing and fertilizer on the productivity of Russian wild ryegrass. Can. J. Plant Sci. 48:75–84.

49. Lehane, J.J.; Staple, W.J. 1965. Influence of soil texture, depth of soil moisture storage and rainfall distribution on wheat yield in southwestern Saskatchewan. Can. J. Soil Sci. 45:207–219.

50. Lindwall, C.W.; Anderson, D.T. 1977. Effects of different seeding machines on spring wheat production under various conditions of stubble residue and soil compaction in no-till rotations. Can. J. Soil Sci. 57:1077–1083.

51. Lindwall, C.W.; Dubetz, S. 1983. Conservation tillage. Pages 121–129 *in* Second Annu. Western Prov. Conf. Ration. Water Soil Res. Manage. Saskatoon.

52. Maas, E.F.; Adamson, R.M. 1978. Soilless culture of commercial greenhouse tomatoes. Can. Dep. Agric. Publ. 1460. 29 p.

53. Maas, E.F.; Webster, G.R.; Gardner, E.H.; Turley, R.H. 1962. Yield response, residual nitrogen and clover content of an irrigated grass–clover pasture as affected by various rates and frequencies of nitrogen application. Agron. J. 54:212–214.

54. MacEwan, Grant. 1983. Charles Noble guardian of the soil. Western Producer Prairie Books, Saskatoon, Saskatchewan. 208 p.

55. MacIntyre, T.M.; Jackson, L.P. 1975. Dikeland drainage and land forming. Agric. Can. Exp. Farm, Nappan, N.S. 19 p.

56. Mack, A.R. 1962. Low-temperature research on nitrate release from soil. Nature (London) 193:803–804.

57. MacKay, D.C.; Langille. W.M.; Chipman, E.W. 1962. Boron deficiency and toxicity in crops grown on sphagnum peat moss. Can. J. Soil Sci. 42:302–310.

58. MacKay, D.C.; MacEachern, C.R.; Bishop, R.F. 1964. The relation of soil test values to fertilizer response by the potato. III. Exchangeable potassium and potassium fertilizer requirements. Can. J. Soil Sci. 44:310–318.

59. MacKay, D.C.; Munro, D.C. 1964. Soil fertility in the Atlantic region. Agric. Inst. Rev. 19(6):16–19.

60. MacLean, A.J.; Bishop, R.F.; Lutwick, L.E. 1953. Fertility studies on soil types. III. Phosphorus supply and requirement as shown by greenhouse studies and laboratory tests. Can. J. Agric. Sci. 33:330–341.

61. MacLean, A.J.; Bishop, R.F.; Lutwick, L.E. 1954. Felrtility studies on soil types. IV. Potassium supply and requirement as shown by greenhouse studies and laboratory tests. Can. J. Agric. Sci. 34:374–384.

62. MacLeod, L.B. 1958. The liming of dikeland soils. Can. J. Soil Sci. 38:69–76.

63. MacLeod, D.J.; Howatt, J.L. 1934. The control of brown heart in turnips. Sci. Agric. 15:435.

64. MacLeod, D.J.; Howatt, J.L. 1934. Magnesium deficiency in potatoes. Sci. Agric. 15:435.

65. MacLeod, L.B. 1965. Effect of nitrogen and potassium on the yield, botanical composition, and competition for nutrients in three alfalfa–grass associations. Agron. J. 57:129–134.

66. MacLeod, L.B.; Bradfield, Richard. 1964. Greenhouse evaluation of the relative importance of lime and potassium for the establishment and maintenance of alfalfa and orchardgrass seedlings. Agron. J. 56:381–386.

67. MacLeod, L.B.; Carson, R.B. 1969. Effects on N, P, and K and their interactions on the nitrogen metabolism of vegetative barley tissue and on the chemical composition of grain in hydroponic culture. Agron. J. 61:275–278.

68. Mason, J.L. 1968. Effect of cultivation and nitrogen on fruit quality, yield and color of McIntosh apples grown in irrigated grass sod cover crop. Can. J. Plant Sci. 49:149–154.

69. Mason, J.L.; Miltimore, J.E. 1964. Effect of nitrogen content of beardless wheatgrass on yield response to nitrogen fertilization. J. Range Manage. 17:145–147.

70. Mason, J.L.; van Ryswyk, A.L.; Miltimore, J.E. 1963. Changes in sulphur, phosphorus and potassium content of alfalfa with soil zone. Can. J. Plant Sci. 43:94–96.

71. McElgunn, J.D.; Lawrence, T. 1973. Salinity tolerance of Alti wild ryegrass and other forage grasses. Can. J. Plant Sci. 53:303–307.

72. McEvoy, E.T. 1954. The relation of ammonium and sulphate ions to magnesium deficiency in tobacco. Can. J. Agric. Sci. 34:281–287.

73. McKeague, J.A.; Stobbe, P.C. 1978. History of soil survey in Canada 1914–1975. Can. Dep. Agric. Hist. Ser. No. 11. 30 p.

74. McLarty, H.R. 1936. Tree injections with boron and other materials as a control for drought spot and corky core of apple. Sci. Agric. 16:625–633.

75. McLarty, H.R.; Wilcox, J.C.; Woodbridge, C.G. 1936. The control of drought spot and corky core of the apple in British Columbia. Pages 142–146 *in* Proc. 32 Annu. Meet. Wash. State Hortic. Assoc., Yakima, Wash.

76. McLarty, H.R.; Wilcox, J.C.; Woodbridge, C.G. 1937. A yellowing of alfalfa due to boron deficiency. Sci. Agric. XVII:515–517.

77. Nielsen, K.F.; Halstead, R.L.; MacLean, A.J. 1967. Ion interactions in oats as affected by additions of nitrogen, phosphorus, potassium, chlorine, and sulfur. Soil Sci. 104:35–39.

78. Ouellette, G.J.; Lachance, R.O. 1954. Soil and plant analysis as means of diagnosing boron deficiency in alfalfa in Quebec. Can. J. Agric. Sci. 34:494–503.

79. Palmer, Asael E. Undated. When the winds came. A.E. Palmer, Lethbridge, Alta. 57 p.

80. Rayment, A.F.; Campbell, J.A. 1980. The influence of different drainage techniques on water outflow, soil aeration and crop growth of a Newfoundland peat. Pages 451–454 *in* Proc. 6th Int. Peat Congr., Duluth, Minn.

81. Rayment, A.F.; Penney, B.G. 1979. The agricultural potential of Newfoundland peat soils. pages 39–50 *in* The diversity of peat. Newfoundland and Labrador Peat Assoc.

82. Read, D.W.L. 1966. Relationship of soil phosphorus to fertilizer response on irrigated forage. Agron. J. 58:225–226.

148

83. Read, D.W.L.; Spratt, E.D.; Bailey, L.D.; Warder, F.G. 1977. Residual effects of phosphorus fertilizer. I. For wheat grown on four chernozermic soil types in Saskatchewan and Manitoba. Can. J. Soil Sci. 57:255–262.
84. Ripley, P.O. 1969. Crop rotation and productivity. Agric. Can. Publ. 1376. 78 p.
85. Shutt, Frank T. 1925. Influence of grain growing on the nitrogen and organic matter content of western prairie soils in Canada. Can. Dep. Agric. Bull. 44, n.s. 15 p.
86. Shutt, Frank T. 1928. Lime in agriculture. Dom. Exp. Farms, Bull. No. 86, n.s. 15 p.
87. Shutt, Frank T.; Charron, A.T. 1905. Recent experiments in the nitrogen-enrichment of soils. Trans. R. Soc. Can. Ser. II, 11(Sect. 3):73–78.
88. Shutt, Frank T.; Wright, L.E. 1933. Peat, muck and mud deposits, their nature, composition and agricultural uses. Dom. Exp. Farms Bull. No. 124, n.s. 27 p.
89. Singh, S. Shah. 1984. Increase in neutral salt extractable cation exchange capacity of some acid soils as affected by $CaSO_4$ applications. Can. J. Soil Sci. 65:153–161.
90. Singh, S.S.; Coote, D.R. 1985. Effects of acid rain on agriculture. Agrologist 14(2):23–24.
91. Sommerfeldt, T.G.; MacKay, D.C. 1982. Dryland salinity in a closed drainage basin at Nobleford, Alberta. J. Hydrol. 55:25–41.
92. Sommerfeldt, T.G.; Rapp, E. 1982. Management of saline soils. Agric. Can. Publ. 1624. 31 p.
93. Sparrow, Herbert O. 1984. Soil at risk, Canada's eroding future. The Senate of Canada, Ottawa. 129 p.
94. Spence, George. 1967. Survival of a vision. Queen's Printer, Ottawa. 167 p.
95. Spratt, E.D.; McCurdy, E.V. 1966. The effect of various long-term soil fertility treatments on the phosphorus status of a clay chernozem. Can. J. Soil Sci. 46:29–35.
96. Taylor, E.M.; Howatt, J.L. 1937. Magnesium in field crop production in New Brunswick. Sci. Agric. XVII:294–298.
97. Timonin, M.I. 1946. Microflora of the rhizosphere in relation to the manganese-deficiency disease of oats. Soil Sci. Soc. Am. Proc. 11:284–292.
98. Turner, R.C.; Clark, J.S. 1966. Lime potential in acid clay and soil suspensions. Trans. Comm. II & IV. Int. Soc. Soil Sci., pp. 208–215.
99. Ward, G.M. 1959. Potassium in plant metabolism. II. Effect of potassium upon the carbohydrate and mineral composition of potato plants. Can. J. Plant Sci. 39:246–252.
100 Ward, Gordon M.; Miller, Marcia J. 1968. Magnesium deficiency in greenhouse tomatoes. Can. J. Plant Sci. 49:53–59.
101. Wilcox, J.C.; Hoy, B.; Palmer, R.C. 1947. Orchard fertilizer tests in the Okanagan valley. Sci. Agric. 27:116–129.
102. Woodbridge, C.G. 1955. Magnesium deficiency in apple in British Columbia. Can. J. Agric. Sci. 35:350–357.
103 Woodbridge, C.G.; McLarty, H.R. 1951. Manganese deficiency in peach and apple in British Columbia. Sci. Agric. 31:435–438.
104. Woods, J.J. 1942. The effect of green crops and manurial treatments in raspberries. Sci. Agric. 23:247–250.
105. Wright, L.E.; Hammond, H.S. 1937. Manures, fertilizers and soil amendments. Their nature, function and use. Dom. Can. Dep. Agric. Publ. 585. 70 p.
106. Wright, L.E.; Pelletier, J.R.; Ripley, P.O. 1949. Soil fertility studies I. Manure, fertilizers and lime for field crops at Ste. Anne de la Pocatière, P.Q. Sci. Agric. 29:128–136.

107. Zentner, R.P.; Read, D.W.L. 1977. Fertilization decisions and soil moisture in the brown soil zone. Can. Farm Econ. 12:8–13.
108. Zentner, R.P.; Campbell, C.A.; Read, D.W.L.; Anderson, C.H. 1984. An economic evaluation of crop rotations in southwestern Saskatchewan. Can. J. Agric. Econ. 32:37–54.

150

Chapter 14
Water Research

Farmers in Canada irrigate more than 600 000 ha of land. About 10 times that area is drained or protected by flood control installations. Even so, the area of irrigated and drained land is only a small portion of the 43 million hectares of improved farmland in Canada. Some areas, such as the Okanagan Valley of British Columbia, must be irrigated to produce any cultivated crop. Other areas such as southern Alberta and Saskatchewan need irrigation to stabilize production and to produce higher-value crops such as sugarbeets and vegetables. Still others, such as southwestern Ontario and the coastal region of British Columbia, require supplemental irrigation, particularly in dry years. Drainage of excess water is required wherever irrigation is practiced, as well as on low-lying ground near natural water courses. Research on irrigation has been done primarily in areas of extreme dryness such as at Swift Current, Saskatchewan; at Lethbridge, Alberta; and at Summerland, British Columbia. Research on drainage has been conducted at a variety of stations throughout Canada.

IRRIGATION

The Great Plains

Much of today's research in agriculture had its beginning with either William Saunders or Frank T. Shutt. This chapter starts with Shutt who, in 1916, assisted with investigations to determine the suitability of certain soils for irrigation. In its examination of the soils in southern Alberta and Saskatchewan for irrigation feasibility, the Reclamation Service of the federal Department of the Interior classified soils as irrigable and nonirrigable on the basis of topography. The reasoning was that to either flood or furrow irrigate soil, level land was needed; undulating land could not be irrigated economically. Surveyors of the Reclamation Service knew that at a depth of about 18 inches (45 cm) there might be alkali which, if irrigated, could rise to the surface. Shutt, as chief of the Division of Chemistry, Experimental Farms Branch, was entrusted with learning what the risks were if these stiff clay loams of southern Alberta and Saskatchewan were irrigated.

In 1923, Shutt and Macoun (30) made their final report on experiments started in 1916 on the Canadian Pacific Railway demonstration farm at Tilley, Alberta. Shutt and Macoun took soil samples at each of four levels to a depth of 5 feet (1.5 m) on two different fields that had been irrigated three times a year since 1915. In each of the following 6 years similar samples were taken from the same two sites and analyzed for salts. In addition to this experiment, they selected several pairs of plots each year, starting in 1917, one producing a good crop and the other, no more than 35 feet (10 m) distant, producing a poor crop or no crop at all. Every year, soil samples were taken from each plot and from a point midway between the two, and analyzed for salts in order to learn the limit of alkali a normal plant could tolerate.

Shutt and Macoun found that the surface soil was free from alkali, the layer between 15 and 45 cm had some alkali but not an excessive amount, and that the layer between 45 and 150 cm had a substantial amount of alkali. Sodium and calcium sulfates were the principal salts present. They did not find any indication of an increase in the concentration of salts due to irrigation in any of the four soil levels examined. If anything, the total salts in the second layer (15 to 45 cm) had decreased. Shutt suggested this could have been an error of sampling. From their second experiment they determined that 0.35 percent sodium sulfate was the limit of plant tolerance to alkali and that as the level reached 1.0 percent, plant growth became stunted. There was no question in Shutt's mind that many of the soils in Alberta and Saskatchewan were suitable for irrigation and that its development in Alberta, under way since 1900, should be encouraged.

While Shutt and Macoun were doing their studies near Tilley, D.H. Bark of the Canadian Pacific Railway and A.E. Palmer were cooperatively seeking methods to reduce the effects of alkali salts on plant growth. They learned that gypsum (calcium sulfate), when applied to alkali soils, decreased the harmful effects of sodium and magnesium sulfates. Marshall and Palmer (24), who took further soil samples in 1937 from the Tilley plots, found that crops seemed to be normal, and that there had been no appreciable accumulation of alkali salts after 20 years of irrigation. They concluded that these heavy clay soils were irrigable, a conclusion confirmed by Chang and Oosterveld (8) after a further 40 years of irrigation. More recently, Chang, Kozub, and MacKay (7) estimated that no more than 9 percent of irrigated land was affected by salinity, much less than that in many districts. They concluded that, although not of immediate concern, the salinity situation requires careful watching.

W.H. Fairfield of Lethbridge, was the first person in the Experimental Farms System to write a farmers' bulletin on irrigation in 1919 (12). He had been preceded in 1914 by S.G. Porter of the Irrigation Branch of the Department of Interior who prepared its circular Number 1 entitled *Practical operation of irrigation works*. A.E. Palmer, in 1929, reported (26) 6 years of intensive research from Lethbridge on methods of irrigating spring wheat, alfalfa, potatoes, sugarbeets, and sunflowers. His research was done with basin irrigation to determine the best stage of plant growth to apply water, the number of irrigations per year needed for each type of crop, and the value of fall irrigation. He learned that wheat required only one irrigation, whereas alfalfa needed at least two and sometimes three irrigations; potatoes reacted poorly if their plants lacked moisture; and sugarbeets and sunflowers required one irrigation during the season but were best when seeded on soil that had an irrigation the previous fall.

In 1930, it was realized that in irrigation districts not all soils, even when irrigated, produced equally, and therefore users of irrigation should pay for their water relative to their soils' production capability. By 1950 Bowser and Moss (6) had developed a seven-factor rating system by which they were able to classify soils from excellent to nonirrigable.

One of the first experiments Fairfield established at Lethbridge was an irrigated 10-year rotation (rotation U). Hill (15) and Bishop and Atkinson (2) reviewed the results after 40 years, concluding that abundant yields were possible provided phosphate fertilizer was supplied and the rotation contained 6 years of alfalfa. Although the response of crops was measurable, no appreciable difference in the phosphorus content of the soil could be detected between fertilized and unfertilized plots. They noted that the percentage of soil organic matter and nitrogen tended to increase over the years and that the physical structure of the soil was maintained. Irrigation in southern Alberta was deemed to be a sound, stable means of increasing productivity.

Sprinkler irrigation started in the late 1930s, although it did not become common farm practice in Canada until 1945 when light, easily moved aluminum pipes were introduced. One of the first problems with this method on the Great Plains was irregular distribution of water caused by wind. Korven (20) at Swift Current studied the problem and concluded that it could be minimized with appropriate types of sprinkler heads spaced on 50-foot (15-m) squares and having the lines across the wind's path.

British Columbia

Irrigation in the Okanagan Valley of British Columbia is essential to produce crops, particularly horticultural crops. Wilcox (38) made intensive studies of irrigation methods, starting with furrow irrigation. He was concerned about the amount of soil erosion that occurred with improper use of furrow irrigation, particularly on light soils with steep slopes. He carefully defined the safe slopes on which furrows could be used. He also learned that the timing of irrigation materially affected the winterhardiness of fruit trees. When summer irrigation was inadequate, trees would mature too early and enter their winter dormancy. When irrigated through August and September, maturity would be too late. In either situation, damage from low winter temperatures could result.

J.C. Wilcox at Summerland was a strong proponent of sprinkler irrigation for most of the Okanagan. In 1947, Wilcox and Swailes (44) found that at water pressures of 100 kPa, and less, distribution from all types of sprinklers without wind distortion, was irregular. Distribution improved as water pressures were increased up to 138 kPa. By 1950, 5 years after the introduction of portable aluminum pipe and quick couplers, he had learned through experimentation that sprinkler irrigation, compared with furrow irrigation, brought about a marked reduction in soil erosion, induced better growth of cover crops, provided improved moisture conditions, reduced difficulties from seepage water, and used both water and labor more economically (37).

By 1955 much had been learned about irrigating grasslands. T.G. Willis, Kamloops, British Columbia, W.L. Jacobson, Vauxhall, Alberta, and J.C. Wilcox, Summerland, British Columbia, concluded (45) that when systems are properly designed and managed, be they surface or sprinkler, they pay reasonable dividends for either grass or hay crops.

Eastern Canada

Experimental stations at both Delhi, Ontario, and L'Assomption, Quebec, demonstrated that supplemental irrigation was of benefit to tobacco crops. Walker and Vickery (36) found that irrigation based on the soil's need for water rather than on a time schedule, improved the yield, quality, and maturity of tobacco in four consecutive years. Either Thornthwaite's evapotranspiration estimates method or the electrical resistance blocks method was suitable for determining the need to irrigate. Allard, Richard, and Bélanger (1) in Quebec, found that irrigating at five specific stages of growth gave yields equal to those from irrigation when soil moisture reached 50 percent of field capacity.

At Harrow, Ontario, in 1964, Fulton and Findlay (13) obtained significant increases in yields with the irrigation of potatoes, provided soil nutrition was adequate. Up to four times the amount of fertilizer was required for optimum yields over the lower yielding nonirrigated crops. MacKay and Eaves (23) at Kentville, Nova Scotia, found the same relationship, using sweet corn and snap beans.

Water requirements

There was little doubt that irrigation of parched soils paid handsomely with increased yields and a capacity to produce a wider variety of crops. Farmers now required a system to determine when to irrigate and how much water to apply. Scientists knew that different crops required different amounts of water for optimum production and that different soils could absorb varying amounts of water and retain it for different lengths of time, depending upon the soil structure and texture. Wilcox was one of the first to study these phenomena on irrigated soil, although Doughty and coworkers (11) at Swift Current had made similar studies on dryland soils in 1943. Working with J.L Mason and J.M. MacDougald, Wilcox found (43) a close and positive correlation between the rate of evaporation from an open pan of water and the consumptive use of water[1] in an orchard. However, there was a negative correlation between the daily consumptive use of water and the time between irrigations. Wilcox (41) then found that the shorter the irrigation interval,[2] the more water was required by the orchard from the irrigation system and therefore the need for greater peak flows in the system. The reason for this is that soils requiring short irrigation intervals are light and often sandy; thus they need more start-ups of the system per unit of water applied.

In Alberta, Krogman and Lutwick (21), and Sonmore (34) showed that alfalfa and grasses required slightly more than 600 mm of water to produce

---

[1]Consumptive use of water is the amount of water used from the soil by plants, plus the loss of water by evaporation from the soil surface.
[2]Irrigation interval is the time between the start of one irrigation and the start of the next.

optimum yields. From similar information for other crops, and using evaporation data from an open pan of water, they were able to calculate the daily consumptive use of water for each of the three major crop types of the area. When they subtracted the water used from the moisture in the soil at the start of the season, they could determine when a soil should be irrigated. The method was tested for several years; it worked, and in 1964 Hobbs and Sonmore (18), in cooperation with the Alberta Extension Service, published a weekly *Irrigation gauge* press release advising which crops on which soils should be irrigated. Heeney, Miller, and Rutherford (14) at Smithfield, Ontario, and Wilcox and Korven (42) in British Columbia, devised similar methods, using evaporimeters. By 1983 Hobbs and Krogman (16) had summarized 30 years' research on water, giving the requirements of the 14 major irrigated crops in southern Alberta, and describing a Lethbridge computer program devised for irrigation scheduling.

Water conservation

There are several ways to conserve water, including efficient irrigation methods, proper transmission from storage to field, and management of snow.

D.S. Stevenson at Summerland, British Columbia, was the first in Canada to experiment with trickle irrigation in 1973. This is a system where plastic pipes are permanently laid on or below the surface of the soil and from which small-diameter tubes continuously drip or trickle water. The system conserves both water and labor but still must be examined each day for malfunction. By 1973, Stevenson concluded the system to be effective in orchards and vineyards (35), provided it was designed properly and operated without any blockage of emitters.

In 1966, Pohjakas (27) at Swift Current, obtained satisfactory moisture distribution in the top 1.2 m of soil from perforated subterranean plastic pipe mechanically laid 40 cm below a grassed surface, providing the delivery pipes were spaced no greater than 1.5 m apart. The system had the advantage of leaving the top 5 cm of soil dry, thereby limiting the loss of water from evaporation. Hobbs and Laliberte (17) at Lethbridge, attempted to distribute water through unlined underground channels, or moles, and found they needed to be spaced about 1.2 m apart at from 22 to 30 cm below the soil surface. Some problems were encountered; these were caused by soil collapsing into the moles when soil moisture was inadequate and by gophers using the moles as runways, adding exits and partitions. The special equipment used to open the subterranean moles and to lay a folded plastic lining when required was subject to damage from underground rocks. Nonetheless, mole irrigation was highly efficient in its use of water.

Cities sometimes are supplied with domestic water from great distances. After it has been used in homes and factories, the water is filtered and cleaned, then returned to the watersheds from whence it came. In 1973 J.B. Bole at Lethbridge and W. Nicholaichuk at Swift Current reasoned this process to be wasteful of both water and plant nutrients. In Oriental countries, and more recently in Europe, waste water from cities is used to irrigate and feed crops. Bole and Bell (3) used nonchlorinated effluent from a municipal lagoon on alfalfa and various grass species. The chance of diseases being transmitted from the sewage to the forage and back to either livestock or humans was known to be remote. However, as an extra measure of precaution, they restricted the irrigation to forages. Forages were chosen because they have a long growing season, use large quantities of water, have a high nutrient uptake, and a capacity to prevent erosion by stabilizing the soil. Following 4 years of sprinkler irrigation, alfalfa produced 45 percent more dry matter than the best grass, using 60 cm of effluent annually without fertilizer. Using 100 cm, or more, of effluent to which a nitrogen–phosphorus fertilizer was added, three of the four grasses produced more dry matter than alfalfa. Jame, Nicholaichuk, and Kilcher (19) compared irrigation with sewage effluent to that with fresh water over a 5-year period. They found that under the conditions at Swift Current the salt content of the upper layers of the root zone increased to nearly that of the irrigation water, but the salt content in the lower soil layers decreased substantially. As expected, yields of alfalfa from the effluent-irrigated soil were markedly higher than those from the freshwater-irrigated soil.

The control of seepage from irrigation canals and dugouts also conserves water and protects arable lands from becoming waterlogged or salinated. In the early 1950s, the Prairie Farm Rehabilitation Administration (PFRA) used a variety of materials to line some of its canals and dugouts. Fifteen years later, Pohjakas and Rapp (28) tested sections of each of these canals and dugouts by sealing off portions, filling the sealed portions with water, and measuring the water loss from the ponded area over a 24-hour period. Following corrections for loss by evaporation, they found polyethylene plastic to be the best lining, followed by concrete and asphalt. Compacted earth linings were the least watertight and the least durable. Plastic linings, except for compacted earth, were the cheapest but were more expensive to maintain than concrete. Following this, extensive research and laboratory testing by Sommerfeldt, McLaughlin, and Allan (31) showed that a mixture of asphalt emulsion incorporated in the ditch with soil, effectively controlled seepage and withstood winter weathering. Herbicides were applied to the soil in advance to prevent weed growth. Time is required to determine the durability of this lining material.

Snow may account for up to half of the prairie precipitation. However, much of it is blown off fields and so is of no use to agricultural crops the following spring. In 1979 F.B. Dyck and W. Nicholaichuk at Swift Current sought to conserve snow on fields by cutting stubble at alternate heights. By 1984 they (25) had devised a system of trapping snow by managing the fallow that increased the soil water intake efficiency from 38 percent to 57 percent. To achieve these results, they swathed grain so that narrow strips of standing stubble had only their heads removed. The remainder was cut to the usual short stubble height. Between these taller barriers, wind velocities were reduced, snow accumulated, and temperatures rose, causing snow to melt and providing increased water for the soil. By subsoiling (slitting the soil with a chisel blade) between the barriers prior to snowfall, overwinter intake of water was doubled. Some of this research is in its preliminary phases but holds great promise for production on the prairies.

DRAINAGE

Historically, drainage of Canadian agricultural soils has followed the pattern of the development of Canada. Commencing about 1700, settlers drained land along the east coast and built mud dikes to reclaim fertile marshlands along the Bay of Fundy (see section of Chapter 13 on dikelands). From 1870, provincial governments led in the construction of outlet drains. These programs continue today. Subsurface drainage began in Ontario with small-diameter randomly placed pipes. Since then, pipe sizes have been increased, materials for pipes have been changed, and equipment for laying them has been revolutionized.

At the Central Experimental Farm, Ottawa, erosion from the runoff of excessive surface water was studied by Cordukes, Turner, Ripley, and Atkinson (9), starting in 1944. They concluded that a good vegetative cover and the use of farm manure materially reduced soil and water losses.

Drainage of irrigated soils in Canada was first studied and reported (39,40) from the Research Branch by J.C. Wilcox at Summerland in the 1950s. Wilcox theorized that a logarithmic relationship existed between time and the rate of drainage on deep soil with no water table. He applied his theory to data from experiments of four other scientists and found it to be valid and therefore probably of universal application.

Scientists at the Research Station, Harrow, were concerned with nutrient losses from fields into water courses several years before environmentalists became concerned with the problem. In 1963 Bolton, Aylesworth, and Hore (4) started a 7-year study that provided information when it was needed. We now take for granted that cultural practices used for row crops such as corn release more nutrients into field drains than do practices used for noncultivated crops such as bluegrass or alfalfa sod. Bolton and colleagues noted this effect and also found that the high level of nitrogen effluent from a 4-year rotation (1-year corn, 1-year oats, and a 2-year alfalfa) cropping system was attributed to the nitrogen produced by the alfalfa as well as to the fertilizer nitrogen applied to the corn. Phosphorus concentration in the effluent from all crops was small. The amount of nutrients flowing from a field was more dependent upon the flow of water than upon the amount of fertilizer applied. Large nutrient losses occurred in seasons of large drain flows. Later, Bolton, Dirks, and McDonnell (5) concluded that the beneficial effects of subsurface drains, fertilizer application, and crop rotation were additive.

At Kapuskasing in northern Ontario, Lévesque and Hamilton (22) back-filled tile drains laid 45 cm deep to 15 cm and 45 cm with either peaty muck or gravel. The drains were on 18-m spacings. Under the conditions of their 1967 experiment none of the treatments consistently lowered the water table more quickly than the untreated drains. On a poorly drained clay soil of eastern Ontario, Culley and Coote (10) in 1984 found that the water table was not

157

affected by an open channel drain beyond a distance of 65 m. With 0.1-m-diameter plastic drain pipes set at 1 m depth about 17 m apart, however, the water table regime was altered dramatically. Although the water table rose rapidly during heavy rain storms, it rarely did so within 0.6 m of the soil surface and drew down at the rate of 1.5 m per day. From the design viewpoint, Culley and Coote selected the Hooghoudt theoretical model as being superior to the Glover model.

Rapp and Laliberte (29) used existing drains in southern Alberta to determine their effectiveness on an irrigated field in reducing both the water table and the salt content. Without drains the water table stood at 0.3 to 2.1 m from the soil surface and the salt content of the upper soil horizon increased. During the period 1955–1965, Rapp and Laliberte found the drains were effective in removing excess groundwater salts, although the delivery of water was highly variable. The water table was maintained at a safe level, but the reclamation effect of the drains was doubtful judged by the salinity of the effluent that showed no significant decrease during the life of the experiment.

Moles similar to those described for irrigation are also used for drainage. Although new to Canada, moles have been used in other parts of the world for more than 100 years. During the past few years Sommerfeldt (32) at Lethbridge found that the clay and moisture contents of the soil were important factors in forming mole drains and in their resultant stability. With suitable soil conditions, moles were less expensive to install than other types of drainage. Sommerfeldt (33) also studied southern Alberta fields that had been mole-drained for 10 years. Half the number of moles were unlined and the other half were lined with perforated plastic. This study proved that the lined moles discharged three to four times the volume of water of the unlined ones. Nonetheless, both were operative over the 10-year period and effective in lowering the water table following each flood irrigation.

Israel is credited with making its deserts bloom. Canada can make a similar claim when we consider areas where sagebrush and cacti grow beside cultivated orchards or along fencelines protecting lush pastures. During the past century irrigation has been refined from flooding, to sprinkling, to trickling. Because computer programs now can schedule complex irrigation systems, farm managers have more time for other important decision-making matters.

Drainage goes hand-in-hand with irrigation. In the 1700s settlers drained coastal areas and marshlands by using whatever methods they knew. Today, the mole drains recently introduced into Canada have proved useful, particularly when lined, for lowering water tables in moist clay soils. Sometime ago Research Branch scientists found that there were nutrient losses, particularly nitrogen, into soil water; this finding since has been espoused by environmentalists.

Producing research results in irrigation and drainage requires patience, persistence, and a breadth of knowledge applied over the long-term, since these practices change the biological and physical soil and water systems slowly.

References

1. Allard, J.; Richard, J.; Bélanger, J.A. 1964. Comparaison de deux méthodes pour déterminer le moment d'irriguer le tabac jaune. Can. J. Plant Sci. 44:276–282.
2. Bishop, R.F.; Atkinson, H.J. 1954. Changes in organic matter and certain other soil constituents in forty years of cropping under irrigation. Can. J. Agric. Sci. 34:177–180.
3. Bole, J.B.; Bell, R.G. 1978. Land application of municipal sewage waste water: Yield and chemical composition of forage crops. J. Environ. Qual. 7:222–226.
4. Bolton, E.F.; Aylesworth, J.W.; Hore, F.R. 1970. Nutrient losses through tile drains under three cropping systems and two fertility levels on a Brookston clay soil. Can. J. Soil Sci. 50:275–279.
5. Bolton, E.F.; Dirks, V.A.; McDonnell, M.M. 1982. The effect of drainage, rotation, and fertilizer on corn yield, plant height, leaf nutrient composition and physical properties of Brookston clay soil in southwestern Ontario. Can. J. Soil Sci. 62:297–309.
6. Bowser, W. Earl; Moss, H.C. 1950. A soil rating and classification for irrigation lands in western Canada. Sci. Agric. 30:165–171.
7. Chang, C.; Kozub, G.C.; MacKay, D.C. 1985. Soil salinity status and its relation to some of the soil and land properties of three irrigation districts in southern Alberta. Can. J. Soil Sci. 65:187–193.
8. Chang, C.; Oosterveld, M. 1981. Effects of long-term irrigation on soil salinity at selected sites in southern Alberta. Can. J. Soil Sci. 61:497–505.
9. Cordukes, W.E.; Turner, R.C.; Ripley, P.O.; Atkinson, H.J. 1951. Water erosion of soil. Sci. Agric. 31:152–161.
10. Culley, J.L.B.; Coote, D.R. 1984. Water table regimes in an eastern Ontario soil with and without pipe drains. Can. Agric. Eng. 26:7–13.
11. Doughty, J.L., et al. 1943. Report of investigations, soil research laboratory, Swift Current, Sask. Dep. Agric. 63 p.
12. Fairfield, W.H. 1919. Irrigation of potatoes. *In* The potato in Canada, its cultivation and varieties. Dom. Can. Dep. Agric. Bull. 90. 16 p.
13. Fulton, J.M.; Findlay, W.I. 1964. Cumulative effects of supplemental irrigation on fertilizer requirement, yield and dry matter content of early potatoes. Am. Potato J. 41:315–318.
14. Heeney, H.B.; Miller, S.R.; Rutherford, W.M. 1961. A meteorological method of calculating the irrigation requirement of the canning tomato crop. Can. J. Plant Sci. 41:31–41.
15. Hill, K.W. 1951. Effects of forty years of cropping under irrigation. Sci. Agric. 31:349–357.
16. Hobbs, E.H.; Krogman, K.K. 1983. Scheduling irrigation to meet crop demands. Agric. Can. Res. Branch Contrib. 1983-10E. 18 p.
17. Hobbs, E.H.; Laliberte, G.E. 1967. Mole irrigation. Can. Agric. Eng. 9:64–65.
18. Hobbs, E.H.; Sonmore, L.G. 1964. Irrigation scheduling. Res. for Farmers. 9(4):4–5.
19. Jame, Y.-W.; Nicholaichuk, W.; Kilcher, M.R. 1981. Soil salinity: Effluent irrigation vs. creek water irrigation. Can. Agric. Eng. 23:31–35.
20. Korven, H.C. 1952. The effect of wind on the uniformity of water distribution by some rotary sprinklers. Sci. Agric. 32:226–240.

21. Krogman, K.K.; Lutwick, L.E. 1961. Consumptive use of water by forage crops in the upper Kootenay river valley. Can. J. Soil Sci. 41:1–4.

22. Lévesque, M.; Hamilton, H.A. 1967. L'efficacité d'un système de drainage souterrain avec remplissage poreux dans un sol argileux de Kapuskasing. Can. J. Soil Sci. 47:55–63.

23. MacKay, D.C.; Eaves, C.A. 1962. The influence of irrigation treatments on yields and on fertilizer utilization by sweet corn and snap beans. Can. J. Soil Sci. 42:219–228.

24. Marshall, J.B.; Palmer, A.E. 1939. Changes in the nature and position of the soluble salts in certain Alberta soils after twenty years of irrigation. Sci. Agric. XIX:271–278.

25. Nicholaichuk, W.; Dyck, F.B.; Steppuhn, H. 1984. Snow management for moisture conservation. Can. Soc. Agric. Eng. Paper 84-304. 6 p.

26. Palmer, A.E. 1929. Use of irrigation water on farm crops. Dom. Can. Dep. Agric. Bull. 125, n.s. 51 p.

27. Pohjakas, K. 1966. Distribution rates and patterns from plastic subirrigation pipe. Can. Agric. Eng. 8:37–38.

28. Pohjakas, K.; Rapp, E. 1967. Performance of some canal and dugout linings on the Canadian prairies. Can. Agric. Eng. 9:58–63.

29. Rapp, E.; Laliberte, G.E. 1968. Performance of tile drains under irrigation in southern Alberta. Can. Agric. Eng. 10:64–69.

30. Shutt, Frank T.; Macoun, John M. 1923. The vertical movement of alkali under irrigation in heavy clay soils. Trans. R. Soc. Can., Ser. III, 17(3):71–74.

31. Sommerfeldt, T.G.; McLaughlin, N.B.; Allan, J.R. 1978. Field evaluation of a soil–asphalt ditch lining. Can. Agric. Eng. 20:13–15.

32. Sommerfeldt, T.G. 1983. Soil moisture and texture effects on mole drain stability and on force requirement for installation—a laboratory study. Can. Agric. Eng. 25:1–4.

33. Sommerfeldt, T.G. 1984. Performance of unlined and lined mole drains in a saline clay loam field. Can. Agric. Eng. 26:1–5.

34. Sonmore, L.G. 1963. Seasonal consumptive use of water by crops grown in southern Alberta and its relationship to evaporation. Can. J. Soil Sci. 41:287–297.

35. Stevenson, D.S. 1981. Responses of 6-year-old diamond grapevines to the change from sprinkler to trickle irrigation and to the time and method of applying nitrogen. Can. J. Plant Sci. 61:571–575.

36. Walker, E.K.; Vickery, L.S. 1959. Some effects of sprinkler irrigation on flue-cured tobacco. Can. J. Plant Sci. 39:164–174.

37. Wilcox, J.C. 1950. Sprinkler irrigation experience in British Columbia orchards. Sci. Agric. 30:418–427.

38. Wilcox, J.C. 1952. Irrigation of horticultural crops. Agric. Inst. Rev. 7(4):20–23.

39. Wilcox, J.C. 1959. Rate of soil drainage following an irrigation. I. Nature of soil drainage curves. Can. J. Soil Sci. 39:107–119.

40. Wilcox, J.C. 1960. Rate of soil drainage following an irrigation. II. Effects on determination of rate of consumptive use. Can. J. Soil Sci. 40:15–27.

41. Wilcox, J.C. 1960. Effect of irrigation interval on peak flow requirements in sprinkler irrigated orchards. Can. J. Soil Sci. 40:99–104.

42. Wilcox, J.C.; Korven, H.C. 1964. Some problems encountered in the use of evaporimeters for scheduling of irrigation. Can. Agric. Eng. 6:29–31.

43. Wilcox, J.C.; Mason, J.L.; McDougald, J.M. 1953. Consumptive use of water in orchard soils. II. Effects of evaporating power of air and of length of irrigation interval. Can. J. Agric. Sci. 33:231–245.

44. Wilcox, J.C.; Swailes, G.E. 1947. Uniformity of water distribution by some undertree orchard sprinklers. Sci. Agric. 27:565–583.

45. Wilcox, J.C.; Willis, T.G.; Jacobson, W.L. 1955. Irrigation of grassland. Agric. Inst. Rev. 10(2):21–23.

161

Chapter 15
Energy Research

The Research Branch objective of learning how to produce food at minimum cost has meant that its research was aimed directly or indirectly at conserving energy—an objective that predated the energy crisis by many years. The agri-food system uses 15 percent of the total energy consumed in Canada. This 15 percent is broken down as follows: agricultural production 4 percent; processing and packaging food 3 percent; distributing food 3.5 percent; preparing food for consumption 4.5 percent. Oil prices have stabilized somewhat following the mid-1970 surge, and there appears to be surplus oil and gas in Canada; however, the fact remains that our fossil fuel supply is finite. Additional energy sources, preferably renewable, may need to be developed to ensure that Canada's food production meets domestic needs and can expand as opportunities arise. For reasons such as climate, land mass, and a high living standard, Canada uses more energy per capita than most other countries (3).

Research Branch now is taking several approaches to the altered energy situation. One initiative is to identify economical sources of renewable energy: solar, wind, and biomass for static uses; vegetable oils and other liquid fuels processed from agricultural crops and residues for use in mobile equipment. Lines of endeavor have included the production of energy efficient crops; the use of energy efficient livestock systems; the reduction of imported off-season fruits and vegetables by developing improved varieties and increasing the number of greenhouse crops grown, as well as by providing advanced storage and preservation techniques for crops already grown locally; the exploitation of biotechnology by improving and extending the nitrogen-fixing capabilities of legume and other crops and by optimizing the use of cellulose; and the production of energy, both direct and indirect, by the adequate treatment of wastes. Another approach is to conserve the energy we have by making food production, manufacture, distribution, and utilization even more efficient. Conservation of energy in the transportation of food from one part of Canada to another could be achieved by lighter packaging, by dehydrating foods, by employing bulk shipments, by growing crops and livestock close to target markets, and by making maximum use of rail and water transport.

Research on many of these subjects has been sponsored by Research Branch through its contracting program. The mechanization program started in 1973 had spent $750 000 on energy research by 1981. By April 1985, $28 million had been used to contract 300 energy research projects. It was expected to reach $36 million and 390 projects in 1986, but new political objectives may result in a reduced contracting program.

This chapter describes some of the work and achievements of the past 10 years. More is to be done; more will be accomplished.

GREENHOUSES

The advent of high oil and gas prices in the 1970s immediately affected the greenhouse vegetable and flower industries. Areas of greenhouse production in Canada include the Maritime Provinces, Quebec, Ontario, southern Alberta, and the coast of British Columbia. Therefore, it was fitting for Research Branch to contract 39 projects to study ways and means of reducing greenhouse energy costs.

Greenhouses are collectors of solar energy, but conventional types have no means of storing excess energy. Indeed, large vents and sometimes cooling devices are used to exhaust heat to the outside and so prevent inside temperature and humidity from becoming excessively high. During cloudy days and cool nights additional heat is required from furnaces to prevent temperatures falling below desirable plant growth levels. It was evident that if heat from the furnaces could be conserved, and heat from the sun stored and then used during cloudy days and cool nights, fuel costs could be reduced.

At the Saanichton Research Station, the Department of Agriculture entered into a contract with the University of British Columbia to test the practicality of using solar energy for heating greenhouses. The principle of storing heat in soil underneath a greenhouse and using it when temperatures in the house dropped below a predetermined minimum was reported first by Japanese scientists as early as 1969. J.M. Molnar, director of the station, had seen the Japanese work in 1980 and believed the same principles could be applied at Saanichton.

To test the system, Molnar contracted with Professor L.M. Staley of the University of British Columbia, to build a new greenhouse that included heat collectors and a heat storage unit. The house had an insulated, perpendicular north wall, whereas the south-sloping, shed-type roof was of standard greenhouse glass. Black shade cloth hung from a duct mounted along the north ridge and acted as a solar collector. A fan, controlled by thermostats, transferred the heated air through ducts into rock storage underneath the greenhouse floor. When temperatures in the house fell below 17°C the flow of air was reversed, bringing the stored heat back into the greenhouse. Auxiliary heating was provided by propane and electricity. A standard greenhouse adjacent to the experimental house and heated with propane fuel was used for comparison.

Both houses were operated from September through April during what was considered to be an average winter with frequent overcast days (1981–1982). The solar-heated house used 30 percent less propane and electricity than the standard house used. Early results encouraged Molnar to continue his experiments. Molnar and Staley built two standard-type, glass-covered greenhouses and two semicircular arching roof-type greenhouses covered with double polyethylene sheeting (4). One house of each design was heated with standard hot-water pipes and the other with solar-produced heat stored in soil under each house floor. The glass solar house also had hot-water pipes for auxiliary heating.

Controls to vent the houses, to store or recycle heat from below ground, and to draw thermal blankets over the crops during cool nights were entirely computerized. The computer also collected all data such as temperature readings, electricity use, and hot water consumption.

Results after the 1st year of operation in 1982–1983 were better than expected. The efficiency of the system was increased by 20 percent, and by using thermal blankets, the demand on the heat source was reduced by 44 percent. Both scientists realize that further improvements to the design will be made, but even at this early stage they are confident that solar heating of greenhouses is a sound commercial investment.

E. Brundrett at the University of Waterloo, Ontario, in his contract with the department, used a commercial greenhouse at Grimsby, Ontario, to test a ventilation air heat recovery system. Warm greenhouse air was exhausted through a series of 23-mm-diameter polyethylene tubes, all of which were inside a 75-mm tube through which cool outside air was drawn. Heat from the inside air was recovered by the inflowing outside air. The proportion of heat recovered depended upon the length of tubes: when they were 7 m long, 50 percent was recovered; when they were 23 m long, 70 percent was recovered.

To conserve heat at night and prevent overheating during the day, Brundrett devised a multilayered bubble pack thermal curtain with an upper surface of chromium, which provided reflectance of light and heat during the day. The accordion-folded curtain is suspended along the greenhouse gutters and pulled to the roof peak when needed. Inflatable tubes seal the curtain edges when in the closed position, further reducing heat loss. Early results indicate that the combined thermal and shade functions provide a superior cost effectiveness for the curtain system.

The problem of heat conservation was approached in a different way by G. Gallagher, at the University of Quebec, Chicoutimi, under another contract with the department. He reasoned that only the air near plants required heating. Therefore, in a commercial greenhouse, Gallagher built a retractable tunnel of clear plastic over the plants so that it could be raised as the plants grew. At night aluminized plastic was overlaid to conserve heat further. The tunnel air was heated by using warm, waste water from a nearby industry.

Warm waste water was also used in greenhouses at Grand Lake, New Brunswick, and Glace Bay, Nova Scotia. One estimate put the savings in fuel costs at 50 percent below oil. At this rate, tomatoes and cucumbers can be grown in Canada during winter months and compete cost-wise with imported products.

The positive results of these experiments have provided vegetable growers and greenhouse manufacturers with information on ways of insulating foundation walls and north walls, of using thermal curtains, of using insulating tunnels over crops within greenhouses, of storing solar heat, of capturing waste energy from outside sources, and of using microprocessors for effectively monitoring and controlling operations. Many of these energy-saving techniques are now cost effective with the increased price of fuel.

ALCOHOL FOR FUEL

R.D. Hayes, E. Manolakis, and J.D. Miller of the Engineering and Statistical Research and the Chemistry and Biology Research institutes did a number of techno-economic and scientific studies (5, 8) starting in 1981 on the production of alcohol for farm fuels. They examined the possibilities of using crop wastes such as cull fruits, vegetables, and spoiled grain as well as growing crops especially for fermenting and producing alcohol. Such crops included Jerusalem artichokes, grown in cooperation with the Research Station, Morden, Manitoba, and fodderbeets, developed in Europe for feeding cattle but which are also high in sugar. They considered the cost of taking land out of food production in order to grow fuel crops.

The technology is available to produce fuel alcohol. The method selected to do so depends upon the type of crop or waste material used. Some material first needs its starch converted to sugars, other material can be fermented directly without conversion. Whichever method is selected requires careful preparation and a full knowledge of the process.

Under 1984 economic conditions in Canada the scientists concluded that the production of alcohols for fuel seemed unlikely. They noted, however, that economies can change rapidly; that with subsidies, some countries are actually producing fuel alcohol; and that new technology being developed both here and elsewhere may yet make the production of fuel alcohol an economic reality.

ANIMAL WASTES

All livestock and crops produce residues that should be returned to the land to improve its tilth and nutrient value. However, with high density feedlots for beef and sheep, and large plants for pig and poultry production, it is not always economically feasible to utilize all their manures as a soil additive, particularly when trucking distances are long. Flushing manure and crop wastes into rivers loads the waters with biodegradable material, consuming the oxygen needed for fish. The extra nutrients result in excessive weed growth in the waterways.

There are some countries that for generations have used biogas from manures for cooking and lighting. In Europe, biogas now is obtained from many municipal sewage systems. Scientists also know that besides biogas, a single-cell protein can be produced, which, when separated from the liquid, contains 30 percent crude protein suitable for feeding livestock. The bacteria used to digest the wastes cling to the digester's surfaces, sluffing off as they grow. P. van Die from the Engineering and Statistical Research Institute inserted rough-surfaced plastic tubes that increased the area-to-volume ratio by more than 100 times. When digestion is complete, the single-cell protein (bacteria) is centrifuged from the liquid and used as animal protein concentrate, a small portion being reserved to start the next digestion cycle. The remaining liquid is spread on fields but is of low nutritional value because most of the nitrogen has been removed. In Canada's cold winters, it is a challenge to maintain the temperature of the digestion chamber at 35°C, where it is most effective. This is done by using adequate insulation and preheating the manure as it is fed to the digester.

The first Canadian contract to build a pilot-scale anaerobic manure-digester was awarded to the University of Manitoba in 1974. It seemed promising, so several full-scale operations on large (5000 head) feedlots have been developed and are under test. In one operation the returns in gas and protein cake make the operation appear economically sound with a possible payback period of 3 years. A second installation has an estimated repay period of 6–10 years. Neither calculation has taken into account environmental benefits, which are considerable but difficult to quantify. Until each digester has run for several years, no final assessment is possible.

It already was known that the potential energy available from crop residues exceeds that from manure, so Research Branch had a German straw burner tested in the Maritimes. It was found that 400 t of straw produced enough energy to dry 1800 t of corn or grain and heat a 260-sow barn; it also provided all the hot water required in the barn. Where straw is not needed for other purposes, such as to control soil erosion, its use as a fuel is profitable.

ICE PRODUCTION

Before methods of mechanical ice making were developed, natural ice blocks were cut from rivers and lakes to be stored in sawdust for future use. With increased labor costs, this system became too expensive, and it was sometimes unreliable. A new scheme, which was started in the fall of 1983 by C. Vigneault of the Engineering and Statistical Research Institute, would again take advantage of our natural winter temperatures to produce ice. He constructed an insulated icehouse 3 m^2 and 3.6 m high. When air temperatures were −5°C, or lower, water was sprinkled on the floor of the building over which outside air was blown to produce a solid ice block.

Following two winters' trials the engineers were able to automate the process and build a 300-t ice block sufficient to cool 1000 t of vegetables harvested throughout the season. The coldness stored in the block is removed by sprinking water on the ice or circulating water around its base, then spraying the cooled water on vegetables to cool them. A second method is to crush ice from the block and use it to pack around the vegetables in the standard manner. One estimate put the energy cost for mechanical refrigeration at $2.02/t and for natural ice at $0.57/t, a 70-percent saving. No estimate of overhead costs has yet been made. One commercial vegetable grower in Quebec has already installed the system.

PLANT ENERGY

Plants must have nitrogen, which is often supplied by ammonia or nitrate fertilizer. Fertilizer, however, is costly, because it requires a high energy input for its manufacture. Nitrogen is present in abundance in the air, but plants cannot use nitrogen in its gaseous state. Certain microorganisms in the soil convert nitrogen gas into ammonia by combining the nitrogen molecule with hydrogen. This conversion is called nitrogen fixation and is the major natural source for replenishing soil nitrogen. Most nitrogen is fixed by the bacteria (rhizobia) associated with legumes. These bacteria invade the roots of legumes, such as alfalfa and clover, develop nodules in the root cortex, and begin to fix nitrogen for use by the plant. In return, the plant provides energy for the bacteria through other nutrients.

As early as 100 years ago, farmers knew that alfalfa, clovers, and other legumes were "good for the soil." The general theory was that their deep roots brought nutrients from the subsoil for both the legume and subsequent crops. In 1886, the year Experimental Farms were started, two German scientists (Hellriegel and Wilfarth) demonstrated that legumes, unlike other plants, harbored symbiotic bacteria in their roots to combine nitrogen with hydrogen for their own growth. Within 10 years F.T. Shutt, chief of the Chemistry Division, had obtained some "nitragin," as the bacteria culture was called (now known to be *Rhizobium* species), and found that with its use legume growth was equal to that experienced in Germany.

Some soils, especially in the prairie region, have few indigenous rhizobia. For this reason legume seeds must be inoculated at the time of planting. New strains of rhizobia were brought into Western Canada in 1914 by W.H. Fairfield, the first superintendent of the Lethbridge Experimental Station. Subsequently, commercial inocula consisting of mixtures of various *Rhizobium* species were imported from the United States. These inocula contain three or four strains, anticipating that one will be of value under any given soil condition. Although of use, it may not necessarily be the best.

L. Bordeleau, the soil microbiologist at Sainte-Foy, Quebec, tested many strains of *Rhizobium* for their capacity to convert or fix nitrogen. He found several strains and selected one that functioned best in wet acid soils containing relatively large amounts of aluminum. He called it Balsac and made it available to producers of inoculum. Today, it is one of the principal strains in Canada and is also used on some soils in the United States, France, Australia, and New Zealand. Because of the increased nitrogen-fixing capacity of Balsac, there has been a marked decrease in the need for energy expensive nitrogen fertilizer in Quebec. A tonne of protein produced with chemical fertilizer costs $90, but a tonne of protein produced with Balsac costs only $6. Some scientists estimate the savings in Quebec from the use of Balsac to be approximately $60 million annually.

At Beaverlodge in the Peace River district, Alberta, Rice, Penney, and Nyborg (14) found *Rhizobium* associated with red clover to be more tolerant of acidic soils (low pH) than those associated with alfalfa. As a consequence Rice (13) selected strains tolerant to low pH from a group of 100 and tested the four best against Balsac and commercial inoculum. One (NRG-185) gave alfalfa yields 8 percent above similar plants inoculated with Balsac. All yielded two- to four-times uninoculated seed. As with the Sainte-Foy Balsac strain, NRG-185 was given to commercial suppliers for inclusion in inoculum being sold in the Peace River district.

In 1978 the Cell Biology Research Institute, Ottawa, started an intensive research program on nitrogen fixation. Five scientists of different disciplines were brought together. By learning about the total soil–bacterium–plant system and searching for places where improvements could be made they hoped to increase the efficacy of nitrogen fixation. By 1982, Macdowall (7) found there tended to be an increase in the nitrogen-fixing activity of alfalfa following its second and third cuttings. He also observed that the nitrogen-fixing activity within alfalfa plants improved when some combined nitrogen was present in the soil. This research led to a better understanding of the photosynthetic capability of alfalfa and was preparatory to improving its efficiency through genetic manipulation. At the same time, Miller and Sirois (9) observed alfalfa plants to be poorly nodulated and stunted in growth when low in calcium and magnesium ions. They saw the importance, therefore, of selecting strains of *Rhizobium* and cultivars of alfalfa under ideal growing conditions. E.S.P. Bromfield studied the relationship between strains of *Rhizobium meliloti* and cultivars of alfalfa so that breeders might better understand the characteristics for which they were looking. Because of the great variability among plants within alfalfa cultivars, Bromfield (2) found the preference of bacteria strains to be more pronounced among plants than among varieties. The fourth line of attack was to develop efficient nitrogen-fixing strains of *Rhizobium*. In this regard Behki (1), in cooperation with two scientists from Carleton University, Ottawa, has moved the hydrogen uptake capacity from one species of *Rhizobium* to another. Using this device he has increased the symbiotic function found in pea bacteria by transferring it to seven strains of alfalfa bacteria.

In 1971 and the years immediately following, several scientists reported other instances of bacteria fixing nitrogen in the root zones of some non-leguminous crops such as rice, sugarcane, and corn. R.I. Larson at Lethbridge, Alberta, crossed the wheat varieties Rescue and Cadet, searching for and finding chromosome substitution lines. When T.G. Atkinson, also of Lethbridge, was ravelling a wheat root rot problem in 1970 he used some of Larson's lines as well as standard wheat varieties. Scattered randomly within the substitution population growing in the field were plants that were greener and stronger than the rest. Atkinson and J.L. Neal, a soil bacteriologist, found that these greener and stronger plants supported a different microflora around their root systems than did the standard wheat plants (10). Later, Neal and Larson (11) were able to consistently isolate acetylene-producing bacilli from their root zones, an indirect indication that the bacilli fixed nitrogen. Twenty cultures of this type of bacilli were isolated (6) for identification and further study.

R.J. Rennie became a member of the Lethbridge team and worked with three scientists in Brazil, using an isotope of radioactive nitrogen to trace its path within the wheat plants that supported the selected bacilli. In 1983, Rennie and colleagues (12) proved that nitrogen fixation did occur when either one of two bacteria was associated with some particular strains of wheat. The association between bacteria and wheat is very specific, often involving the substitution of only a single wheat chromosome. The practical application of this discovery will require much more intensive research, but the potential return to the farmer should more than repay the effort by providing a crop in addition to legumes capable of supplying energy in the form of fixed nitrogen.

PEAT

Canada ranks second in the world in peat resources, with an estimated 89 billion tonnes. The USSR is first, with 143 billion tonnes, and Finland is third with 16 billion tonnes. The Canadian resource represents an energy equivalent of 14 trillion cubic metres of gas and is equal to 1000 years of coal supply at the present rate of consumption. The quality of Canadian fuel peat is equal to or better than the quality of fuel used for energy in other countries. The development of peat resources in Canada has lagged because we have few experts who know how to exploit it, and because of the high cost of transport to centers of population.

Peatlands cover about 10 percent of Canada's land mass. A large proportion is in northern regions where 60 percent is frozen permanently. There are, however, some deposits close to urban centers that represent appreciable reserves of highly productive agricultural land, especially for horticultural crops. In addition to their use for food production, peats are valuable natural sources of fuel, they are raw material for industry, and they serve as sponges to control the flow of water in drainage basins.

In May 1981, Research Branch cooperated with the National Research Council (NRC), the Department of Energy, Mines, and Resources, and the Department of the Environment in studying the best way to use peat as an energy source. Contracts for studying the peatland resources on the Pacific Coast and the lowlands of the St. Lawrence Valley are being proposed.

The first inventory of Canada's peat resources and their characteristics was completed in 1984 by surveyors of the Land Resource Research Institute. It provides information regarding the extent, volume, and weight of peat for each province and territory. The inventory, consisting of a report and maps, has been published by NRC and has attracted interest nationally and internationally. It will be included as part of the reports on the Pacific Coast and St. Lawrence Valley lowlands peatland surveys. In addition, the National Wetland Data Bank has been developed to facilitate the use of peatland information. The *Wetland Regions of Canada* and the *Wetland Distribution of Canada* maps became available in 1984 as part of the National Atlas published by the Department of Energy, Mines, and Resources.

References

1. Behki, R.M.; Selvaraj, G.; Iyer, V.N. 1985. Derivatives of *Rhizobium meliloti* strains carrying a plasmid of *Rhizobium leguminosarum* specifying hydrogen uptake and pea-specific symbiotic functions. Arch. Microbiol. 140:352–357.
2. Bromfield, E.S.P. 1984. The preference for strains of *Rhizobium meliloti* by cultivars of *Medicago sativa* grown on agar. Can. J. Microbiol. 30:1179–1183.
3. Downing, C.G.E.; Powrie, W.D. 1983. Strategy for energy research, development and demonstration in the agriculture and food system. Engineering and Statistical Research Institute Report I-526. 84 p.
4. Green, James L. 1984. Two solar greenhouses and a computer promise future efficiency and savings. Am. Nurseryman 159(l):101–106.
5. Hayes, R.D. 1981. Farm scale alcohol: Fuel for thought. Engineering and Statistical Research Institute Report I-265. 8 p.
6. Larson, R.I.; Neal, J.L., Jr. 1978. Selective colonization of the rhizosphere of wheat by nitrogen-fixing bacteria. Ecol. Bull. 26:331–342.
7. Macdowall, F.D.H. 1983. Kinetics of first-cutting regrowth of alfalfa plants and nitrogenase activity in a controlled environment with and without added nitrate. Can. J. Bot. 61:2405–2409.
8. Manolakis, E.; Miller, J.D. 1984. Agriculture Canada's bioenergy/biotechnology programs. Engineering and Statistical Research Institute Report I-656. 16 p.
9. Miller, Richard W.; Sirois, Jean Claude. 1983. Calcium and magnesium effects on symbiotic nitrogen fixation in the alfalfa *(M. sativa)–Rhizobium meliloti* system. Physiol. Plant. 58:464–470.
10. Neal, J.L., Jr.; Atkinson, T.G.; Larson, Ruby I. 1970. Changes in the rhizosphere microflora of spring wheat induced by disomic substitution of a chromosome. Can. J. Microbiol. 16:153–158.
11. Neal, J.L. Jr.; Larson, R.I. 1976. Acetylene reduction by bacteria isolated from the rhizosphere of wheat. Soil Biol. Biochem. 8:151–155.
12. Rennie, R.J.; deFreitas, J.R.; Ruschel, A.P.; Vose, P.V. 1983. ^{15}N isotope dilution to quantify dinitrogen (N_2) fixation associated with Canadian and Brazilian wheat. Can. J. Bot. 61:1667–1671.
13. Rice, W.A.; 1982. Performance of *Rhizobium meliloti* strains selected for low-pH tolerance. Can. J. Plant Sci. 62:941–948.
14. Rice, W.A.; Penney, D.C.; Nyborg, M. 1977. Effects of soil acidity on Rhizobia numbers, nodulation and nitrogen fixation by alfalfa and red clover. Can. J. Soil Sci. 57:197–203.

Chapter 16
Animal Research

Domestically, livestock and animal products account for 60 percent of farm cash receipts, nearly one-half coming from the sale of cattle, calves, and pigs. In relation to our human population, the Canadian livestock population is large, as substantiated by data in Table 16.1. We have twice as many cattle, swine, and chickens as most other countries. There are several reasons for the importance of animal production to our national economy and for the superiority of Canada's animal production capabilities.

Table 16.1 Animal populations in 1982[1]

Species	Total number		Number per 100 inhabitants	
	Canada	World	Canada	World
	(millions)			
Cattle	13	1226	52	27
Sheep	0.5	1158	2	27
Swine	9	764	36	17
Chickens	82	6578	328	143

[1]From FAO-WHO-OIE, 1982 Animal Health Yearbook FAO Animal Production and Health Series, No. 19.

Canada has immense tracts of grazing land upon which many thousands of cows rear their calves annually. There are abundant feed grains for swine and poultry to convert into proteins. For all livestock classes, the Canadian federal and provincial departments of agriculture have encouraged the organization of assertive breed associations. Experimental Farms played a vital role in the early stages of animal production and improvement to make today's success possible.

This chapter highlights some of the contributions made by the branch to the Canadian livestock industry through its research in breeding, feeding, and managing of livestock.

DAIRY CATTLE

The first dairy cattle breeds arrived on the Central Experimental Farm in 1889 with the purchase of Ayrshire, Holstein, Jersey, Polled Angus, and milking Shorthorn. Some were destined for shipment to other experimental farms. A breed called The Canadian was added the following year. As with other livestock, the main purpose was to improve the national herd by making its purebred registered bulls and cows available to local breeders. The Department of Agriculture and the Experimental Farms were strong supporters of the various breed associations and urged dairymen to upgrade their herds with the objective of registering their stock in the national herd books. Each experimental farm and station selected animals on the basis of conformity to type, high milk production, and butterfat. Superintendents of the old school who were also animal husbandmen, took pride in accompanying their livestock to major North American agricultural shows. They often returned with a suitcase full of ribbons among which they hoped to have at least either one best of breed or one grand champion.

In 1890 J.W. Robertson, Chief, Agriculture Division, recognized the differences between animals within each breed and realized the futility of comparing breeds with only a few representative animals. He therefore drew no conclusions from the first feeding trial with milk cows except to say that the value of ensilage was surprisingly high.

Over the years, nutritional and management experiments became more sophisticated. G.B. Rothwell in the 1920s compared returns from commercial dairy meals with well-balanced, home-mixed ones. He found that although there was little difference in the animals' performance, the home-mixed meals were invariably the most profitable because their cost was about 66 percent of those commercially prepared. Rothwell also concluded that it was advantageous for farmers to raise their own dairy animals rather than to buy them. He demonstrated that Holsteins were more profitable than any of the other breeds.

By the 1930s, when G.W. Muir was Dominion Animal Husbandman and C.D. MacKenzie was responsible for dairy cattle and dairying, feeding experiments were refined with the introduction of "double reversal" designs. Test animals were divided into as many groups as there were diets to be tested. Each group was fed one diet for 4 weeks. Data were collected on only the last 2 weeks of the period. At the beginning of each ensuing 4-week period, diets were changed among groups. The cycle was run twice. Additives such as cod oil, linseed oil meal, and soybean oil meal were also tested in this manner. Little difference was found among them; their usefulness depending upon price.

Artificial insemination

A. Deakin was the first animal geneticist on the staff of experimental farms. He recognized the limitation each bull had for contributing desirable characteristics to the national herd because of the relatively few offspring produced by natural breeding. Russian workers developed a method to artificially inseminate dairy cattle to overcome this limitation. The method was then used by the School of Agriculture, Cambridge University, England, in 1933. In 1935 Deakin imported the necessary equipment from England, used it immediately, and the first calf to be born on an experimental farm as a result of artificial insemination occurred at Sainte-Anne-de-la-Pocatière in February 1936 (8). Deakin studied methods of collecting semen, inseminating cows, sterilizing equipment, and storing semen. He shipped semen in glass vacuum bottles by rail to other experimental farms but had little success coordinating its arrival with a cow's readiness to receive it. The inauguration of air express in 1938, moderated these difficulties. By 1951, scientists learned how to freeze semen; thus its usefulness became more flexible. Since Deakin made his original studies, techniques for handling semen and testing bulls for their transmission of desirable characteristics have been greatly improved. One of the first routine uses of frozen semen was made by C.G. Hickman. In a dairy cattle breeding research project started in 1954, experimental stations from coast-to-coast were supplied with frozen semen from a common group of bulls. The resulting controlled and geographically widespread experiment would not have been practical had fresh semen or natural mating been used.

Breeding

Following World War II when H.K.C.A. Rasmussen was appointed Chief, Animal Husbandry Division, many other well-educated scientists were hired, and research with large animals rapidly developed. Breeding work with livestock is both slow and expensive. Unlike most plant crops where generation times are 1 or 2 years and often can be truncated to a few months, cattle have a generation time of at least 5 years where the possibility of truncation using embryo transfer methods is still experimental. More important, however, is the difference in the practical size of populations. Cereal plants can be readily grown by the millions, whereas large animals can be handled only in hundreds, or at most thousands. On the positive side, livestock generally are not as sensitive to the environment as are field crops. Their food and housing can be adjusted and their management systems, at least within any major climatic zone, are reasonably constant. The limitations restrict the rate of progress in livestock breeding and govern the way in which livestock research must be managed.

To compensate, Hickman reorganized the breeding program for dairy cattle on experimental farms and stations by considering all animals of any one breed (Holstein, Jersey, Guernsey, or Ayrshire) regardless of location, to be part of one herd. This increased the population size within each breed from which selections could be made. In the mid-1960s the number of breeds was reduced to two to maximize their numbers. Holsteins were retained by Fredericton, New Brunswick; Lennoxville, Quebec; Ottawa, Ontario; Lethbridge, Alberta; and Agassiz and Prince George, British Columbia. Ayrshires were at Charlottetown, Prince Edward Island; Normandin and Sainte-Anne-de-la-Pocatière, Quebec; Ottawa and Kapuskasing, Ontario; Morden, Manitoba; and Prince George, British Columbia. The size of each herd varied from 50 to 80 cows.

Hickman (32) published results from a 12-year selection experiment, using these Holstein and Ayrshire herds. Initiated in 1954, this project evaluated selection for 180-day milk-solids yield in the first two lactation periods. Average annual rates of genetic improvement were 1.06 and 1.30 percent of average annual milk-solids yield for the Holsteins and Ayrshires, respectively. Cows in third and later lactation periods were available for other nutrition and management research. Hickman and Bowden (33) related genetic changes in feed efficiency, growth, and body size to selection for milk-solids yield. Animals

175

producing the most milk were the heaviest at calving, gained more weight during each lactation, and consumed more feed than those producing less milk.

In 1968, a team of scientists representing all cooperating research stations (56) organized the next dairy cattle breeding program. This time they hoped to improve the lifetime productivity in Canadian dairy cattle by using a design developed by Hickman and Freeman (34). Both pure line selection and crossbreeding were planned. The selection criterion was high yield of milk protein. Nine hundred foundation Ayrshire and Holstein cattle were maintained from the previous two populations at five research institutions: Charlottetown, Normandin, Lennoxville, Ottawa, and Lethbridge. Each breed is at three locations, Ottawa having both breeds. The genetic bases of both populations were broadened by using frozen semen from proven bulls outside of the Research Branch herd. The crossbreeding program uses first-generation crossbred males from elite H- and A-line males and females mated to a continuous-breeding crossbred female population begun with reciprocal crosses of the A- and H-lines. The program is large and complicated and would be difficult to manage without optimum employment of the computing capabilities of the department.

Comparisons of the parental lines indicate that the H-line cattle produce heavier calves sooner, have more difficulty calving, and are superior in milk and protein production to the A-line cattle. To date, the study has confirmed that genetic factors play a minor role in influencing cow reproduction, the major role in this regard is played by management. Progeny from crosses of the two breeds exhibit hybrid vigor in all heifer body weight and dimension traits, the greatest being heifer body weight (53). Reproductively, crossbred heifers have a shorter gestation period and calf survival is 5–9 percent greater than for either parental line. Crossbred and pure line cattle have similar conception rates and calving intervals, but crossbred and A-line cows produce more live calves than H-line cows. A more complete assessment of the advantages of purebreeds and crossbreeds for lifetime productivity awaits completion of multiple lactations.

Research by departmental scientists, using the ROP (record of performance) dairy cattle data, was fundamental in establishing methods of obtaining accurate national sire and cow evaluations.

BEEF CATTLE

Breeding

The great central plains of North America, stretching from Texas into Canada, were once the domain of the bison, the only bovine native to North America. Through this region the climate varied from extreme dry heat in summer to frigid blizzard winds in winter. Summer pastures varied from arid to semi-forested areas. During the winter, parts were buried in snow, often heavily crusted from periodic chinook winds. Through centuries of natural selection, the bison adapted to these conditions. Herds in uncounted millions, followed the grazing, the water supplies, and the seasons. When necessary, they pawed through snow to feed on the fall-ripened forage. If caught in a blizzard, they faced into the wind until the storm subsided.

The cattle brought from Eastern Canada did not possess these attributes. They performed adequately if intensively managed with particular attention to winter feed and shelter, but they lacked the capacity to survive and reproduce under the extensive management appropriate to range conditions. Settlers quickly learned that more reliable production could be obtained from cattle imported from the western United States. These cattle, many of which had originated as far south as Texas, were extremely hardy, a quality that compensated for their deficiency in conformation and growth characteristics traditional of British breeds. They were not purebreds. Their ancestors were of many different breeds and their hardiness was a direct result of unprogrammed, natural selection, and hybrid vigor.

It was predictable that winterhardiness would be the first issue to attract research attention. Improved winterhardiness resulting from natural bison $\times$ cattle crosses had been observed as early as 1750 through practical experience in the south central United States. By 1885, deliberate matings were made by ranchers in Canada and the northern United States. Among the recorded successes was that of an Ontario breeder, M.M. Boyd, who began his crossing program in 1894. He called the hybrid, cattalo. When his herd was dispersed in 1916, E.S. Archibald, Dominion Animal Husbandman, acquired 16 females and four males for experimental evaluation at Scott, Saskatchewan. In 1919, in cooperation with the Department of the Interior, the herd was moved to Buffalo National Park, Wainwright, Alberta, and was augmented by the introduction of cattle, bison, and yak. Deakin, Muir, and Smith (9), the scientists responsible for this work, reported that none of the original Boyd cattalo produced offspring, and all results subsequently obtained in the experiment were from the animals introduced after 1919. The final move was to Manyberries, Alberta, in 1950, where the herd came under the direction of H.J. Hargrave and H.F. Peters.

Hybridization with the yak was discontinued in 1928, its progeny having been judged as inferior in winterhardiness to those incorporating bison. The bison crosses with domestic cows resulted in 77 percent calf mortality, and males of this and the reciprocal cross were invariably sterile. Fertility of the hybrid female was also low but, through backcrosses with domestic sires, a cattalo line was produced. Functional males carrying more than one-eighth bison never were observed. This established the maximum limit to the genetic contribution of bison. Subsequent studies indicated that, compared with Herefords, the cattalo had no advantage in growth or carcass performance. The work was discontinued in 1965.

The 1916 decision to purchase the cattalo herd marked the advent of cattle breeding experimentation by the Experimental Farms. No further breeding research was undertaken, however, until 1949 when J.G. Stothart and H.T. Fredeen at the Experimental Station, Lacombe, Alberta, initiated a selection study with Shorthorns that involved two half-sibling bulls of distinctly different feedlot performance. One had an average daily gain of 1.11 kg/day, the other

0.77 kg/day. Each was mated to representative cow herds. The resulting progeny from the high performance bull were heavier both at birth and when weaned than those from the low performance bull. Furthermore, the average daily gain, feed required per kilogram of gain, and age to finish were each in favor of progeny of the high performance bull. The results formed the basis for the national beef ROP testing program adopted in 1957. The study evolved into a large-scale Shorthorn selection project with herds at Brandon, Manitoba; Indian Head and Scott, Saskatchewan; and Lacombe, Alberta. From 1957 to 1970, Newman, Rahnefeld, and Fredeen (62) achieved annual genetic gains in yearling weight of 4.5 kg per animal and established the practice of selecting for performance in beef cattle.

Shortly after the cattalo herd was transferred to Manyberries, Hargrave, Peters, and J.E. Lawson initiated studies to evaluate the reproductive potential under range conditions of crossbred females other than cattalo. They mated Highland and Brahman sires with Hereford cows. Highland cattle, native to the damp mountainous regions of northern Scotland possess long, heavy hair, and gave promise of transmitting winterhardiness to their offspring. The appeal of the Brahman did not stem from its potential hardiness. It had originated in the southern United States from zebu strains imported from India with little promise of adapting to winter climates. What it did possess was a genetic background uniquely different from the British breeds and the scientists reasoned that this difference might provide additional vigor in their female hybrids.

This expectation was fulfilled. The Brahman crosses proved well adapted to the Manyberries range. Though somewhat temperamental, they conceived and calved easily, were good mothers, and consistently produced calves with high weaning weights. The most successful cross was the Brahman × Shorthorn, a prime example of hybrid vigor, since the offspring were substantially more hardy than either of the parent breeds. Brahman crosses, compared with Herefords, suffered less attrition, lived longer, and during their lifetime produced 25 percent more calves in number and 52 percent more calf weight at weaning.

In 1963, Lawson initiated an experiment with the Angus and Hereford breeds to evaluate their response to selection under a high-energy diet (60 percent barley) and a low-energy diet (100 percent chopped alfalfa). Though not complete yet, the initial results from three generations indicate that the lines selected on the low-energy diet have superior milk production, greater fat, protein, and solids-not-fat content in the milk, and increased calf weights at weaning than those selected under the high-energy diet. These results are expected to influence the way beef animals will be chosen in future.

A new era in Canadian beef production began in 1951 when some enterprising cattlemen imported a few Charolais cattle from the United States. Other cattlemen, impressed with the outstanding carcass and growth potential of Charolais crosses, began to switch their allegiance from the traditional breeds to hybrids. Pressure mounted to open the Canadian border, long closed for reasons of animal health security, to importation of breeds from continental Europe. The Honourable Alvin Hamilton, Minister of Agriculture in 1960, instructed K.F. Wells, Dominion Veterinarian, to arrange some means of safely importing European cattle. By the time all arrangements were complete, the Honourable Harry Hays was Minister of Agriculture. He inaugurated an import scheme in 1964 by establishing quarantine facilities and rigorous health inspection procedures for cattle entering the country.

Charolais and Simmental breeds were the first European imports to clear quarantine in 1967. They were augmented in 1968 by the Limousin, Maine Anjou, and Brown Swiss breeds. Although many others have entered Canada since 1968, these initial breeds have had the greatest impact on cattle production throughout Canada.

Growth rate, muscular conformation, and mature size were the attributes that attracted the interest of western Canadian cattlemen. Thus, the initial emphasis was to produce first-generation hybrid calves for slaughter. Some attention was given to selecting superior hybrid females required for breed propagation. It was evident, however, that the difference between these breeds and the British breeds might confer greater heterosis for reproductive traits than had been attained when crossing among the British breeds. This had been the practical experience of those who had worked with Charolais crosses since the early 1950s.

H.T. Fredeen, J.A. Newman, and J.G. Stothart, of the Research Station, Lacombe, Alberta, initiated a study in 1967 to measure the effects of crossing European and British breeds. Their objective was to evaluate the lifetime reproductive performance of hybrid females from matings of Charolais, Limousin, and Simmental sires with Angus, Hereford, and Shorthorn cows. In 1969 J.E. Lawson at Lethbridge and G.W. Rahnefeld at Brandon joined the team, making additional facilities available to evaluate the hybrid females under diverse environmental conditions. To the Lacombe group came G.M. Weiss, a statistician, and A.H. Martin, a meats specialist.

The program developed into four interrelated phases. Phase I produced hybrid females to evaluate their lifetime reproduction capabilities and hybrid males for feed efficiency and carcass evaluation. Phase II measured the reproductive performance of hybrid females under intensive pasture production at Brandon and extensive semiarid range production at Manyberries. Phase III examined both cow herds for winter feed requirements and summer lactation. Phase IV, still in progress, evaluates the lifetime reproductive performance of females from backcrossing the hybrid females to sires of their parental breeds. Phases I and II now are completed and are reported (17) by Fredeen and coworkers. The size and complexity of the experiment can be judged by realizing that 10 crossing groups resulted in 5053 pregnancies.

The findings of Phase II validated the original assumption that heterosis of the hybrid females would be related to the genetic diversity of the parent breeds. The lifetime productivity of animals from the Hereford × Angus cross, the breed combination long favored by western cattlemen, was inferior to the productivity of animals from all other crosses. Weaning weights of Hereford × Angus progeny were 10 to 15 percent less than those of the progeny of hybrid dams sired by Charolais and Simmental. Differences between progeny from at least one European breed varied depending upon environment. For instance, the most productive progeny under the conditions at Manyberries were from Charolais × Angus. At Brandon, progeny from Simmental × Shorthorn were superior.

An integrated evaluation of all the evidence revealed that these differences (interactions) weré related to the milking potential of the hybrid females. Progeny of the Simmental × Shorthorn cross were superior in weaning weight at both locations, an indication of greater milk production. This superiority was reflected in a longer reproductive cycle and a reduced conception rate under the harsh environment at Manyberries.

These results clearly have demonstrated that heterosis is not a panacea for beef producers in Western Canada. Genetic superiority in the reproductive potential of beef cattle can be detrimental to lifetime productivity unless matched with environmental or management practices adequate to sustain the biological demands imposed by that potential.

Some insight into this problem has been provided by the results of Phase III. Subcutaneous fat represents insulation as well as an important energy source for the brood cow. Experience at Manyberries indicates that cows carrying less than 5 mm of fat when weaned in October do poorly during the winter and have substandard conception rates the following summer. Except for the Hereford × Angus cows, which average 8 to 9 mm of fat at weaning time, all nursing cows of the other crosses at Manyberries range from 0 to 5 mm of fat in the fall. To restore them to an acceptable level of fatness for winter has required up to 50 percent more feed than amounts commonly recommended for the winter period. Phase IV will not be completed for several years.

The beef breeding program has shown the value of careful selection for economic traits, of maintaining high-quality purebred lines, and of using these purebreds in hybrid programs designed to match genetic characteristics with environmental constraints. As the advantages of advanced technologies increase, a fuller understanding of the biology becomes essential.

Feeds for cattle

Cattle (and sheep) are ruminant animals and have the advantage over other stock and poultry of being able to convert cellulose into a digestible form of carbohydrate. This is done by bacteria in the rumen, the second portion of the stomach. Before this happens, however, ruminants regurgitate food from the reticulum, the first portion, and remacerate it with their broad molars, mixing it with their digestive juices (this action is called chewing the cud) before swallowing again when it reenters their digestive tract. Ruminants can also use concentrated foods in the form of grains and minerals to balance their diets.

Forages (grass and legumes) eaten by the ruminants are consumed directly from pastures or are fed to them either as preserved hay or as fermented silage. The silage is made from grasses, legumes, and cereals such as whole oat or corn plants. V.S. Logan, in summarizing (54) the research done with grass silage at a number of experimental farms and stations such as Charlottetown, Nappan, Sainte-Anne-de-la-Pocatière, Lennoxville, Ottawa, and Lethbridge emphasized the fact that high-quality roughage is frequently more easily kept in many parts of Canada as silage than as hay. Silage has the advantage over hay because it takes less storage room and there is less wastage when feeding. It has the disadvantage of freezing in extremely cold weather and may develop objectionable odors. The many experiments during the period from the mid-1920s through 1955 at numerous experimental stations comparing various types of feed did not reveal many consistent differences. In the Maritime Provinces and Quebec, where corn was difficult to grow until shorter season varieties became available, root crops such as mangels, turnips, and rutabagas, were favored over oat and pea silage but not always over grass silage.

Sylvestre and Mercier (73) from Ottawa and Lennoxville summarized the use of grass silage in maintaining and fattening beef cattle under feedlot conditions. They concluded that grass silage was as good as or better than corn silage. For both types, depending upon the objectives of feeding (maintenance, fattening, or reproduction), concentrated feeds such as grains are needed with both silage and hay roughages.

Melfort, Saskatchewan, is in the aspen parkland, a region that includes all the arable land in Manitoba and those areas of arable land in Saskatchewan and Alberta to the north and west of the Palliser Triangle. There, the climate and land generally are suited to growing grass and legumes as well as cereals and canola. The inferior land is used for pasture and hay, the superior for cereals and oilseeds. Scientists at the Melfort Experimental Station believed that more beef could be raised if they learned how best to manage forage crops. They knew that forage yields, and hence beef yields, reduce on a per hectare basis as pastures age over years. They also knew that dry matter yields are dependent on precipitation in May and June with 41 percent of seasonal live weight gains of cattle occurring in the first quarter of the pasture season. Beginning in 1954 they started a series of beef–pasture experiments, the results of which have led to improved use of pasture and hay crops.

In their first 7-year grazing study from 1955 to 1961 at Melfort, Cooke, Beacom, and Dawley (6) increased beef yields by 20 kg/ha when they fertilized a bromegrass–alfalfa pasture with 22 kg of nitrogen. Adding 45 kg of phosphorus to the nitrogen increased beef yields by 36 kg. By increasing the applied nitrogen to 84 and 168 kg/ha, beef yields were greater by 61 and 98 percent, respectively. However, the higher rates of nitrogen reduced and nearly eliminated the alfalfa because of the greater competition from the more vigorous growing bromegrass. Later (1961–1966), Beacom (2), and Cooke, Beacom, and Robertson (7) increased beef yields by 5 percent, the grazing season by 15 days, and feed efficiency by 30 percent when they adjusted stocking rates to the amount of forage available. Robertson, Cooke, and Beacom (65) expanded on these results with the objective of optimizing beef yields and minimizing management problems from insufficient feed during mid-season. By converting surplus spring

181

grass to silage and feeding the silage during the summer, they were able to increase beef yields by 3 percent and retain the same number of animals on pasture throughout the season. When they produced extra oat soilage (immature oat plants) for supplemental stock feed during low summer pasture production, they increased animal gain per hectare by 35 percent relative to the control. Supplementing with barley grain instead of oat soilage resulted in a further 10 percent live weight gain per hectare.

Beacom and the Melfort scientists went the next step with grass-fed beef by developing a finishing system, using ground high-quality hay with little grain. The hay is partly field dried, chopped, then mechanically loaded into a hay tower, developed in cooperation with the Engineering and Statistical Research Institute. The tower has a movable roof and skirt, and a blower to circulate ambient air for final drying. Fatty acids from canola oil and antibiotics are added when fed to cattle implanted with growth-promoting substances. By using this good-quality hay in high-forage rations supplemented with acidulated fatty acids from canola oil and with growth promotants, rates of gain and feed-to-gain ratios equal those of stock finished on grain.

A critical period in finishing cattle in a feedlot occurs when they are put on full-feed. This is done in southern Alberta and elsewhere by gradually increasing their daily grain allotments over a 4- to 6-week period without causing digestive disturbances. Hironaka (42) at Lethbridge, found he could bring yearling steers to full-feed in 10 days without any digestive disorders. To do so he used a starter ration containing 25 percent dried molasses beet pulp or dried brewer's grain. The remaining portion of the ration contained alfalfa, oats, barley, minerals, and vitamins. As a result of this research, feedlot operators saved 2- to 4-weeks of finishing time, and a market for all the beet pulp produced in southern Alberta was developed. The system did not prevent feedlot bloat, the solution to which is discussed in the next section. At Melfort, S.E. Beacom and colleagues achieved a similar result by feeding high-quality ground grass hay in half the ration and grain in the other half. After 3 days, the grain portion was gradually increased over a 2-week period to 90 percent of the ration. Under the conditions at Melfort, no grain poisoning or bloat was experienced.

This has been only a small sampling of the many hundreds of feeding experiments with cattle throughout the past 100 years. It provides a quick overview of the type of research done and the information that has been supplied to livestock farmers.

BLOAT, THE UNPREDICTABLE KILLER

The following dilemma is a typical one many dairymen have to face: "Last night was cool but today will be hot and sunny. Following an hour in the milking parlor, the dairy herd of one farm is moved to a lush legume–grass pasture essential for high production. The pasture is still wet from last night's rain."

"Two cows bloated yesterday but only one was saved. Perhaps that good legume–grass pasture should be plowed down and seeded with grass only. The national average for bloat loss is less than 1 percent, but three good cows out of 60 have died on this farm in the past month."

Such thoughts torment the minds of many dairymen each spring and summer. National averages are acceptable to them provided their cattle are part of the lucky 99 percent.

Bloat is of two kinds—pasture and feedlot. Both are caused by the accumulation of gases in the rumen. If the gas cannot escape from the rumen, it swells until pressure prevents breathing and blood circulation. If caught in time, herdsmen can relieve the pressure by one of several methods, the most common of which is drenching with mineral oil or other foam-dispersing agents. In severe cases a trocar (a metal tube with a pointed, removable rod) or a knife must be used to puncture the rumen through the left flank to release the gas and prevent death.

Some of the first experiments in Canada on the prevention of bloat were started in 1957 and reported by J.E. Miltimore (61) from the Summerland Experimental Station in 1963. He found that penicillin, mineral oil, or tallow given either as a drench or mixed with feed immediately following the morning milking, appreciably reduced the incidence of bloat but did not prevent it.

The following year J.M. McArthur, the chemist who worked with Miltimore (57), found that the protein 18S, also known as Fraction 1, was probably the causal agent for pasture bloat. This particular protein had a high sedimentation velocity measured as 18 Svedberg units, hence its name. At the 10th International Grassland Congress in Finland in 1966, W.J. Pigden reported for McArthur and Miltimore that 18S protein trapped the rumen gas by forming foams with a high shear strength—the bubbles would not break easily. He reported that forages which caused bloat had about 5 percent 18S, whereas those that did not, contained less than 1 percent. Bloat was described in terms of the speed of release of 18S protein, as well as in terms of fermentation, gas production, and acidity of the rumen contents. Because of the significant contribution made by Miltimore and McArthur toward solving the bloat problem, their names became well known throughout the world. How could they use their information to help the dairyman?

By 1968 Miltimore and McArthur had shown that bloating forage plants such as alfalfa and clover had about one-third of their proteins in the form of 18S, whereas non-bloating forages such as trefoil, sainfoin, vetch, and several grasses had one-sixth, or less, of this type of protein. They established a minimum level of 18S protein for which to look when selecting new varieties of alfalfa and clover. Since then, other scientists have shown that the analytical methods used, although accepted at the time, produce questionable results.

The previous year, D.H. Heinrichs, the alfalfa breeder at Swift Current, Saskatchewan, became interested in the amount of 18S protein in various alfalfa species. Heinrichs had bred several varieties of alfalfa suitable for use on dryland pastures in the Prairie Provinces, but their adoption had been restricted because producers were concerned about their potential bloat-causing properties. He and Miltimore found that the yellow-flowered Siberian alfalfa (*Medicago falcata*) contained about half the 18S protein of standard alfalfa (*M. sativa*). Because the two species could be crossed, they suggested that Siberian alfalfa should be used in breeding programs to produce a variety with a low content of 18S.

In the meantime, Miltimore had been appointed director of the Research Station at Kamloops, about 250 km north of Summerland. Kamloops is in the beef country of British Columbia, so he transferred his interest in bloat from dairy cattle to beef cattle. Beef usually is finished for slaughter in feedlots, where high concentrate diets are fed. Bloat is a constant concern to feedlot operators, even though they may not feed alfalfa. Most research in the Research Branch on feedlot finishing of beef was done by R. Hironaka at Lethbridge. Hironaka pelleted whole barley in which the barley kernels were broken during the pelleting process. He found that animals fed these pellets had almost no bloat and gained weight more efficiently than those fed pelleted ground barley. Furthermore, many of the animals fed ground barley bloated. On examining the two kinds of pellets, Hironaka found the ground barley pellets had smaller particles than the whole grain pellets. When diets were changed to whole grain pellets, bloating stopped almost immediately. Recognizing the need for scientific confirmation of this observation, he joined with the Summerland and Kamloops team to find the causes of feedlot bloat. They used identical twin cattle, which made their experiments much more valuable than had they used unrelated pairs of animals of the same age. By 1972 they reported (43) that animals fed diets of a large particle size (715 microns, 1 micron is 0.001 mm) produced much less foam, less stomach gas, and less bloat than animals fed diets of a small particle size (388 microns). With the addition of K.J. Cheng at Lethbridge, they were able to provide a microbiological and biochemical basis for bloat. They found that by adding salt to the diet, feedlot bloat tended to be controlled. Although the large-sized feed did not completely eliminate feedlot bloat, its use substantially reduced the number of occurrences; thus bloat was no longer the serious problem it had been.

More scientists joined the team. In 1971 B.P. Goplen, the forage breeder from the Saskatoon Research Station, initiated the world's first major program to breed bloat-safe alfalfa. He spent a year at Summerland studying bloat. R.E. Howarth, a zestful, young chemist, was recruited in 1972 to develop practical laboratory methods for selection of bloat-resistant legumes. Miltimore moved again, this time to become director at Agassiz, but the chemist, W. Majak, continued working on this problem at Kamloops. The Saskatoon, Lethbridge, and Kamloops scientists started a series of cooperative studies, sharing their resources and expertise to resolve this difficult and complex problem. One of the first things Goplen wanted to know was the heritability of soluble protein in different species of alfalfa. If the level of 18S was readily inherited from one generation to the next, then a reasonable approach to the problem was to breed a non-bloating alfalfa. If the heritability was low, then to breed such a variety was not a viable option. The team analyzed more than 1300 alfalfa plants of diverse origin and found that even though the heritability was not high (28 percent), it was high enough to make a breeding program worthwhile.

Most scientific theories are tested continually and challenged. So it was with the 18S theory. Goplen found that progress was slow in selecting alfalfa for low 18S protein. Other workers doubted that 18S was the only causal agent of bloat, and therefore Howarth, by now also in Saskatoon, reinvestigated the foaming properties of alfalfa proteins and found that two proteins were implicated. For 5 years Howarth and Goplen used total soluble proteins as their selection criteria, but by 1977, after two cycles of selection, they had made little headway. Was something else involved?

In 1978, Howarth and coworkers (47) discovered that the structure of cell walls in forages was associated with their bloat-potency. With a simple but effective experiment they showed that cells in leaves of bloat-safe legumes were 10 to 20 times more resistant to mechanical rupture than those in bloat-causing legumes. This changed the direction of the research program to a search for plants that released bloat-causing substances slowly. G.L. Lees was employed at Saskatoon to identify the structural features in plants associated with bloat. He found that the rate of microbial digestion and the strength of cell walls were probably the two most important features.

In 1981 the team of seven scientists (45) from the Research Branch augmented by J.W. Costerton from the University of Calgary built on Lees' work and demonstrated that bloat-causing legumes were digested more rapidly than bloat-safe legumes, although the total digestibility of each type was about the same. Therefore Goplen and Howarth began to select for low initial rate of digestion in the Saskatoon alfalfa breeding programs. The laboratory method for making these tests is to place samples of legumes in nylon bags and immerse them in an artificial rumen for 2 to 3 hours. At Lethbridge, Cheng has found a positive correlation between this laboratory method and actual digestion rates in rumens of both sheep and cattle.

While Canadian scientists sought solutions to the bloat problem by adjusting the diet of the cow, New Zealand scientists were selecting animals with high and low susceptibility to bloat. Their intention is to breed a bloat-resistant animal. They have found that animals with low susceptibility have 25 percent less rumen fluid than the highly susceptible ones. This variation is thought to be controlled genetically. In Canada, Majak examined rumens on a daily basis for frothiness. The cattle then were fed fresh, chopped alfalfa with the intention of producing bloat. Bloated animals tended to be the ones with froth prior to feeding, having rumen fluid high in chlorophyll and many microorganisms. This provided an active inoculum for the fermentation of incoming feedstuffs. Based upon this and other research Howarth and colleagues (46) concluded that microorganisms adhering to chlorophyll membranes prevent the bubbles in the rumen from breaking, thus causing froth. Bloat-causing legume species are more susceptible to microbial digestion than bloat-safe species, the specific reasons for which are understood and are being used in breeding bloat-safe legumes.

Following the presentation of this work at the Sixth International Symposium on Ruminant Physiology in September 1984, cochairman R.W. Dougherty congratulated Agriculture Canada scientists on the bold approach to the

185

problem of pasture bloat. He made special comment of the unique achievements made possible by the close cooperation of plant and animal scientists.

The problem of bloat is not completely solved, but Canadian scientists are much closer to a solution than they were in 1957 when Miltimore first started his investigations. Farmers and feedlot operators now know how to reduce the incidence of bloat, thus saving several thousand animals each year.

NUTRITION

Trace elements

Livestock need a number of trace elements to grow properly and to function. When the plant material they consume grows on soils containing adequate concentrations of these elements, the livestock normally do not require supplements to their diets. When soils contain an excess of one or more trace elements toxicity, and in extreme cases death, can occur in livestock. Sometimes elements interact within the body, the availability of one affecting that of another. At other times they interact with vitamins. Many trace elements exist. The following three examples illustrate their importance.

Manganese

The proper functioning of both male and female reproductive systems is dependent upon adequate supplies of manganese. The element occurs in its highest concentration in the glandular organs, particularly the liver (35). Deficiency of manganese frequently occurs when dietary calcium levels are high. In 1975 Hidiroglou and Shearer (41) found that manganese was probably physiologically significant in the normal reproductive cycling of ewes. This was confirmed 3 years later by Hidiroglou, Ho, and Standish (38) when they found that ewes fed diets low in manganese required nearly twice as many services per conception as those fed diets containing normal amounts of manganese. The cause of the difference was not determined conclusively. Ivan (50) found that 30 percent of the rumen manganese in sheep was in its protozoa and bacteria fractions, indicating that the sheep themselves did not utilize all the manganese.

Copper

Canadian soils in many areas are deficient in copper. In 1964 J.E. Miltimore at Summerland, British Columbia, injected copper into cattle grazing the pastures of low-lying, poorly drained organic soils. Their average daily gains increased by more than 200 g. Later, with a number of colleagues (60), he repeated the treatments on two ranches, obtaining an increased daily gain per animal of 118 g (22 percent). Forage on the poorly drained soils had less copper than molybdenum (Cu/Mo ratio of 0.7), whereas on well-drained soils copper was in excess of molybdenum (ratio of 1.9). The amount of copper in the hair of injected animals was 50 percent higher than that of the control animals. It was evident that a deficiency of copper was a limiting factor in growth. In 1967 A.G. Castell, at Melfort, Saskatchewan, working with swine, suspected their diets were deficient in copper because they were not gaining weight as expected. He therefore added 125 ppm of copper (0.5 kg copper sulfate per tonne) to their diet. This improved the growth rate and feed conversion efficiency of the gilts by 9 percent.

There also are situations where copper toxicity in livestock can arise. Copper was found by Hidiroglou, Heaney, and Hartin (36) to be in excess and therefore poisonous to sheep reared in total confinement. The level of copper in the liver of the sheep was considered to be normal but knowing that chronic copper poisoning occurs under conditions of moderate copper intake along with low dietary levels of molybdenum and sulfur, the Animal Research Centre scientists suspected that the intake of these two elements would be low. By adding minute quantities of molybdenum and sulfur to the daily feed of each animal, the rate of copper excreted by the sheep increased and the mortality rate decreased rapidly. This slight change in feed demonstrated the importance of mineral interaction in livestock.

W.T. Buckley at Agassiz, British Columbia, wishing to study copper in the nutrition of cattle, found there was not a satisfactory radioisotope tracer for the element. However, there was a stable copper isotope (^{65}Cu). Because it could be used for experiments of long duration and did not require the elaborate human or animal facilities that were needed with radioisotopes, Buckley and coworkers decided to use the stable isotope. The major disadvantage was its relative difficulty of determination. Buckley and coworkers (3) devised two analytical methods, testing them on dairy cattle being treated with a stable isotope of copper. The same rate of dietary copper absorption was established with each method, and therefore the reliability of the methods was established and will be used in future nutritional studies.

Selenium

Vitamin E and selenium deficiencies are closely related and are often corrected together. Calves born in the winter or early spring frequently appear healthy, then one day they are no longer frisky, seem less eager to feed, and within a few days indifference gives way to a stiff gait. Tremors may affect their entire bodies, they move with pain, cannot feed themselves, and after 10–14 days die of starvation if nothing is done to correct the situation. These are the symptoms of nutritional muscular dystrophy (NMD). During winter months when cows are fed a dry diet, there is little vitamin E present and therefore young animals on a milk-only diet have a deficiency in vitamin E. Selenium also plays a part.

It is only in the past two decades that animal nutritionists and veterinarians have known the cause of and cure for NMD, also known as white muscle disease. It was in 1958 that an American scientist found selenium to be an element essential to growth. During the past few years there has been increasing evidence that NMD is prevalent in regions with granite soils that are low in selenium. The disease never has been reported on farms where the trace mineral selenium in forage is at levels greater than 0.1 ppm.

In the northern Ontario district of Kapuskasing, in 1960, beef cattle at the experimental station and in commercial herds showed NMD symptoms. M. Hidiroglou injected selenium and vitamin E into the calves. He was successful in curing the symptoms but could not study the problem further because there was no known method of chemically analyzing for selenium in tissues of either plants or animals. Therefore, he and R.B. Carson, Chief, Analytical Chemistry Research Service, devised a method (44) for determining selenium levels. By 1967, following Hidiroglou's move to the Animal Research Institute, Ottawa, he, with Heaney and Jenkins (37), demonstrated that the rumen bacteria of a sheep metabolized inorganic selenium and incorporated it into its own protein. This fixed selenium then became available to the sheep.

Meanwhile at Lethbridge, Alberta, Slen, Demiruren, and Smith (69) showed that range sheep supplied with selenium produced more clean fleece with an increased fiber thickness. In addition, the body weights of the selenium-treated animals were 8–11 percent greater than the control animals, indicating a selenium deficiency in southern Alberta soils.

Before selenium could be used for livestock that might reach the commercial market the scientists had to learn if there was any danger of selenium being deposited in the meat or in the milk. K.J. Jenkins and H. Hidiroglou knew that selenium, in more than trace amounts, could be poisonous to livestock and humans. They made hundreds of tests on various animals and found that at dosages sufficient to cure NMD there was no danger of residues reaching either milk or meat. The health departments in both Canada and the United States agreed to its use based upon the findings of Jenkins and Hidiroglou.

Hidiroglou, Jenkins, and Ivan (39) found that calves whose dams had been fed oat silage during the winter months did not suffer from NMD, whereas three of four calves from cows fed on barley silage did die from NMD. Calves from cows injected with selenium and vitamin E, regardless of the type of winter feed, remained healthy.

Injection is a costly, unsatisfactory method of dispensing selenium. In 1974 Jenkins and colleagues from Kapuskasing (51) found that incorporating selenium into salt licks provided sufficient selenium to keep users healthy. However, some animals either ignore salt licks or use them sparingly. Recently (1985) Hidiroglou, Proulx, and Jolette (40) employed a simplified Australian method of delivering selenium to cattle. It consists of administering two 30-g pellets of 10 percent selenium and 90 percent iron in the stomach of each cow. Because of their iron content, the pellets attached themselves to the hardware magnet often placed in the stomach of a cow to prevent nails or other sharp iron objects from puncturing the rumen wall. Stomach bacteria interact with the elemental selenium, making it available to the cow's digestive system. After 2 years of testing at Kapuskasing, no calves born to cows with selenium–iron pellets suffered from NMD.

The problem of nutritional muscular dystrophy is solved for sheep and cattle. Unfortunately, muscular dystrophy in humans is genetic in origin and the treatments suitable for livestock do not apply.

Mycotoxins

Secondary metabolites of filamentous fungi have beneficial and deleterious biological effects in animals and humans. The most widely known secondary metabolite is penicillic acid, a powerful antibiotic obtained from green bread mold. It was initially isolated in 1913. The first recognized mycotoxins, deleterious secondary metabolites harmful to mammals, were alkaloids obtained in 1875 from ergot, a poisonous fungus that grows on the heads of various grains. There are numerous mycotoxins, each with its own designation. Here, only the general term mycotoxin is used.

The first fatal outbreak of mycotoxin poisoning occurred in 1962 in England where thousands of turkeys died. In Canada, the Department of Health and Welfare consistently screens cereal products for minute amounts of mycotoxins to ensure the purity of the food chain. A crisis developed in Eastern Canada in 1980, when Health and Welfare Canada and Agriculture Canada detected quantities of deoxynivalenol, a mycotoxin in wheat infected with *Fusarium graminearum*. The Research Branch started an intensive program at the Harrow, Fredericton, Ottawa, and Charlottetown research stations, the Chemistry and Biology Research Institute (CBRI), the Biosystematics Research Institute (BRI), and the Animal Research Centre (ARC), Ottawa, on the biology of field-produced mycotoxins in grains. The research station at Winnipeg had been working on mycotoxins produced in stored grains for some years.

I.A. de la Roche, director, CBRI, brought together a team of chemists and a fungal physiologist to study the chemistry, biology, and physiological reactions of *Fusarium* mycotoxins. The preliminary need was a reliable method of producing the mycotoxin in a laboratory. By September 1982, Miller, Taylor, and Greenhaugh (58) were successful and quantities were supplied to ARC for research with farm animals and to Health and Welfare Canada for research with laboratory animals.

R.S. Gowe, Director, ARC, assembled another multidisciplinary team in the relevant animal sciences with a mandate to establish safe levels of mycotoxin in feed and to identify residues in animal products such as meat, milk, and eggs. Tremholm and colleagues (75) determined the precise toxicity of mycotoxin in cooperation with the Department of Health and Welfare. As a result, the allowable limit of mycotoxin in uncleaned wheat was raised from 0.3 ppm to 2 ppm.

Research in Canada on *Fusarium* mycotoxins involves scientists at the National Research Council Atlantic Regional Laboratory, the Department of Health and Welfare, the universities of Quebec, McGill, Carleton, Guelph, and Saskatchewan, several private companies, the Food Production and Inspection Branch of Agriculture Canada, as well as the original team from Research Branch. Canada now has a commanding lead on the subject; however, the problem is far from being solved. ARC trials showed that naturally contaminated wheat is usually more toxic than the equivalent amount of pure toxin. This means that the many new secondary toxins produced by *F. graminearum* identified by the CBRI team are found in grains under as yet unknown circumstances. A number of encouraging discoveries, however, have been made by Miller, Young, and Sampson (59). For example, certain cultivars of wheat and possibly corn have enzymes that can degrade the toxin. There is only moderate resistance by wheat and corn to head blight, a symptom of root rot disease caused by *Fusarium* spp. The possibility of capitalizing on their degradative enzyme system affords an opportunity of finding other solutions to the problem for Eastern Canadian farmers. Research is under way on toxigenic *Fusarium* species from Western Canada to understand their potential mycotoxin problems.

SWINE

The Lacombe pig

The first breed of pig developed and registered in Canada was introduced by Fredeen and Stothart (16) of the Experimental Station, Lacombe, in 1957. In 1944 R.M. Hopper, superintendent, Experimental Farm, Brandon, suggested that pig production should follow the example of corn in the use of heterosis. R.D. Sinclair, Dean of Agriculture, University of Alberta, agreed. Lacombe was the location chosen for the development of a white, productive breed, that, when crossed with the Yorkshire, would bear vigorous offspring with good carcass characteristics.

In 1947, H.T. Fredeen and J.G. Stothart selected Berkshire, Danish–Landrace, and Chester White as the foundation breeds from which they hoped to develop a new breed that would cross well with Yorkshires. Berkshire, even though black, was chosen as the female because of its high milking capacity, uniformity of back, and fullness of ham. The Danish–Landrace and Chester White breeds were chosen because of their desirable white color and bacon characteristics. Five of the seven foundation boars were hybrids of the two selected breeds. The remaining two, used in the final foundation matings of 1951, were pure-bred Danish–Landrace boars.

First-generation females of the original cross were mated with their sire breed, not their sires. The progeny of this backcross population then were mated *inter se*. Selection of breeding stock in each generation was based upon a combination of own, sib, and half-sib performance of a variety of economically important traits, including the number of pigs per litter and their weight at weaning and at 6 months; 14 well-developed, uniformly-spaced teats; vigor; physical soundness; strength of feet and legs; freedom from defective conditions; and carcass merit, growth rate, and feed requirements in accordance with Canadian ROP testing policy. As an indication of the care in and pressure of selection, only 6 percent of males and 24 percent of females weaned were used in further breeding. These were descendants of all seven foundation males, six foundation females, and involved 716 litters.

It was a requirement that the new breed have white hair. Black pigmentation was associated with "seedy belly," a condition necessitating the trimming of pigmented mammary tissue prior to curing the belly bacon. Genetically, white pigmentation is dominant to black. Therefore, all progeny from the first cross and the backcross to the Landrace–Chester boars were white in appearance. Some, however, contained black-haired genes that showed for the first time in 1950 when progeny from three of the six backcross boars had black hair. All black-haired pigs were assigned to ROP testing. By 1955 the low incidence of black hair was removed by mating each Lacombe pig with registered Berkshire pigs. If no black-haired pigs resulted, the Lacombe male or female was known to have no recessive black-hair genes and was used for further breed development. Since 1959 the Lacombe breed has produced only white pigs.

To develop a new breed that would cross well with the Yorkshire and result in progeny superior to either parent, 392 litters of Yorkshire pigs were raised. These served as a control herd against which the Lacombe was measured. Between 1954 and 1956, the Lacombe breed was tested against the control herd and two other herds of Yorkshires at the Experimental Station, Scott, and the Experimental Farm, Indian Head, Saskatchewan. In all instances the Lacombe was superior to the Yorkshire in litter size, average weight, and age to 90 kg. Lacombes were exceeded by Yorkshires only in carcass scores. When the two breeds were crossed to produce hybrid pigs, they, too, were superior to Yorkshires in each trait.

The new Lacombe breed was registered in 1957 and released to the public in 1958. Registration was based upon performance and restricted to those animals genetically pure for white hair color. The Canadian Lacombe Breeders Association was incorporated in 1959 and by 1968, 12 percent of all pedigree certificates issued for pigs in Canada were for purebred Lacombes. They were exported to 13 countries but because of their similarity to Danish–Landrace and because they have some deficiencies, the Lacombes are now of only minor importance in the production of hybrid pigs.

Hog grading

Canada introduced national grading of live market hogs in 1922. The objectives were to establish uniform quality standards as a basis for payment and to encourage improvement of hogs moving to market (15). In 1937, carcass grading began. Firm, mandatory standards were established by 1944. This system remained unchanged until 1968 when it was replaced by an index system devised by Fredeen to reflect the value of a commercially trimmed carcass. The system gave preference to carcasses between 56 and 81 kg with less than 50 mm thickness of fat on the lighter carcasses and 85 mm thickness of fat on the heavier carcasses. In 1969, 48 percent of the Canadian hogs slaughtered met the standard. Because producers were paid on the basis of a meaningful grade, within 6 years, the proportion increased to 60 percent. Hog producers recognized the advantages of using genetically superior breeding stock, managing their herd to reduce the amount of fat, and marketing at appropriate live weights. Fredeen (15) estimated that the change increased the amount of commercially trimmed retail pork by 0.47 kg per carcass, totaling 3.7 million kilograms in 1974. A. Fortin, Animal Research Centre, Ottawa, working with D.W. Sim, Food Production and Inspection Branch, Ottawa, and A.H. Martin, H.T. Fredeen, and G.M. Weiss, Research Station, Lacombe, continued studies on grading and found (14) that rather than two, only one measurement of fat taken at the loin was needed to accurately judge its amount on a carcass. The type of carcass delivered to Canadian abattoirs by hog producers improved and therefore in 1981 Fortin, Hoskins, and Sim, (13) recalculated the relationship between fat measurement and grade to bring the system in line with what was actually happening in the industry. For developing the Lacombe breed, introducing the improved grading system, and being a leader in livestock breeding, Fredeen received a Public Service of Canada Merit Award in 1969.

SHEEP

R. Robertson, Superintendent, Experimental Farm, Nappan, Nova Scotia, was the first to introduce sheep to experimental farms. In the spring of 1898 he bought 24 sheep and grazed them on 10 acres (4 ha) of inferior-quality land to improve it without adding more manure or fertilizer. Robertson reported that although the soil was enriched somewhat, the sheep did not do well. The following year he seeded 2 acres to rape for feeding later in the season and sold spring lambs to the value of feed consumed the previous winter. Wool was exchanged for additional summer feed. Robertson's final results were not reported. He left in 1914 before the experiment was concluded. Later reports indicate no appreciable improvement in the pasture was evident.

The first sheep upon the Central Experimental Farm, introduced by J.H. Grisdale in April 1899, were four ewes, some lambs, and one ram of each of purebred Leicester and Shropshire breeds. He bought 18 grade sheep from a city butcher as well. Shropshire and Leicester represented short- and long-wooled breeds. Grisdale hoped to show farmers the advantages of using purebred sires to improve their flocks, to gain data on the cost of raising lambs for market, and to determine the value of sheep as enrichers of soil and destroyers of weeds.

By 1915, sheep were kept at Charlottetown, Prince Edward Island; Cap Rouge and Lennoxville, Quebec; Brandon, Manitoba; Indian Head and Scott, Saskatchewan; Lethbridge and Lacombe, Alberta; and Agassiz, British Columbia, as well as at Nappan and Ottawa. In each case, as with other large animals, one objective was to provide vigorous breeding stock to livestock producers within the geographic region served by each experimental farm or station. In addition, some conducted simple feeding experiments, often aimed at determining the cost of placing a lamb on the market or maintaining a ewe throughout the year.

Breeding

S.J. Chagnon of the Animal Husbandry Division, Ottawa, reported in 1925 that the practice of crossbreeding Shropshire with Leicester and vice versa resulted in early maturity and improved weight in comparison with purebred lambs. Under western Canadian conditions 6-year-old ewes could not withstand the severe range conditions, whereas under more favorable eastern Canadian conditions, these ewes could produce two or three more lamb crops. In 1936, sheep research in the Animal Husbandry Division became more sophisticated. S.B. Williams and P.E. Sylvestre compared western and eastern Canadian straight-bred and crossbred 4- and 5-year-old ewes, concluding that older western range ewes when mated with Shropshire rams could compete successfully with eastern domestic-type ewes in the production of lambs. Williams and Sylvestre tested many other crosses, management, and feeding systems with sheep.

Because of the poor conformation of the Rambouillet breed of sheep in Alberta and Saskatchewan, L.B. Thomson, superintendent, Experimental Range Station, Manyberries, Alberta, started a breeding program to develop a hardy, well-conformed range ewe. In 1935 he bought from the United States 15 purebred Romney–Marsh rams, which have a desirable conformation and, from the best range flocks in southeastern Alberta, 520 Rambouillet ewes. He crossed these two breeds, then closed the flock, so that all further matings were between unrelated hybrids. After 12 years of consistent and rigid selection for wool and mutton, the new Canadian breed called Romnelet was released. It could withstand the rigors of the western Canadian climate, herded well, had long stapled wool, and had a deep body with thick conformation to produce top grade lamb carcasses. Tests of the new breed under range conditions gave highly satisfactory results. The breed became established and was used by a number of ranchers.

At Lethbridge, Rasmussen (64) was the first Canadian to study the heritability of fleece weight. Thomson at Manyberries culled on this basis with some success, which Rasmussen confirmed by showing that the heritability of fleece weight was 56 percent. Rasmussen concluded that distinct improvement in fleece weight could be attained through selection. The only wool laboratory in the Department of Agriculture was built at the Lethbridge Experimental Station in 1946 by Rasmussen to better study the genetics, nutrition, and management of sheep. That same year Rasmussen arranged for the appointment of S.B. Slen, a wool technologist, to operate the laboratory. Slen, with Vesely and Peters (80), studied the relationships between the age and breed of sheep to the production of wool. With Banky (68), Slen included the type of rearing (single, twin, or triplet) and the effect of year (climate) to the study. Age of dam and type of rearing had little effect on the production of wool. Contrarily, breed and seasonal weather from year to year were the greatest source of variation. This meant breeders of sheep could confidently select for wool production provided they did so within, and not between, years.

193

Despite this research input, in the mid-1940s, the Canadian sheep population commenced a steady decline (see Table 16.2) except for a slight upsurge between 1942 and 1946. This decline was primarily a result of the difficulty of attracting herders to live under range conditions. Other factors included losses from predators, lack of high performance breeding stock, low levels of management, inadequate nutritional programs, suboptimal treatment of diseases, little product development, and a low retailer and consumer awareness of lamb as an alternate meat source. Some of these problems result from the seasonal nature of the industry, the birth of lambs being only in the spring. Slen, Whiting, and Rasmussen (70) of Lethbridge prepared an informative textbook-like publication for the use of sheep ranchers, but even this did not prevent the continued reduction in sheep populations.

Table 16.2 Canadian mid-year sheep populations[1]

Year	Sheep (thousands)	Year	Sheep (thousands)
1871	3156	1973	779
1881	3049	1974	730
1891	2564	1975	650
1901	2510	1976	577
1911	2174	1977	559
1921	3204	1978	587
1931	3627	1979	648
1941	2840	1980	734
1951	1968	1981	830
1961	1773	1982	822
1971	851	1983	809

[1]From Statistics Canada.

Interactions occur when two or more kinds of animals or plants react differently under varying situations. Frequently, interactions are noted when some breeds of livestock or varieties of plants perform better than others under various climatic conditions. G.M. Carman at the Animal Husbandry Division, Ottawa, and R.C. Carter of the Virginia Polytechnic Institute, Blacksburg, Virginia, tested the interaction of various breeds of sheep at these two locations, nearly 9 degrees of latitude (about 870 km) apart. The experiment, using five lamb crops between 1961 and 1965, compared hybrid ewes and lambs from the cross of North Country Cheviot by Leicester ewes from the Lac-Saint-Jean area of Quebec with Hampshire by Hampshire–Rambouillet backcross ewes from the Virginia Agricultural Experiment Station. Carter, Carman, McClaugherty, and Haydon (5) found that when half of each flock was raised at Ottawa and half at Blacksburg, the most important factors of adaptation concerned the breeding performance of the ewes. The Hampshire-based hybrids were 18 days earlier in lambing than the Cheviot-based hybrids in Blacksburg. At Ottawa there was a difference of only one day. At Blacksburg, 91 percent of the Hampshire-based ewes lambed and only 80 percent of the Cheviot-based ewes lambed. At Ottawa there was no variation between breeds, both lambing at 93 percent. Lambs from the two types of sheep graded best when raised in the area of selection— Hampshire-based in Virginia and Cheviot-based in Ontario. From these experiments Carter and Carman concluded that local adaptation of ewe breeds and breed crosses was economically important to total production.

Production systems

Under natural conditions sheep are seasonal breeders, with the traditional management system based upon a single lamb crop each year. Multiple births were discouraged because orphan lambs resulting from such births were difficult to rear. The production system was well below its potential. Even so, in 1977 Canadian consumers enjoyed imported frozen lamb valued at $37.6 million, and other sheep products worth $111 million. Returns to Canadian sheep producers during 1977 were only $21.4 million. This indicated there was considerable potential for the development and expansion of Canada's sheep and lamb industry.

In 1967, in an attempt to extend the breeding season of sheep, J.J. Dufour of Lennoxville, Quebec, crossed Australian Dorsets with Canadian Leicester and Suffolk sheep. For the next 4 years Dorset–Leicester rams were mated with Dorset–Suffolk ewes (as well as the reverse cross); then the Dorset–Leicester–Suffolk (DLS) sheep were intermated. By 1974 Dufour (10) reported that the DLS sheep had longer breeding seasons than any of the three parent breeds. By 1983 Fahmy (11) mated the DLS with Finnish–Landrace rams to increase litter size and added 55 lambs weaned to each 100 litters. In addition, he improved the weaned litter weight by 3 kg (11 percent) over the DLS. Although the sheep-breeding program at Lennoxville is in its infancy, it is already contributing to the productivity of the Canadian sheep industry.

H.F. Peters at the Animal Research Institute, Ottawa, knew from working with J.A. Vesely and G. Kozub (79) at Lethbridge and Manyberries, Alberta, that single crosses surpassed purebreds at market weight by 6 percent, and three-breed-crosses by 11 percent. He also knew from the work of others that the sheep's reproductive system responded to day length. For instance, Vesely and Bowden (78) at Lethbridge had shown Rambouillet and Suffolk ewes, exposed to 16 hours of light per day for 106 days, then 8 hours of light per day for 106 days, produced, respectively, 24 and 39 percent more lambs than the controls, which were exposed to normal day lengths. Peters also knew some breeds tended toward multiple births, others to high growth rates and lean meat yield. Such results with Finnish–Landrace sheep were reported by Vesely (77).

The program started by Peters at the Animal Research Institute in 1968 and advanced by many others including D.P. Heaney, L. Ainsworth, A. Fortin, A.J. Hackett, K.E. Hartin, G.A. Langford, J.N.B. Shrestha, and P.S. Fiser, has as its objective, the development of an intensive, totally confined sheep production system similar to the advanced methods used to rear poultry. Among the

195

methods developed are year-round mating, decreased lambing intervals, early mating of ewe lambs, increased litter size, artificial rearing, and increased rate and efficiency of growth. To achieve these goals required the use of advanced breeding and management techniques. How it was done is outlined by Heaney and coauthors (31). An economic analysis of the results based on the first 12 years was prepared by Smith, Howell, Lee, and Shrestha (71), the first three of whom are from the University of Saskatchewan.

Research on this new system of rearing sheep continues, but results to date are encouraging. University of Saskatchewan consultants believe commercial lamb producers can capitalize on the reproductive potential of the ewe if they apply an intensive total confinement system. They could rear 300 lambs to market each year for every 100 ewes mated. This is possible because each ewe produces two lamb crops in 3 years, with each crop being 280 percent (2.8 lambs per ewe lambing, or approximately 40 percent with twins and 25 to 30 percent with three or more lambs). The improved productivity is the result of crossbreeding, selection, and advanced management systems.

The intensive system, using year-round housing and dividing the flock into two alternate mating groups, was noted to have several advantages: losses from predators and parasites were eliminated; fences, except for crop production, were unnecessary; the use of labor and buildings was optimized because they were used year round; and the size of buildings was minimized on a per ewe basis because buildings for both lambing and rearing were in continuous rather than seasonal use, as is the case with a single flock. From the marketing and consumer viewpoints, fresh Canadian lamb now is available on a 12-month basis.

Commercial sheep ranchers are starting to use some of the technology developed in this program. The Canadian sheep population has increased over 40 percent (see Table 16.2) since its low in 1977 because of the commercial use of Finnish–Landrace to improve litter size and Dorset Shorthorn to lengthen the breeding season. Reliable estimates place the number of ewes on controlled reproduction at 10 000 per year to supply out-of-season markets. The use of artificial rearing is increasing, although still with only a small proportion of the lamb crop. No commercial sheep enterprises are yet using total confinement.

Feeds

Intensive production of sheep requires skillful feeding of ewes and lambs. As with cattle, also ruminants, sheep require roughage such as grass, hay, and silage in their diets. By the 1950s, dairymen had established that silage made from grasses and legumes was suitable feed for cattle. Between 1955 and 1957 G.M. Carman and colleagues of the Animal and Poultry Science Division, Ottawa, examined the diet of sheep between 1955 and 1957. Carman knew from research in Kentucky that alfalfa and grass silage were often of less value than alfalfa and grass hay when fed to sheep. Results varied. So Carman, Pigden, Haskell, and Winter (4), in a 3-year study with ewes, found that legume–grass silage fed at various proportions with hay had no effect upon birth weights,

196

weaning weights, or ewe fleece weights. However, the quality of lambs from ewes fed hay was superior to that of lambs from ewes fed silage. Furthermore, only 70 percent of the lambs from silage-fed ewes survived, whereas 88 percent from the hay-fed ewes survived. The reason for the difference was that ewes on silage produced insufficient milk to support their lambs for a period of 2 weeks after lambing. These results proved that pregnant ewes required at least 66 percent of their roughage as hay.

POULTRY

Egg production

Within 2 years of establishing Experimental Farms A.G. Gilbert, a successful commercial poultryman, was chosen as Chief of the Poultry Husbandry Division. Part I refers to his search for high egg producing strains of several breeds of chickens and to egg-laying contests started in 1919 by F.C. Elford, the new Dominion Poultryman, using lines from breeders' flocks. Elford and his successor, G. Robertson (1937–1946), were both practical poultrymen but without the advantage of any formal education in science. Robertson appointed three assistants, A.G. Taylor (waterfowl), H.S. Gutteridge (nutrition), and S.S. Munroe (genetics). In 1946 Gutteridge became Dominion Poultryman.

Initially, commercial poultry production was directed principally to improving egg production; meat production was a by-product of the dual-type birds. The barred Plymouth Rock breed was most popular. The notion of crossing breeds was first tried at the Central Experimental Farm when A.G. Gilbert crossed Brahma cockerels with Plymouth Rocks in 1888. He hoped to obtain improved meat qualities. He regarded the experiment as a success because the resulting birds gained 1.5 lb (0.68 kg) per month for a total gain of 6 lb 2 oz (2.8 kg) in 18 weeks. This cross was made repeatedly and proved better than several others.

In 1927 the Third World Poultry Congress was held in Ottawa, Ontario, under the patronage of His Excellency the Governor General, Viscount Willingdon, with F.C. Elford, Dominion Poultryman, the General Director. It was a large congress even by today's standards, for there were 1932 delegates from 28 countries. This congress gave impetus to commercial poultry production and poultry research in Canada.

In 1948 some poultrymen moved their hens from floor pens containing several hundred birds to individual cages, each containing one to five birds. Cages could be stacked as batteries and the number of birds per square unit of floor area increased severalfold. The industry was based upon pure breeds until 1945, when the production advantage of crossbreeding and the use of hybrid birds started to be realized by some commercial poultrymen.

As early as 1941, Gutteridge and O'Neil (28) at the Poultry Division, Ottawa, in cooperation with the Experimental Stations at Harrow, Ontario, and Charlottetown, Prince Edward Island, found that egg production and days to first

egg were influenced more by environment than by heredity, egg weight equally
by both, and maximum body weight more by heredity than by the environmental differences experienced at the three locations. Genetic differences, however,
were not large or well characterized. Two years later, Gutteridge, Pratt, and
O'Neil (29) working with barred Plymouth Rocks found that this strain of bird
housed in batteries produced 30 eggs more per bird than similar ones housed in
floor pens.

From 1950 to 1973, Gowe et al. (27) conducted an extensive series of
investigations on genotype–environment interactions in poultry. Research
establishments from Agassiz to Charlottetown participated in these early studies
up until 1963. One such study (23) compared the performance of seven strains
of poultry housed in floor pens and in cages at a time when the industry was
starting to house laying birds in cages. It was clearly shown that some strains
performed better in cages than did other strains. By 1985 most laying birds in
Canada were housed in cages. Restricted versus full-feeding effects and farm
(location) effects also were investigated thoroughly.

The long-term, multi-trait selection project (24), utilizing several Leghorn
strains, that was started by Gowe in 1950 and continues today was the first
animal poultry study to use random breed controls. It developed a means by
which the genetic trend could be accurately separated from any environment
trend. Theoretical work on the design of control populations (25, 26) as well as
the actual demonstration of their value were both important parts of these
selection studies. Egg production was increased by over 70 eggs in a 51-week
laying period at a rate of 2.4 eggs per bird per year over the 30-year study. This
improvement in production was achieved while simultaneously increasing egg
size, eggshell strength, and albumen quality and decreasing mortality, body size,
age to sexual maturity, and the average incidence of blood spots. These highly
improved strains have been made available to Canadian poultry breeders for
further development and incorporation into their improvement programs; the
unique control strains are important gene pools.

The large-scale diallel cross of the six selected Leghorn strains reported by
Fairfull et al. (12) in 1983 is the definitive study on general levels of heterosis,
reciprocal effects, and general and specific combining ability in egg strains.
Heterotic and reciprocal effects were substantial for all the key traits such as egg
production, viability, and size. For his outstanding research in support of the
Canadian poultry industry, R.S. Gowe, Director, Animal Research Centre, was
honored with a Public Service Merit Award in 1980.

Breeding for disease resistance

Two virus diseases of poultry have been the subject of a successful, mutually
beneficial cooperative program between the Animal Research Centre and the
Animal Disease Research Institute. The latter, although part of Agriculture
Canada, is in the Food Production and Inspection Branch. From 1937 to 1952
the institute and its several laboratories were part of Science Service.

Marek's disease of chickens was first identified in 1907 by Professor Marek in Hungary. It is caused by a virus that produces tumors on nerves, muscles, and viscera, resulting in debilitation and eventual death. The virus attaches to feather and skin scurf of infected birds and is transmitted on these particles through the air. The interior of hatching eggs is not infected; thus newly born chicks are clean provided they are kept free from contaminated dust particles. In Canada during the 1960s Marek's disease resulted in losses of about $20 million annually. The only means of control was to destroy infected flocks, thoroughly clean buildings and equipment, and start again with noninfected day-old chicks.

The major break came in 1969 when, in the United Kingdom, three scientists, Churchill, Payne, and Chubb learned how to immunize against Marek's disease by injecting day-old chicks with a live attenuated virus. The vaccine dramatically reduced losses from Marek's disease but did not eliminate them. One experiment in Ottawa that exposed vaccinated chicks to the disease resulted in 11 percent mortality with a strain of egg-laying White Leghorns and 27 percent mortality with one strain of meat-type breeders. Gavora, Gowe, and McAllister (18) in 1977 found differences in mortality from Marek's disease among strains of both selected and control strains of White Leghorns, even though the birds were vaccinated against the disease. This led Gavora and colleagues (19) to study two strains of White Leghorns selected for high egg production and other related economically important characters for their heritability of resistance to the disease. They were delighted to find resistance to have 61 percent heritability. In addition, birds of low body weight, early sexual maturity, and high rate of egg production were' those most resistant to the disease. From these observations Gavora and Spencer (21), in a cooperative program with the Animal Disease Research Institute, were able to design and execute breeding programs for the production of poultry strains resistant to Marek's disease.

The second viral disease of poultry studied by Gavora and Spencer was lymphoid leukosis. Unlike Marek's disease, lymphoid leukosis affects the reproductive tract and is transmitted through eggs as well as by direct contact. In a study with Gowe and Harris (20) in 1980 they found that strains of chickens selected for high egg production had a lower incidence (3.9 percent) of lymphoid leukosis than unselected strains (18.5 percent) had. Mortality from infection is generally low. For this reason, Gavora and Spencer wondered if the difference between the two strains of chickens could be attributed to disease resistance rather than to greater propensity of increased egg production. Unlike Marek's disease where the solution was breeding for resistance as well as inoculation, Gavora and Spencer (22) concluded that eradication of the lymphoid leukosis pathogen from breeding flocks is the best permanent prevention.

This research is an example of a cooperative effort in which neither partner could otherwise realize success alone. It is highly regarded worldwide. For their research achievements J.S. Gavora and J.L. Spencer received a Public Service Merit Award in 1982.

Feeds

Management and breeding systems for poultry have improved over the years. The point now has been reached where poultry are more efficient in converting plant food into meat than any other animal. A conversion rate of 2 kg of food to produce 1 kg of weight gain is not uncommon in commercial broiler plants. Good swine enterprises require 3.5 kg and beef feedlots use 9 kg of feed to produce 1 kg of gain.

Feeding trials with poultry started in 1890 when A.G. Gilbert recommended that chicks should be pushed from the start by providing milk and bread every 2 hours, then gradually adding cracked grain. By 1896 he recommended well-balanced rations, including freshly ground green bones.

In 1914 A.G. Gilbert and V. Fortier of the Poultry Division forced a pen of 14 White Leghorn hens to prematurely molt by restricting their feed for 4 weeks in July, then providing full-feed again. In so doing, they increased the number of eggs laid from November through December in comparison to similar birds full-fed. The starved birds laid 47 percent more eggs than birds on full-feed. In the 1920s F.C. Elford, concerned over the high cost of commercial poultry feeds, compared four commercial feeds with two home-mixed feeds during November through April. He found that home-mixed feeds provided a profit of $2.68 per bird, whereas the best of the four commercial feeds gave only $2.02 profit per bird. No allowance was made for the time spent in preparing feeds from homegrown and mixed grains.

Poultry nutrition in the 1930s involved vitamins A and D. Codliver oil was the standard source of vitamin A but pilchard oil was also readily available. Through comparative tests Elford showed that pilchard oil was equal to or slightly better than codliver oil based upon the rate of chicken growth.

MacIntyre and Aitken (55) at Nappan, Nova Scotia, in 1957, subjected both barred Plymouth Rocks and White Leghorns to high-energy rations in comparison with low-energy rations at both normal and high protein levels. At the time there was uncertainty about the claim that high energy reduced feed intake and increased egg production. After 2 years of testing they confirmed that high energy decreased feed consumption but did not increase egg production with White Leghorn birds.

The amount of feed used to produce eggs materially affects profits in egg production. Walter and Aitken (76) at Brandon, Manitoba, in 1961, found that restricting feed during the rearing period caused birds to commence laying later, but when they did start to lay they laid more and larger eggs than those birds on full-feed. Reducing feed by 12 percent during the laying period effected a considerable saving in feed but was offset by a drop in egg production.

The diet fed meat-type birds influences their progeny and their meat production. At Ottawa and at Fredericton, New Brunswick, Aitken, Merritt, and Curtis (1) showed that hens fed a low-density diet laid at a significantly higher rate, gained less weight, and suffered less mortality than hens fed on a high-density diet. However, hens on the high-density diet laid heavier hatching eggs that produced heavier chicks than those on low-density diets. In maritime areas fish meal is often a cheaper form of protein than soybean or other oilseed meal. In 1971 Proudfoot, Lamoreux, and Aitken (63) increased the fish meal component from 4 percent to 10 percent of the diet fed to broiler birds, which resulted in a marked positive growth response. Increasing the fish meal component to 15 percent made no change and reduced bird weights when increased to 20 percent.

In 1975 Sibbald and Price (67) of the Animal Research Institute and the Statistical Research Service, Ottawa, studied the variation of apparent metabolizable energy (AME[1]), using poultry as their test animal, and found a "sawtooth" effect in the data. Observations for a particular bird, on successive days, tended to be alternatively higher and lower than average. When I.R. Sibbald investigated the cause of the variation, he learned that AME values varied with feed intake.

During the next several years Sibbald developed a new bioassay to measure true metabolizable energy (TME[2]). He improved upon the old method by precision feeding and by removing body wastes from the excreta. The first improvement prevented feed selection on the part of test birds, provided each bird with exactly the same amount of food, and avoided the need to recover waste feed. The second improvement was to keep one bird in each test without food for the 48-hour test in order to estimate losses other than undigested food in the excreta. A third improvement was the introduction of a correction for nitrogen retained in the body. The new system required only 2 days of feeding compared with 2 weeks for the conventional AME method and it used only 200 to 300 g of food sample instead of 10 kg. The chemical analyses of food and excreta were the same for each method. Details of the system are given by Sibbald (66) in a bulletin now in its second printing.

TME was designed to determine the true available energy in feedstuffs, but Sibbald has found it to be useful also in assessing the available amino acids, lipids, and minerals in food samples. The method may be useful in monogastric nutrition generally but is inapplicable to cattle and sheep. The system is reported as being used by other scientists and commercial companies in more than 100 laboratories in 51 countries. R.E. Salmon has used it successfully with turkeys at Swift Current, Saskatchewan, in his development of management and nutrition programs for large institute-type turkeys. Feed companies in both Canada and the United States employ the technique in assessing ration formulations in order to supply poultry producers with feeds of known energy content.

201

---

[1]AME is the energy in the food less the energy lost in feces, urine, and combustible gases. In poultry work the combustible gases are ignored. The AME underestimates the energy actually available to the bird because excreta contain materials that are not derived directly from the feed.

[2]TME differs from AME inasmuch as correction is made for excreta energy not derived directly from the food. The assay involves precision feeding of fasted birds to ensure that a known quantity of feed enters the bird at a known time; the excreta voided during the subsequent 48 hours are collected for analyses. Control birds receive no feed and their excreta are used to estimate the excreta energy from sources other than food.

Recently, plant breeders wishing to obtain a nutritive profile of cereal breeding material have used the technique. Under the old system about 10 kg of precious breeders' seed was sacrificed for the test. This might have taken 2 or 3 years to produce. With TME only 200 g (one-fiftieth) is needed, meaning that breeders can now test and discard their breeding lines earlier in their program, saving valuable space in field nurseries, greenhouses, and growth chambers. They need not sacrifice a significant portion of their breeders' seed. This adds yet another tool that enables plant breeders to release new cultivars without delay.

At the Research Station, Kentville, Nova Scotia, F.G. Proudfoot and H.W. Hulan have developed novel low-protein diets for roasting birds that improve feed efficiency and reduce mortality, leg weakness, and days to market. In 1981, they (48) showed that lifetime rations should be divided into three stages for starting, growing, and finishing. When the metabolizable energy of the diet was increased by 7 percent during all three stages, the greatest body weights, best feed conversions, and the highest monetary returns were realized. The Kentville team has identified new sources of high-quality protein, such as squid meal and oat groats, for poultry. Hulan has found that the sudden death syndrome in broilers and roasters is caused by insufficient B vitamin in poultry diets. This syndrome causes up to $25 million loss annually in the Canadian poultry industry. By adding biotin to the diet and by using pelleted soybean meal, which alters its soluble protein fraction, the incidence of the syndrome in commercial flocks is reduced from 4.5 percent to 0.5 percent.

Eggshell strength

The eggshell is one of the miracles of nature, a perfect package that man cannot duplicate. The eggshell contains the embryo and all the nutrients needed for its development from oviposition to hatching. It must be strong enough to resist breaking and yet weak enough to permit the chick to peck its way out. The shell and its inner membranes must allow the exchange of air but insulate the embryo from bacteria. The hen's productivity has been improved through genetic and nutrition research. Concomitantly, production systems have become more mechanized and marketing systems have become more organized; thus the increased volume of eggshell a hen produces must resist greater insults without breaking. Hamilton and coworkers (30) noted that the importance of eggshell strength is indicated by losses up to the farm gate alone of $66 million annually in North America.

In 1960, J.R. Hunt, a poultry nutritionist at the Animal Research Institute, needed a means of measuring eggshell strength while studying how diet affected it. Since the late 1880s, direct methods of measuring had included crushing or dropping objects on the shell. Indirect measurements of strength such as shell thickness and specific gravity of the whole egg were determined by floating it in brine solutions. Hunt and later R.M.G. Hamilton collaborated with P.W. Voisey of the Engineering Research Service on the problem.

Voisey and Hunt (81) used high-speed photography for the first proof that a crack in an eggshell starts at the point where an insult is applied and propagates outward. They then introduced materials testing machines (49) that precisely compared direct and indirect methods of measuring shell strength without breaking the shell. At the same time, Voisey and Hunt made some basic stress analyses on the complex loading because all insults are in the form of a concentrated force due to the shell curvature. It was found that the fracture mechanism depended on the tensile strength of the shell at its inner surface.

Voisey hypothesized that because eggshells are brittle, like cast iron, they should exhibit a property called strain rate sensitivity. Experiments proved (82) that the faster an insulting force was applied, the greater the force required to fracture the shell. This finding was important because measuring devices previously used did not control the rate of application of force. Other implications involved the design of egg-handling equipment.

The research on measuring the complex interrelationships between an egg's physical characteristics and its shell strength continued. Indirect measures of shell strength have proved impossible to achieve. They account for up to only 40 percent of the variations found in shell strength. The instrument developed by Voisey and MacDonald (83) measures shell strength precisely but destructively, and provides a standard for comparison. Because the range of shell characteristics is extremely small and the variations of shell strength are numerous, statistical analyses of the data by B.K. Thompson (74) have been crucial.

Emanating from this research is the clear understanding that the mechanisms governing shell strength rest in the shell's microstructure, subsequently investigated by Stevenson, Voisey, and Hamilton (72). If means can be found to study the effects of breeding and nutrition on the microstructure of the shell material, then it will be possible to control eggshell strength.

The work of the Ottawa scientists has made the branch an international leader in eggshell strength studies. The techniques that they developed are applied worldwide.

SILVER FOXES

Foxes were one of the first wild animals in Canada to be raised in captivity. Jones (52) describes sporadic attempts to raise foxes in Ontario and Quebec between 1898 and 1905, but the real beginning of commercial fox ranching was in Prince Edward Island, in 1887, by Charles Dalton of Tignish. He started with red foxes but soon bought a pair of the more highly valued silver foxes. In the wild, the silver fox is a rare mutation of the red one.

By 1891 others noted Dalton's success and followed his lead. They found that the major problem was a lack of good fencing materials such as woven wire. Fox ranchers had not yet recognized the monogamous nature of foxes and kept several pairs in one pen. This interference with the foxes' innate behavior caused young to be killed by their parents. The price of fur did not provide sufficient funds to permit farmers to experiment. Nonetheless, in Prince Edward Island the industry thrived. In 1913 there were just over 3000 foxes on 277 farms. Ten years later the number of farms with foxes rose to 450 and the number of foxes rose to more than 13 000.

The health of foxes on Prince Edward Island concerned ranchers as well as the Canada Department of Agriculture. In 1919 the Health of Animals Branch established a small Fox Research Station at Charlottetown. Their resident animal pathologist inspected foxes being imported onto the island and advised fox ranchers on disease problems, particularly distemper and parasites. By 1924, hookworms were controlled with carbon tetrachloride but no control for lungworms had been found. At the Animal Disease Research Station in Hull, Quebec, the Health of Animals Branch also commenced nutrition experiments with foxes in collaboration with their animal pathologists, the National Research Council of Canada, and Prince Edward Island fox breeders.

Within the first few years it became obvious that there were many more problems to solve than had been contemplated. Therefore the superintendent, G. Ennis Smith, recommended the ranch be moved from Hull to Prince Edward Island. In 1925 an experimental fox ranch was established at Summerside as a cooperative venture among provincial business people, the Canadian National Silver Fox Breeders' Association, the Silver Fox Breeders' Association of Prince Edward Island, and the Dominion Experimental Farms. Smith moved from Hull and became superintendent of the ranch. The Fox Research Station under the Health of Animals Branch at Charlottetown closed when Summerside became operative.

Fox ranchers were acutely aware of their lack of knowledge of the nutritional requirements for foxes, of methods to control both external and internal parasites, and of an understanding of the way in which various important fur characteristics were inherited. Smith's task was to develop sound information on these problems for the benefit of those who had provided funds to build and stock the experimental ranch.

Within 10 years Smith had many of the most important answers. He had determined the correct amount of feed for each season, he knew the protein requirements to produce the best pelt, he had learned the amount of feed needed by pups at each stage of growth, and he had discovered the importance of feeding vitamin C to prevent a spontaneous fracture of the tail with the resultant loss of its desirable white tip. The average litter size on commercial ranches increased from less than one pup per pair per year to slightly more than three pups per pair per year.

Both external and internal parasites were a serious menace in the early days of fox ranching. Smith learned that applying flea powders to foxes provided only temporary control. His unfailing solution was to control external parasites in the fox nests, kennels, and woodwork of the pens by spraying with ordinary fuel oil. Internal parasites such as lungworm and bladderworm were controlled by keeping the foxes on boards during summer months to prevent infestation from the soil. Board floors drained poorly, however, and pelts became stained with urine. Shortly after 1935, therefore, boards were replaced with wire mesh floors covered with bedding straw, eliminating both problems.

The successes at the Experimental Fox Ranch led to the establishment of Fox Illustration Stations in 1938. These were privately owned fox ranches with which Summerside had agreements to test, under practical commercial conditions, new management methods devised through research. Three stations were opened in each of New Brunswick and Nova Scotia and two on Prince Edward Island. There was a marked improvement in the quality of fur shipped by each station resulting from the use of wire floors, improved feeding and management practices, and the introduction of outstanding sires combined with better selection and breeding methods. Each year illustration stations held field days for all fox ranchers, providing an opportunity for them to learn of recent improvements.

In the late 1930s C.K. Gunn, by now the superintendent, and Alan Deakin of the Animal Husbandry Division, enquired into the possibility of using artificial insemination with foxes. They found, however, that fox semen had a relatively low sperm count, was not readily collected, and did not store easily. They also discovered a few fox males to be polygamous in their mating habits with up to 70 foxes being sired by one male in a season. For these reasons further work on artificial insemination was abandoned. Gunn found that smear tests of vaginas following each mating to determine the fertility of males was an essential part of ranch management. The practice became common among fox ranchers.

Between 1939 and 1941 the fox pelt market collapsed. Wartime economics and transportation problems, the vagaries of fashion, new chemical dyeing techniques, overproduction, and growing humanitarianism all coalesced to defeat the fox industry. Furthermore, short-haired furs such as mink and muskrat grew in popularity. Although fox ranchers on Prince Edward Island did not make the shift, the Experimental Fox Ranch added mink, changing its name to Experimental Fur Farm; this happened sometime before 1952.

Research work with mink included a search for substitutes for meat proteins. Gunn found that he could substitute up to 50 percent of meat protein with soybean meal in both fox and mink diets without any discernible difference in their rate of growth or quality of pelt. He demonstrated that rations low in vitamin B caused retardation of growth, a reduced fur density, and a loss of pigment in mink.

Gunn retired in 1968 and the Experimental Fur Farm came under the jurisdiction of Research Station, Charlottetown. The farm was closed in 1969.

References

1. Aitken, J.R.; Merritt, E.S.; Curtis, R.J. 1969. The influence of maternal diet on egg size and progeny performance in meat-type hens. Poult. Sci. 48:596–601.
2. Beacom, S.E. 1970. Symposium on pasture methods for maximum production in beef cattle: Finishing steers on pasture in northeastern Saskatchewan. J. Anim. Sci. 30:148–152.
3. Buckley, W.T.; Huckin, S.N.; Budac, J.J.; Eigendorf, G.K. 1982. Mass spectrometric determination of a stable isotope tracer for copper in biological materials. Anal. Chem. 54:504–510.
4. Carman, G.M.; Pigden, W.J.; Haskell, S.R.; Winter, K.A. 1958. Legume–grass silage as a roughage for the pregnant ewe. Can. J. Anim. Sci. 38:200–207.
5. Carter, R.C.; Carman, G.M.; McClaugherty, F.S.; Haydon, P.S. 1971. Genotype–environment interaction in sheep. I. Ewe productivity. II. Lamb performance traits. J. Anim. Sci. 33:556–562; 732–735.
6. Cooke, D.A.; Beacom, S.E.; Dawley, W.K. 1968. Response of six-year-old grass–alfalfa pastures to nitrogen fertilizer in northeastern Saskatchewan. Can. J. Plant Sci. 48:167–173.
7. Cooke, D.A.; Beacom, S.E.; Robertson, J.A. 1973. Comparison of continuously grazed bromegrass–alfalfa with rotationally grazed crested wheatgrass–alfalfa, bromegrass–alfalfa, and Russian wild ryegrass. Can. J. Anim. Sci. 53:423–429.
8. Deakin, Alan; Muir, G.W.; Davies, W.D. 1942. Artificial insemination of dairy cattle. Dom. Can. Dep. Agric. Exp. Farms Service. Mimeo. Publ. 15 p.
9. Deakin, Alan; Muir, G.W.; Smith, A.G. 1935. Hybridization of domestic cattle, bison and yak. Dom. Can. Dep. Agric. Publ. 479. 30 p.
10. Dufour, J.J. 1974. The duration of the breeding season of four breeds of sheep. Can. J. Anim. Sci. 54:389–392.
11. Fahmy, M.H. 1983. Maternal performance of DLS sheep and its crosses with the Finn Landrace breed. Proc. Fifth World Conf. Anim. Prod. 2:55–56.
12. Fairfull, R.W.; Gowe, R.S.; Emsley, J.A.B. 1983. Diallel cross of six long-term selected Leghorn strains with emphasis on heterosis and reciprocal effects. Br. Poult. Sci. 24:133–158.
13. Fortin, A.; Hoskins, V.E.; Sim, D.W. 1983. Regional characteristics of hog carcasses in Canada: A survey conducted in the spring of 1981. Can. J. Anim. Sci. 63:551–561.
14. Fortin, A.; Martin, A.H.; Sim, D.W.; Fredeen, H.T.; Weiss, G.M. 1981. Evaluation of different ruler and ultrasonic backfat measurements as indices of commercial and lean yield of hog carcasses for commercial grading purposes. Can. J. Anim. Sci. 61:893–905.
15. Fredeen, H.T. 1976. Recent trends in carcass performance of the commercial hog population in Canada. J. Anim. Sci. 42:342–351.
16. Fredeen, H.T.; Stothart, J.G. 1969. Development of a new breed of pigs: The Lacombe. I. Foundation and developmental procedures. II. Evaluation. III. Public distribution and propagation. Can. J. Anim. Sci. 49:237–273.
17. Fredeen, H.T.; Weiss, G.M.; Lawson, J.E.; Newman, J.A.; Rahnefeld, G.W. 1981. Lifetime reproductive efficiency of first-cross beef cows under contrasting environments. Can. J. Anim. Sci. 61:539–554.
18. Gavora, J.S.; Gowe, R.S.; McAllister, A.J. 1977. Vaccination against Marek's disease: Efficacy of cell-associated and lyophilized herpesvirus of turkeys in nine strains of Leghorns. Poult. Sci. 56:846–853.

19. Gavora, J.S.; Grunder, A.A.; Spencer, J.L.; Gowe, R.S.; Robertson, A.; Speckmann, G.W. 1974. An assessment of effects of vaccination on genetic resistance to Marek's disease. Poult. Sci. 53:889-897.

20. Gavora, J.S.; Spencer, J.L.; Gowe, R.S.; Harris, D.L. 1980. Lymphoid leukosis virus infection: Effects on production and mortality and consequences in selection for high egg production. Poult. Sci. 59:2165–2178.

21. Gavora, Jan S.; Spencer, J. Lloyd. 1983. Breeding for immune responsiveness and disease resistance. Anim. Blood Groups Biochem. Genet. 14:159–180.

22. Gavora, Jan S.; Spencer, J. Lloyd. 1985. Effects of lymphoid leukosis virus infection on response to selection. Pages 17–24 in Hill, W.G.; Manson, J.M.; Hewitt, D., editors. Poultry genetics and breeding. British Poultry Science Ltd.

23. Gowe, R.S. 1955. Environment and poultry breeding problems. 2. A comparison of the egg production of 7 s.c. White Leghorn strains housed in laying batteries and floor pens. Poult. Sci. XXXV:430–435.

24. Gowe, R.S.; Fairfull, R.W. 1980. Performance of six long-term multi-trait selected leghorn strains and three control strains, and a strain cross evaluation of the selected strain. Proc. S. Pac. Poult. Sci. Conv., Auckland, New Zealand, pp. 141–162.

25. Gowe, R.S.; Fairfull, R.W. 1985. The direct response to long-term selection for multiple traits in egg stocks and changes in genetic parameters with selection. Pages 125–146 in Hill, W.G.; Manson, J.M.; Hewitt, D., editors. Poultry genetics and breeding. British Poultry Science Ltd.

26. Gowe, R.S.; Lentz, W.E.; Strain, J.H. 1973. Long-term selection for egg production in several strains of White Leghorns: Performance of selected and control strains including genetic parameters of two control strains. 4th. Eur. Poult. Conf., London, England, pp. 225–245.

27. Gowe, R.S.; Robertson, A.; Latter, B.D.H. 1959. Environment and poultry breeding problems. Poult. Sci. 38:462–471.

28. Gutteridge, H.S.; O'Neil, J.B. 1942. The relative effect of environment and heredity upon body measurements and production characteristics in poultry II. Period of egg production. Sci. Agric. 22:482–491.

29. Gutteridge, H.S.; Pratt, Jean M.; O'Neil, J.B. 1944. Level and source of protein in poultry production. II. As related to economical production of eggs. Sci. Agric. 24:240–250.

30. Hamilton, R.M.G.; Hollands, K.G.; Voisey, P.W.; Grunder, A.A. 1979. Relationship between egg shell quality and shell breakage and factors that affect shell breakage in the field—a review. World's Poult. Sci. J. 35:177–190.

31. Heaney, D.P.; Ainsworth, L.; Batra, T.R.; Fiser, P.S.; Langford, G.A.; Lee, A.J.; Hackett, A.J. 1980. Research for an intensive total confinement sheep production system. Agric. Can. Anim. Res. Inst. Tech. Bull. 2. 56 p.

32. Hickman, C.G. 1971. Response to selection for 180-day milk solids. J. Dairy Sci. 54:191–198.

33. Hickman, C.G.; Bowden, D.M. 1971. Correlated genetic responses of feed efficiency, growth, and body size in cattle selected for milk solids yield. J. Dairy Sci. 54:1848–1855.

34. Hickman, C.G.; Freeman, A.E. 1969. New approach to experimental designs for selection studies in dairy cattle and other species. J. Dairy Sci. 52:1044–1054.

35. Hidiroglou, M. 1979. Manganese in ruminant nutrition. Can. J. Anim. Sci. 59:217–236.

36. Hidiroglou, M.; Heaney, D.P.; Hartin, K.E. 1984. Copper poisoning in a flock of sheep. Copper excretion patterns after treatment with molybdenum and sulfur or penicillamine. Can. Vet. J. 25:377–382.
37. Hidiroglou, M.; Heaney, D.P.; Jenkins, K.J. 1967. Metabolism of inorganic selenium in rumen bacteria. Can. J. Physiol. Pharmacol. 46:229–232.
38. Hidiroglou, M.; Ho, S.K.; Standish, J.F. 1978. Effects of dietary manganese levels on reproductive performance of ewes and on tissue mineral composition of ewes and day-old lambs. Can. J. Anim. Sci. 58:35–41.
39. Hidiroglou, M.; Jenkins, K.J.; Ivan, M. 1977. Influences of barley and oat silages for beef cows on occurrence of myopathy in their calves. J. Dairy Sci. 60:1905–1909.
40. Hidiroglou, H.; Proulx, J.; Jolette, J. 1985. Intruminal selenium pellet for control of nutritional muscular dystrophy in cattle. J. Dairy Sci. 68:57–66.
41. Hidiroglou, M.; Shearer, D.A. 1975. Concentration of manganese in the tissues of cycling and anestrous ewes. Can. J. Comp. Med. 40:306–309.
42. Hironaka, R. 1969. Starter rations for beef cattle in feedlots. Can. J. Anim. Sci. 49:181–188.
43. Hironaka, R.; Miltimore, J.E.; McArthur, J.M.; McGregor, D.R.; Smith, E.S. 1973. Influence of particle size of concentrate on rumen conditions associated with feedlot bloat. Can. J. Anim. Sci. 53:75–80.
44. Hoffman, I.; Westerby, R.J.; Hidiroglou, M. 1968. Precise fluorometric microdetermination of selenium in agricultural material. J. Assoc. Off. Agric. Chem. 51:1039–1042.
45. Howarth, R.E.; Cheng, K.-J.; Fay, J.P.; Majak, W.; Lees, G.L.; Goplen, B.P.; Costerton, J.W. 1981. Initial rate of digestion and legume pasture bloat. Proc. XIV Int. Grassl. Congr. pp. 719–722.
46. Howarth, R.E.; Cheng, K.-J.; Majak, W.; Costerton, J.W. 1985. Ruminant bloat. Proc. VI Int. Symp. Ruminant Physiol. Reston Publishing Co. Chapter 27.
47. Howarth, R.E.; Goplen, B.P.; Fesser, A.C.; Brandt, S.A. 1978. A possible role for leaf cell rupture in legume pasture bloat. Crop Sci. 18:129–133.
48. Hulan, W.H.; Proudfoot, F.G. 1982. The effect of different dietary energy regimes on the performance of two commercial chicken broiler genotypes reared to roaster weight. Poult. Sci. 61:510–515.
49. Hunt, J.R.; Voisey, P.W. 1966. Physical properties of egg shells. 1. Relationship of resistance to compression and force at failure of egg shells. Poult. Sci. XLV:1398–1404.
50. Ivan, M. 1981. Distribution of radiomanganese in the rumen of sheep. Can. J. Physiol. Pharmacol. 59:76–83.
51. Jenkins, K.J.; Hidiroglou, M.; Wauthy, J.M.; Proulx, J.E. 1974. Prevention of nutritional muscular dystrophy in calves and lambs by selenium and vitamin E additions to the maternal mineral supplement. Can. J. Anim. Sci. 54:49–60.
52. Jones, J. Walter. 1913. Fur-farming in Canada. Commission of Conservation, Canada. Gazette Printing Co., Montreal. pp. 13–15.
53. Lin, C.Y.; McAllister, A.J.; Batra, T.R.; Lee, A.J.; Roy, G.L.; Vesely, J.A.; Winter, K.A. 1984. Reproductive performance of crossline and pureline dairy heifers. J. Dairy Sci. 67:2420–2428.
54. Logan. V.S. 1955. Grass silage in dairy cattle rations. Can. Dep. Agric. Publ. 929. 11 p.
55. MacIntyre, T.M.; Aitken, J.R. 1957. The effect of high levels of dietary energy and protein on the performance of laying hens. Poult. Sci. 36:1211–1216.

56. McAllister, A.J., et al. 1978. The national cooperative dairy cattle breeding project. Agric. Can. Anim. Res. Inst. Tech. Bull. 1. 46 p.
57. McArthur, J.M.; Miltimore, J.E.; Pratt, M.J. 1964. Bloat investigations. The foam stabilizing protein of alfalfa. Can. J. Anim. Sci. 44:200–206.
58. Miller, J.D.; Taylor, A.; Greenhalgh, R. 1983. Production of deoxynivalenol and related compounds in liquid culture by *Fusarium graminearum*. Can. J. Microbiol. 29:1171–1178.
59. Miller, J.D.; Young, J.C.; Sampson, D.R. 1985. Deoxynivalenol and *Fusarium* head blight resistance in spring cereals. Phytopathol. Z. (In press.)
60. Miltimore, J.E.; Mason, J.L.; McArthur, J.M.; Strachan, C.C.; Clapp, J.B. 1973. Response from copper and selenium with vitamin E injections to cattle pastured on mineral and organic groundwater soils. Can. J. Anim. Sci. 53:237–244.
61. Miltimore, J.E.; Pigden, W.J.; McArthur, J.M.; Anstey, T.H. 1964. Bloat investigations. I Reduction of bloat incidence in dairy cattle with penicillin and antifoaming agents. Can. J. Anim. Sci. 44:96–101.
62. Newman, J.A.; Rahnefeld, G.; Fredeen, H.T. 1973. Selection intensity and response to selection for yearling weight in beef cattle. Can. J. Anim. Sci. 53:1–12.
63. Proudfoot, F.G.; Lamoreux, W.F.; Aitken, J.R. 1971. Performance of commercial broiler genotypes on fish meal diets with a charcoal supplement. Poult. Sci. 50:1124–1130.
64. Rasmussen, K. 1942. The inheritance of fleece weights in range sheep. Sci. Agric. 23:104–116.
65. Robertson, J.A.; Cooke, D.A.; Beacom, S.E. 1979. Comparison of four systems of managing yearling beef steers on rotationally grazed, bromegrass–alfalfa pasture. Can. J. Anim. Sci. 59:519–529.
66. Sibbald, I.R. 1983. The T.M.E. system of feed evaluation. Agric. Can. Res. Branch Contrib. 1983-20E. 89 p.
67. Sibbald, I.R.; Price, K. 1975. Variation in the metabolizable energy values of diets and dietary components fed to adult roosters. Poult. Sci. 54:448–456.
68. Slen, S.B.; Banky, E.C. 1958. The relationship of clean fleece weight to age in three breeds of range sheep. Can. J. Anim. Sci. 38:61–64.
69. Slen, S.B.; Demiruren, A.S.; Smith, A.D. 1961. Note on the effects of selenium on wool growth and body gains in sheep. Can. J. Anim. Sci. 41:263–265.
70. Slen, S.B.; Whiting, F.; Rasmussen, K. 1953. Range sheep production in western Canada. Can. Dep. Agric. Publ. 886. 47 p.
71. Smith, E.G.; Howell, W.E.; Lee, G.E.; Shrestha, J.N.B. 1982. The economics of intensive sheep production. Agric. Can. Anim. Res. Inst. Tech Bull. 4. 73 p.
72. Stevenson, I.L.; Voisey, Peter W.; Hamilton, R.M.G. 1981. Scanning electron microscopy of fractures in eggshells subjected to the puncture test. Poult. Sci. 60:89–97.
73. Sylvestre, P.E.; Mercier, E. 1960. Grass silage for beef cattle. Can. Dep. Agric. Publ. 955. 13 p.
74. Thompson, B.K.; Hamilton, R.M.G.; Voisey, Peter W. 1981. Relationships among various egg traits relating to shell strength among and within five avian species. Poult. Sci. 60:2388–2394.
75. Trenholm, H.L.; Hamilton, R.M.G.; Frierd, D.W.; Thompson, B.U.; Hartin, K.E. 1984. Feeding trials with vomitoxin (deoxynivalenol)—contaminated wheat. Effects of swine, poultry and dairy cattle. J. Am. Vet. Med. Assoc. 185:527–531.
76. Walter, E.D.; Aitken, J.R. 1961. Performance of laying hens subjected to restricted feeding during rearing and laying periods. Poult. Sci. 40:345–354.

209

77. Vesely, J.A. 1978. Performance of progeny of Finnish Landrace and Dorset Horn rams mated to ewes of various breeds and crosses. Can. J. Anim. Sci. 58:399–408.
78. Vesely, J.A.; Bowden, D.M. 1980. Effect of various light regimes on lamb production by Rambouillet and Suffolk ewes. Anim. Prod. 31:163–169.
79. Vesely, J.A.; Kozub, G.C.; Peters, H.F. 1977. Additive and nonadditive genetic effects on growth traits in matings among Romnelet, Columbia, Suffolk, and North Country Cheviot breeds. Can. J. Anim. Sci. 57:233–238.
80. Vesely, J.A.; Peters, H.F.; Slen, S.B. 1965. The effects of breed and certain environmental factors on wool traits of range sheep. Can. J. Anim. Sci. 45:91–97.
81. Voisey, Peter W.; Hunt, J.R. 1964. A technique for determining approximate fracture propagation rates of egg shells. Can. J. Anim. Sci. 44:347–350.
82. Voisey, P.W.; Hunt, J.R. 1969. Effect of compression speed on the behaviour of eggshells. J. Agric. Eng. Res. 14:40–46.
83. Voisey, Peter W.; MacDonald, D.C. 1978. Laboratory measurements of eggshell strength. 1. An instrument for measuring shell strength by quasi-static compression, puncture, and non-destructive deformation. Poult. Sci. 57:860–869.

210

Chapter 17
Crops Research

Crops research has always been an important function of the Research Branch. Some of the research done during the past century is highlighted here, with emphasis on plant breeding and development.

CEREALS

Cereals are the most universally consumed of the cultivated crops: rice, millet, sorghum, and corn in the tropics and subtropics; wheat, barley, oats, and corn in the temperate zones. Canada is on the northern fringe of the temperate zone in which all these cereals, except rice and millet can be grown. A few sorghum varieties mature in some parts of Canada. Modern corn is a national Canadian crop, but the corn of two decades ago was not. Winter wheat can survive in areas of ample winter snow or moderate winter temperatures. Today's spring wheat matures in most agricultural areas of Canada, but we were not always so fortunate. Spring varieties of barley, oats, and rye are completely adapted to Canadian agricultural climates.

How did these changes occur? The climate has remained constant during the past century; therefore, the plants and their cultures must have been altered. The teams of research scientists responsible for these changes have been led by plant breeders, entomologists, plant pathologists, plant nutritionists, or agronomists. Farmers who take advantage of these improved plant characteristics need to be aware of innovative field practices.

Anderson and Morrison (1), when speaking to the Canadian Centennial Wheat Symposium in 1967, reminded scientists that there are no final solutions to agricultural problems. They emphasized that the resolution of one difficulty often just leads to a puzzle elsewhere.

Wheat

One reason Members of the House of Commons favored the passage of the Experimental Farm Station Act in 1886 was that Canada, particularly Manitoba and the North-West Territories, needed an early maturing hard, red, spring wheat. Wheat was first produced in Canada in 1605 at Port Royal, Nova Scotia, during the French settlement there, and a decade later it was produced at Quebec City. The next 200 years of wheat production in Eastern Canada and, after 1812, in the Red River Valley of Manitoba was based on varieties from western Europe, New England, and New York that were not suited to the cold winters and short growing seasons of Western Canada.

The first significant progress in the development of suitable Canadian varieties occurred in the spring of 1841 or 1842 with the introduction of totally new germ plasm from the steppes of the Ukraine. David Fife, a farmer near Peterborough, Ontario, received some wheat from a friend in Glasgow, Scotland, who had obtained it from a cargo direct from Danzig, now called Gdansk, Poland. Fife seeded the sample that same spring. All but one plant proved to be a winter wheat variety and therefore did not produce flowers or grain. That one plant was different. It was a spring variety and did produce seed, which Fife kept and multiplied. From the Canadian perspective the important thing was that Red Fife, as the variety came to be known, was earlier and yielded more than other spring wheat varieties. It remained free from rust, had harder kernels than most wheats then grown, and produced flour of good quality. Newman (62), Archibald (3), and Buller (8) credit Red Fife with advancing Canadian wheat production far enough to create an export market. By 1886, Red Fife was the dominant variety on the one million hectares of wheat grown in Eastern Canada. In the west, however, Red Fife was too late maturing to assure consistent crops.

212

William Saunders, the first director of the Experimental Farms System, from personal visits to Manitoba and the North-West Territories (Brandon and Indian Head), saw the need for an early maturing spring wheat. With support from western parliamentarians he started the Experimental Farm wheat improvement program in 1888 by crossing early maturing varieties with high-quality varieties. At his side was W.T. Macoun, later to become Dominion Horticulturist, and two of Saunders' five sons, A.P. and C.E. Saunders. They toured the four experimental farms from Nappan, Nova Scotia, to Agassiz, British Columbia, when wheat and other cereals were in flower. They crossed early maturing and high-quality varieties, selecting the best from the progeny grown in succeeding years. One variety, Markham, originated from a cross between Red Fife and Hard Red Calcutta. William Saunders had imported Hard Red Calcutta from India, because it ripened 2–3 weeks earlier than Red Fife.

C.E. Saunders was appointed Experimentalist, Central Experimental Farm in 1903, relieving his father of the arduous details of the cereal breeding program. One of the first things he did was to examine the progeny of earlier crosses. From the progeny of Markham, which was not a true breeding variety, he was able to select Marquis. Morrison (60) describes the details of the way Saunders chewed samples of grain to produce "gum" (gluten) in order to judge the baking quality of its flour, and of how he ground the grain and baked experimental bread to assure that the selections were of high quality. In 1909 the first samples of Marquis were sent to prairie farmers for final test. Marquis proved to be such a superior variety that by 1920, 90 percent of the 6.9 million ha seeded to hard, red, spring wheat on the Canadian prairies was Marquis.

Marquis initially was resistant to wheat stem rust but later succumbed to new races, and in northern regions it frequently was damaged by fall frosts. Saunders continued to hybridize wheat in his search for earlier varieties resistant to stem rust. He introduced Preston, Huron, and Stanley, all excellent varieties except for quality. Later introductions were deficient in other ways. Saunders' task was difficult, complicated by the fact that early maturity, usually associated with low yield and quality, is a genetically complex characteristic, dependent for expression in part upon a variable climate. Although a firm believer in the value of Mendelian Laws of Heredity, some frustration crept into his 1910 report as he wondered whether "... the discovery of Mendelian unit characters is sometimes due to the unhappy combination of a great deal of enthusiasm with very few facts". His search for early maturing wheat was continued at Ottawa by J.G.C. Fraser with the release of Garnet in 1925. Garnet matured 5–7 days earlier than Marquis but was of lower quality. It was intended for the Peace River District of northern Alberta where, to a limited extent, it is still grown. In 1948 Fraser and his colleague F. Gfeller released the variety, Saunders, which had the earliness of Garnet and the quality of Marquis. It was useful in northern areas of the Prairie Provinces during the 1950s. Today, the breeding of such early maturing wheat is centered at the Beaverlodge Research Station, Alberta.

Resistance to rust

Wheat stem rust, caused by the fungus *Puccinia graminis*, was well known to cereal farmers on the Great Plains, but it was not until 1916 that it reduced the yield of spring wheat by over 100 million bushels (2.7 million t)! It was evident to everyone that rust-resistant varieties were required if the western wheat industry was to flourish.

Chapter 5 mentions the discovery by E.C. Stakman at the University of Minnesota that wheat rust included many distinct forms to which different varieties of wheat were susceptible in different ways. This discovery launched the modern era of wheat development for specific Canadian needs. The Rust Research Laboratory, Winnipeg, Manitoba, was established in 1924 with D.L. Bailey and C.H. Goulden as plant pathologist and wheat breeder, respectively. Goulden hired a second plant breeder, K.W. Neatby, in 1926. Their goal was to produce varieties of wheat resistant to stem rust. The first rust-resistant variety grown in Canada was Thatcher bred by H.K. Hayes of the University of Minnesota and released in 1935. It was almost identical to Marquis with the added advantage of resistance to rust. The following year Goulden, Neatby, and Bailey had their first rust-resistant variety, Renown, available from Winnipeg. It was selected from a cross between Reward, developed by the Cereal Division in 1928, and H.44 from South Dakota, a cross between Marquis and a variety of Emmer wheat. R.F. Peterson transferred from the Experimental Farm, Brandon, to replace Neatby, and released Regent in 1939 and Redman in 1946, each with improved disease resistance and agronomic characteristics. New races of rust continued to develop and move northward along the Mississippi Valley and eventually they attacked Canadian wheats. One of the more serious was race 15B. A.B. Campbell was ready with Selkirk in 1953 and even today he continues to be prepared, as new races of rust develop naturally by hybridization. So successful has been the breeding of rust-resistant wheat varieties that there has not been an epidemic of stem rust on wheat in Western Canada since 1954.

Starting in 1969, varieties of wheat from the Winnipeg program have had other desirable characteristics in addition to wheat stem rust resistance combined with outstanding milling and baking quality. That year, Neepawa, which could withstand root rot, loose smut, and bunt infection and still maintain its bright kernel color became available to the seed trade. Other varieties were given awns, which enabled them to form a better swath when harvested, and were designed so that sprouting would not occur in the swath, even during a prolonged, wet harvest. By 1984, varieties developed at Winnipeg were seeded on 87 percent of the Canadian hard, red, spring wheat area.

For their continued outstanding performance, plant pathologists and plant breeders at Winnipeg have received many awards, including the following:

1937—Gold Medal, Professional Institute of the Public Service of Canada (PIPS), the first recipient:
    J.H. Craigie, plant pathologist.
1953—Gold Medal, PIPS:
    C.H. Goulden, plant breeder.
1962—Gold Medal, PIPS:
    T. Johnson, plant pathologist.
1967—Order of Canada:
    J.H. Craigie, plant pathologist.
1971—Order of Canada:
    T. Johnson, plant pathologist.
1976—Public Service of Canada Merit Award:
    A.B. Campbell, plant breeder; and
    G.J. Green, plant pathologist.
1983—Gold Medal, PIPS: Winnipeg Bread Wheat Research Team—
    A.B. Campbell, plant breeder;
    P.C. Dyck, plant geneticist;
    E.R. Kerber, plant cytogeneticist;
    J.J. Neilsen, plant pathologist; and
    D.J. Samborski, plant pathologist.

During this period smaller programs to breed hard, red, spring wheat for specific climatic conditions were under way simultaneously at three other stations. At Scott, A.G. Kusch bred and released the variety Lake in 1954. Lake had the drought resistance needed for central Saskatchewan. However, it was susceptible to race 15B stem rust and therefore was never grown extensively. The high parkland belt in western Alberta centering around Lacombe frequently experiences earlier fall frosts than the true prairies. Recognizing the need for an early maturing bread wheat, A.D. McFadden and M.L. Kaufmann at the Experimental Station, Lacombe, bred and released Park in 1963. Park matures early, yields well, and has satisfactory baking quality. It proved popular in central Alberta and, in 1984, was grown on 3.8 percent of the prairie hectarage usually seeded to hard, red, spring wheat. Returning to the dry area of the prairies, E.A. Hurd at the Research Station, Regina, started to breed a drought-resistant variety, completing the program following his 1970 move to Swift Current. In

1975 he released Sinton, a variety that yields well under low moisture conditions and is also resistant to leaf rust. By 1984 it occupied just over 2 percent of the prairie hard, red, spring wheat seeded area.

Resistance to wheat stem sawfly

The heavy-headed stems of wheat started lodging (falling over) on 10 August 1926, in southern Saskatchewan. The kernels were yet to ripen. Neither wind nor rain had caused the lodging. The culprit was a tiny wasp, commonly called a sawfly. The solution to this serious problem required the development of entirely new varieties of wheat. The story of how this was done will be told in Chapter 18.

Durum wheat

Pasta is made from varieties of durum wheat. The kernels of durum wheat are much harder than those of bread wheat; hence the name of the wheat. Durum kernels are also larger, and the flour, called semolina, which tends to be yellowish, has more protein. The gluten fraction is less extensible than flour from bread wheat. Genetically, durum has two genomes (two times seven pairs of chromosomes), whereas bread wheat has three genomes (21 pairs). They are well adapted to the semiarid regions of the prairies because of superior tolerance to drought. Durum was introduced into Canada about 1918, using two varieties, Golden Ball and Mindum, from the United States. Subsequently a succession of United States varieties was brought to Canada until 1969.

In the mid-1950s durum wheat yields were severely restricted by race 15B stem rust. C.H. Goulden, newly appointed Dominion Cerealist, decided in 1949 that the Rust Laboratory at Winnipeg should initiate a new durum wheat breeding program to incorporate resistance to rust into otherwise acceptable durum varieties. A.B. Masson started the program. When he took charge of the seed increase and distribution program in 1955, E.R. Kerber replaced him. D. Leisle assumed responsibility for durums in 1961 and released Hercules, the first Canadian durum variety, in 1969. Hercules was resistant to stem and leaf rusts and to loose smut. It was also superior in yield, maturity, straw strength, and kernel size to the United States varieties available at that time. Since 1969, Coulter and Medora have both come from Leisle's breeding. They are equal to or better than Hercules, are resistant to bunt, and have excellent pasta qualities. Because both Coulter and Medora lack drought tolerance they are suited only to the eastern prairies.

In 1961 E.A. Hurd at the Research Station, Regina, organized a durum program that had the same objectives plus tolerance to drought. Hurd moved to Swift Current in 1970 and was joined by T.F. Townley-Smith. The varieties Wascana and Wakooma were released in 1971 and 1973 because their yield was higher than the Winnipeg varieties under drought conditions. At that time the Swift Current and Winnipeg programs were integrated, resulting in the introduction of Macoun in 1974 and Kyle in 1984. Ninety percent of the 1.7 million ha of durum wheat grown in Canada in 1984 was seeded to Winnipeg and Swift Current varieties.

Other spring wheat varieties used for pastry and biscuit flours and for feed have been developed by Experimental Farms and Research Branch scientists. In 1947 J.G.C. Fraser and F. Gfeller of the Cereal Division, Ottawa, released Cascade, a semihard, white, spring, feed wheat. Cascade had resistance to stem rust and powdery mildew, with moderate resistance to leaf rust. It was grown in Ontario and the irrigated areas of Alberta. In 1951 Fraser and Gfeller also released the hard, red wheat, Acadia, which was popular in Eastern Canada for some years. H.G. Nass joined the staff of the Charlottetown Research Station in 1971 to breed red, feed wheats. He was quickly successful, using advanced breeding material available from the program started by J.D.E. Sterling several years prior. Dundas, his first variety, was released in 1979. It is early, awned, but susceptible to powdery mildew. Vernon, released in 1981, is awnless with resistance to powdery mildew. Milton, made available in 1982, has a wider adaptation than the first two varieties and seems to be suitable for culture throughout the Atlantic Provinces and parts of Quebec.

Winter wheat

A winter wheat must be seeded in late summer; it then germinates and grows 10–15 cm. During the cool, short days of fall, each plant undergoes a physiological process called vernalization that enables it to flower the following spring. Provided the wheat survives winter's low temperatures, normal spring temperatures bring rapid growth, the development of flowers, and seed that ripens in July or early August. Winter wheats use winter and spring moisture effectively and therefore yield more than spring-seeded wheats.

The first type of wheat sown by the Selkirk settlers in 1812, on land where the city of Winnipeg now stands, was winter wheat. It was not hardy enough to withstand that year's severe winter; consequently, from 1813 on the Selkirk settlers sowed spring wheat. Spring wheat predominated in Acadia and New France under the French regime for the same reason it does today in Atlantic Canada and Quebec. Winter wheat thrives in the milder climate of southwestern Ontario, where it was introduced in the 1780s by United Empire Loyalists from upstate New York. The production of winter wheat in Alberta began in the late 1880s, using Ukrainian varieties obtained from the United States. In 1888 slightly under one million hectares of winter wheat were grown in Ontario. Today only half that amount is cultivated, 250 thousand ha in Ontario and a similar area in Alberta and Saskatchewan. The soft, white winters are grown in Ontario for the pastry and biscuit trade and over 60 percent of the crop is exported; in Alberta and Saskatchewan red winters are grown for the pastry trade and export. Hard, red, winter varieties, usually from Alberta, are used to make crackers. Other winter wheats, including those grown in the Maritime Provinces, Ontario, and British Columbia, are used for feed.

The initial breeding program in winter wheat was a cooperative effort among the Ontario Agricultural College, Guelph, the Experimental Station, Harrow, and the Cereal Division, Central Experimental Farm. It resulted in the release of Rideau in 1941. The Cereal Division continued the program under A.G.O. Whiteside with the objective of improving soft, white, winter wheats for the pastry trade. Richmond was released in 1953 and Talbot in 1962, but the most successful variety was Fredrick, released in 1971 and named after F. Gfeller, a wheat breeder at Ottawa from 1936 until his retirement in 1969. Fredrick soon accounted for 90 percent of the Ontario crop. Gordon, bred by D.R. Sampson at Ottawa, was licensed in 1980 and Harus, bred by A.H. Teich at Harrow, was licensed in 1985. Both have better pastry quality than Fredrick. Winter feed wheats developed by H.G. Nass at Charlottetown, primarily for Atlantic Canada, include Lennox (1975), Valor (1981, in cooperation with the Ottawa Research Station), and Borden (1984).

At Lethbridge, Alberta, J.E. Andrews sought a high-quality, hardy, winter wheat for southern Alberta and southwestern Saskatchewan. Andrews started the program in 1951 by using progeny from a cross of Minter × Wichita made at Lethbridge in 1949. Andrews' first success was Winalta (2) released with M.N. Grant in 1961. This high-yielding, hardy variety had a kernel quality approaching that of hard, red, spring wheat. It was widely sought by millers of bread flour and quickly replaced the previously grown varieties. When Andrews moved to Brandon, Grant continued the program, developing Sundance in 1971 and Norstar 7 years later. Sundance yielded about 19 percent more than Winalta with the added advantage of resistance to shattering. Norstar brought together all the desirable characteristics of the two previous varieties plus exceptional winterhardiness.

The tolerance of winter wheats to low winter temperatures has been a subject of study by D.W.A. Roberts for many years. At Lethbridge, Roberts (68) has found that resistance in winter wheats to low temperatures (called hardening) starts to develop when air temperatures fall below 10°C. The process requires light, carbon dioxide, and proper plant nutrition. Fully hardened plants of suitable varieties can withstand temperatures as low as −20°C. The cultural practices used in preparing seedbeds, the amount of snow cover, the frequency and duration of mild periods, the amount of moisture in the soil, and the incidence of disease such as wheat streak mosaic all influence the survival of winter wheat plants.

Oats

Settlers, in the 1600s, brought oats with them to North America as feed for themselves and fodder for their livestock. Because of the large horse population, oats soon became second only to wheat in market value. Prior to the formation of Experimental Farms, several people selected genetically different imported varieties, then named and propagated them.

C.E. Saunders commenced breeding oats at the Central Experimental Farm in 1906. His first variety, Legacy, released in 1920, was medium early, resistant to halo and Victoria blights, and yielded well, particularly in central Alberta. Several other varieties followed; one, Abegweit, was developed in cooperation with Charlottetown.

Like wheat, the oat plant is attacked by leaf and stem rusts. J.N. Welsh joined the Cereal Breeding Laboratory, Winnipeg, to assist C.H. Goulden in breeding wheat. His attention was soon redirected to oats because of the need for rust-resistant varieties. With W.F. Hanna, the plant pathologist, Welsh in 1936 introduced Vanguard, the first rust-resistant oat variety. Vanguard was selected from a cross made in 1926 between Hajira and Banner. It had resistance to seven races of stem rust as well as to halo and Victoria blights. Among Welsh's most significant achievements was the introduction of the varieties Garry and Rodney. Garry, in particular, was characterized by wide adaptability, high agronomic performance, and disease resistance. The partnership of Welsh and T. Johnson, a plant pathologist, was profitable in the production of oat varieties and in the collection of information about rust organisms themselves.

R.I.H. McKenzie transferred from Indian Head to Winnipeg in 1956 and, with plant pathologists J.W. Martens and D.E. Harder, provided Canadian farmers with an array of rust-resistant oat varieties to satisfy nearly all their needs. The variety Harmon gained widespread acceptance because of its good performance and its attractive large kernel. In the late 1970s and early 1980s varieties, such as Dumont and Riel, produced by McKenzie possessed high levels of resistance to stem and crown rusts as well as resistance to smut. In 1984, one-half of the land seeded to oats in the Prairie Provinces grew varieties developed by Winnipeg scientists.

Lodging of cereal plants before and at time of ripening causes crop losses because the grain fails to ripen and harvesting is difficult. Some varieties, particularly of oats, lodge more readily than others. In 1947 D.G. Hamilton of the Cereal Division studied (26) the factors affecting resistance to lodging in oats with the objective of identifying specific characteristics for plant breeders to select. He found that varieties resistant to lodging had culms (stems) of greater diameter and larger, more rigid, and more widely spreading root systems than those of susceptible varieties. Fortunately, these characteristics occurred simultaneously, making selection for resistance to lodging possible.

A.D. McFadden at Lacombe, Alberta, selected early maturing lines of oat crosses received from Ottawa; he chose Larain and released it in 1947. When M.L. Kaufmann joined the Lacombe staff in 1956 he used the pedigree method in his breeding work. This required that hundreds of lines be subjected to expensive yield trials. The highest yielding varieties were selected as parents, but he found that his best progeny were no more productive than their parents and concluded that a production plateau may have been reached. To reduce his costs and still obtain high-yielding new varieties he devised a "random" method of selection by harvesting one seed from each plant in the segregating generations (F_2 to F_6) and seeding them in bulk the following year. Seed from the F_6 generation, much of which was expected to be homozygous (true breeding), was planted in head rows and selections made from among these in the normal way.

Using more than 800 lines, Kaufmann (42) assessed the method to be efficient and from the program he introduced Random oats in 1971. Random is moderately early with short, strong straw and has a good yield potential. Kaufmann assumed responsibility for the barley program in 1971. At the same time, H.T. Allen took over the oat program and introduced three varieties between 1975 and 1979. Cascade, the last, was the most productive and is used extensively throughout the Prairie Provinces.

On both coasts, D.K. Taylor at Agassiz, and J.D.E. Sterling and R.B. MacLaren at Charlottetown had been unobtrusively developing oat varieties suitable for their particular climates and soil conditions. In 1967 both met with success: Taylor (80) introduced Fraser, a strong-strawed, high-yielding variety; Sterling and MacLaren (77) had their new high-yielding variety, Cabot, ready for Prince Edward Island.

At the time Kaufmann made his studies at Lacombe, Sampson (71), of the Ottawa Research Station, investigated methods of selecting parents, hybrid lines, and individual plants within lines. He developed an index system that combined data from both the second and third generations. Rajhathy and Thomas (65) added further to the technical understanding of the genus *Avena* by providing a monograph on oat cytogenetics.

Over the years Ottawa has built a strong tradition of oat improvement starting with C.E. Saunders and progressing through L.H. Newman, R.A. Derek, F.J. Zillinsky, and V.D. Burrows. Up to the early 1950s the methods used to breed oats were conventional. Zillinsky changed this by creating a vast new genetic pool of variation through an extensive program of interspecific hybridization involving many oat species. Burrows joined Zillinsky in 1958 as a plant physiologist and investigated these genetic resources. When Zillinsky accepted a position with the Rockefeller Foundation in 1969, Burrows became the oat breeder. Out of the interspecific program Burrows licensed the varieties Gemini, Foothill, and Hinoat. Foothill was Canada's first dual-purpose grain and forage-type oat. Hinoat (high nitrogen oat) was the first high-protein oat released for consumption by humans.

Canadian oat varieties are daylength sensitive, meaning that the plants require exposure to long daylengths to flower normally. Burrows bred the first daylength-insensitive variety, Donald, in 1982. The gene that permitted the variety to flower normally under shorter daylengths was derived from an *Avena byzantina* specimen collected from Turkey in 1964. The transfer of this gene to Canadian oat breeding material led to the establishment of winter oat nurseries in California, making it possible to grow two generations in the field per year. Like other daylength-insensitive cultivars of wheat, barley, and rice, Canadian varieties of daylength-insensitive oats flower normally under either long or short photoperiods.

Burrows has worked many years to develop a different class of oats called dormoats for growing in northern climates. The seed dormancy genes from *A. fatua* have been blended with those governing superior agronomic performance in *A. sativa*. Dormoat seeds are sown in autumn. They remain dormant over

219

winter but germinate in early spring to take advantage of cool, moist conditions by producing plants with many tillers and large panicles containing numerous seeds. The crop remains experimental and awaits the development of suitable seed management techniques to condition the seed prior to planting in autumn in order to obtain uniform germination the following spring.

Although Canadian oat breeders have released seed of several naked oat varieties (Laurel, Brighton, Vican, Terra) that are high in energy and rich in protein, they have yet to be accepted by growers. Burrows, in 1985, released a fifth variety, Tibor, that has the potential of being a nutritious feed for poultry and swine. The energy content of the grain approaches that of corn and the protein content is such that supplemental soybean meal is not required.

The Mediterranean is where many plants were first domesticated and it is probably where *Avena* species originated. There, the greatest variation in plant type occurs. In 1970, plant explorers, B.R. Baum and T. Rajhathy, Ottawa, and J.W. Martens and G. Fleischmann, Winnipeg, spent about 8 weeks in areas bordering on the Mediterranean Sea searching for native oat species of use to the Canadian breeding program. They returned with over 15 000 samples representing 10 oat species! These samples were added to the collection from a similar exploration in 1964. This collection, which provides Canadian oat breeders with the best and most extensive original assemblage of specimens, is maintained by the Plant Gene Resources of Canada Office, Ottawa Research Station, headed by R. Loiselle. There are 78 500 stocks in the collection stored for either a short period of time at 4°C or for a long period of time at −20°C. Canada is responsible for the world collections of oats and barley, and stores a duplicate of the world collection of millet. In addition, the Plant Gene Resources Office maintains working collections of many other crops such as wheat, alfalfa, grasses, and vegetables. The collections have been used to obtain new genes for disease resistance and early maturity. Donald oat was bred from its material.

Seven years following the opening of the Research Station, Sainte-Foy, Quebec, J.P. Dubuc joined the staff to continue the oat breeding program started earlier by F. Gautier and C.A. St.-Pierre. The latter spent only 2 years at Sainte-Foy, but he laid a firm foundation, for Dubuc was able to release his first variety, Alma, the same year he started work. Alma has short, strong straw, and came from a cross made at the Experimental Station, Sainte-Anne-de-la-Pocatière. Four other short, strong-strawed varieties have come from the Sainte-Foy program; the most recent of which, Kamouraska, was bred in cooperation with Charlottetown. Kamouraska has large kernels, good protein content, and superior resistance to lodging.

Barley

Champlain introduced barley into Canada in 1605 to meet the needs of his brewers. Two centuries later a two-rowed variety, Bay of Quinte, became so popular with brewers in the United States that the McKinley tariff of 1890 was imposed to restrict its import (28). With this loss of market, consideration was given to using barley for livestock feed. However, because of low yields from two-row varieties and no organized market for feed barley, the crop declined relative to other cereals. Such was the situation when Experimental Farms started and when barley began its move into Western Canada.

There are two kinds of barley—two-rowed and six-rowed, both *Hordeum vulgare*. The number of rows refers to the arrangement of kernels in the spike. A century ago, maltsters used two-rowed barley because of its yellow aleurone.[1] The six-rowed barley, which has blue aleurone, has been used only for feed.

Three events returned barley to a position of importance in Canada. In 1910 the Ontario Agricultural College introduced a new variety, O.A.C. 21, which was a selection from seed brought from Manchuria in 1889. In 1910 the Canada Malting Company was incorporated and encouraged the production of malting barley. In 1918 J.H. Grisdale, Deputy Minister, Canada Department of Agriculture, established the National Barley Committee with the objectives of increasing barley production, improving barley quality, and finding wider markets for the cereal. With funding supplied by the barley industry, he brought representatives of the grain trade, the maltsters, the universities, and the provincial and federal departments of agriculture together. The group later formed an Expert Committee of the National Research Council.

The two-rowed European barley varieties tested by Ontario farmers in 1888 proved to be weak-strawed and late-maturing. In 1889 Wm. Saunders crossed (13) six-rowed with two-rowed types in an attempt to combine the best characteristics of each. He also made selections from the Manchurian material, releasing Mensury Ottawa 60, a blue-aleuroned variety, and Manchurian Ottawa 50, a yellow-aleuroned variety. The former gained some prominence in Manitoba, but both were overshawdowed by O.A.C. 21. The program at Ottawa continued under P.R. Cowan until 1950. From his breeding program came the varieties Fort and Nord.

In 1912 J.A. Clark, superintendent, Experimental Station, Charlottetown, Prince Edward Island, selected (7) from within a variety of two-rowed barley grown by local farmers and named the selection Charlottetown No. 80. The variety became popular throughout the Maritime Provinces because its awns were shed at harvesttime, making it more easily handled than other awned types. In addition, it outyielded other varieties and was tolerant to acid soils.

A most successful Canadian barley breeding program was started in 1923 by S.J. Sigfussion at the Experimental Farm, Brandon, Manitoba. The productive period, building upon the early work of Sigfussion and R.F. Peterson (1933–1936), started with W.H. Johnston in 1936. Johnston sought high-yielding, disease-resistant two- and six-rowed barleys suitable for malting and for feed. Sigfussion had crossed Lion with Beaver and produced Plush, which was made available in 1936. However Plush, which gave a good yield, was susceptible to smuts, rusts, and most head and leaf diseases. By 1947 Johnston had

221

---

[1]The aleurone is the outer cell layer of the endosperm of cereal grains and contains small, colored, protein granules in some varieties.

developed a new variety, Vantage, that resisted attack from stem rust. During the next quarter century Johnston bred and introduced ever better varieties of both two- and six-rowed barleys, 13 in all. He was the first to combine broad resistance to diseases with earliness and high yield. The last variety he introduced, Bonanza, is now the standard for malting and brewing quality. In recognition of his contributions to Canadian agriculture, the University of Manitoba awarded Johnston an honorary doctor of science degree in 1968 and the federal government presented him with a Public Service of Canada Merit Award in 1969.

D.G. Hamilton assumed responsibility for the national barley breeding program in 1950. One of his prime interests was the selection of barley (and oat) varieties resistant to root rot caused by the fungus *Helminthosporium sativum*. Breeding programs were hampered because no satisfactory techniques were available with which to select resistant seedlings. Hamilton (27) together with R.V. Clark of the Botany and Plant Pathology Division, were able to obtain reliable differential readings in 21 days by modifying existing techniques and using sterilized seed, sterilized sand, and a sand–cornmeal mix containing the inoculum, at appropriate temperatures and humidities. This technique made the selection of seedlings resistant to *H. sativum* rapid and simple. Over the past 30 years R. Loiselle, G. Fedak, S.O. Fejer, and K.M. Ho have successively been barley breeders at the Ottawa Research Station. From their program three outstanding feed barleys were developed—Massey, Vanier, and Léger—named after three former governors-general.

Support of Johnston's breeding was provided by scientists at the Laboratory of Plant Pathology, Winnipeg. To begin, Johnston (59) and D.R. Metcalfe, both at Brandon, studied the inheritance of resistance to loose smut. In 1966 Metcalfe transferred to Winnipeg to initiate a two-rowed barley breeding program, pursuing his studies on loose smut with K.W Buchannon, W.C. McDonald, and E. Reinbergs (58).

R.I. Wolfe, who had assisted Johnston since 1968, continued with the program after Johnston's retirement in 1971. He introduced four more varieties (one named in honor of Johnston), each of which provided additional yield, improved agronomic characteristics, or better disease resistance than earlier varieties. M.C. Therrien has had responsibility for the program since 1981. An indication of the impact Johnston and his colleagues have had on barley production was reflected in 1984, when 60 percent of the area in Western Canada seeded to barley was from the Brandon breeding program.

Scientists at other research stations, both east and west, have produced barley varieties adapted to their own environmental conditions. The first release, Wolfe, was by A.D. McFadden at Lacombe, Alberta, in 1954 from a cross made in Ottawa. It was early with strong straw but lacked disease resistance. S.A. Wells at Lethbridge, Alberta, in cooperation with D.S. McBean of Swift Current, Saskatchewan, produced Galt in 1966. This high-yielding, six-rowed variety is widely adapted to irrigated as well as to dryland conditions. On Prince Edward Island in 1974, J.D.E. Sterling developed Kinkora, the first variety with resistance to barley jointworm. Kinkora also has some tolerance to acid soils. J.P. Dubuc, at Sainte-Foy, Quebec, working in cooperation with plant breeders at the University of Laval, in 1980, provided Sophie, a barley well adapted to the eastern parts of Canada. M.L. Kaufmann, at Lacombe, using a random breeding method that he developed for oats, produced Diamond in 1982. In central Alberta it yields more and matures earlier than Galt.

Cereal quality

The quality of cereals, particularly that of bread wheats, is of vital importance to the maintenance of Canada's international grain trade. For this reason, by law, no variety of bread wheat can be licensed for sale in Canada without its bread-making quality equaling that of Marquis. The Expert Committee on Grain Quality supplies information to the Seeds Division, Food Production and Inspection Branch of Agriculture Canada, which advises the Minister on the quality characteristics for selections to be licensed as new varieties. All cereal breeders in Canada recognize the need for maintaining the quality of Canadian cereals, even though it places an added burden on the selection process. A potential variety may be outstanding in all its agronomic characteristics, lacking only acceptable characteristics for baking. In such a case the breeders must retrace their steps and try again.

Sir Charles Saunders was the first cereal breeder in Canada to recognize the need for good bread-making characteristics (60). P.R Cowan followed Saunder's lead. When A.G.O. Whiteside joined the staff in 1924 Experimental Farms had a complete cereal quality laboratory. Whiteside, and his technician H. Miller, ground small samples of breeders' seed, tested the flour and dough for strength of gluten, and baked bread from each sample. He received support from the Dominion Chemist, C.H. Robinson, in determining each sample's protein content, one of the factors affecting baking qualities of wheat.

Samples from breeders across Canada were sent to the Cereal Division laboratory in Ottawa until 1951. At that time V.M. Bendelow was appointed to Winnipeg to handle the western work and by 1959 he had completely taken over the testing of bread wheats from Ottawa. The contributions made by both the Winnipeg and Ottawa laboratories to all cereal breeding programs have increased in both quantity and value over the years. Bendelow added durum, red winter, and soft, white, spring wheats to the Winnipeg tests. He also studied the malting characteristics of barley. At Ottawa, Whiteside, and his successors, I. de la Roche and then R.G. Fulcher, tested the pastry-baking quality of all candidate soft, white, winter wheats and the protein contents in feed wheats. In 1981 a third laboratory was added, when food technologist, R. Stark, at Kentville, Nova Scotia, provided quality assessments for maritime breeders.

Original research is still being done by scientists at both the Winnipeg and Ottawa laboratories, much of it on the improvement of methods for assessing cereal quality. As an example, Fulcher, Wood, and Yiu (21) at Ottawa have investigated the use of fluorescence microscopy in the detection of specific carbohydrates and enzymes in cereal grains and other foods. With the use of a high-intensity light source and the insertion of two filters to adjust the light passing through the microscope, appropriately stained samples dramatically reveal the presence of fluorescent material. The system is used to observe differences in concentration and distribution of specific carbohydrates in barley and oat cultivars, and to detect structural characteristics of digestive systems and component changes during germination or malting of seed.

CANOLA

Canadian margarines, cooking oils and salad dressings are generally all made from canola. Canola is one of the most recent in a series of success stories of agricultural research and production in Canada.

The precursor of canola is rapeseed. The word "rape" is derived from the Latin word rapum, meaning turnip (19), of which the rape plant is a close relative. Its seed is crushed for oil and its leaves are used as a forage to feed livestock. In Asian countries 3000–4000 years ago rapeseed was crushed and used as a cooking and illuminating oil (6). It was introduced into Japan from China about 2000 years ago and later into Europe along the Mediterranean, where it was also used as a lighting oil. Rapeseed oil was favored over whale oil because it burned with a smokeless flame. It was improved upon only when petroleum-based oils became available in the nineteenth century.

When steam engines were developed in the eighteenth century, engineers found that rapeseed oil was the sole oil known which would cling to metal and not be washed off by hot water and steam. It therefore became the favored lubricant for marine use. North American supplies were imported from eastern Europe. In the 1930s (6) T.M. Stevenson of the Forage Crops Division introduced seed of forage and oil rapes from Europe for testing by experimental farms across Canada. He found that rape grew vigorously. It could be seeded late, and because of the many hours of daylight during Canadian summers, it matured before severe autumn frosts occurred. There was, however, no commercial production in Canada until 1943.

Early in 1942, after supplies of rapeseed from Europe and Asia were cut off because of World War II, Mrs. Phyliss Turner, fats and oils administrator of the Wartime Prices and Trade Board (85), asked Stevenson, now Dominion Agrostologist, if Canada could produce rapeseed. Stevenson knew from his 1930 experiments that rapeseed could be grown. About 1000 kg of seed were produced on experimental farms that year. A further 18 000 kg were purchased from seed companies in the United States and sown by Canadian farmers in 1943, most under contract. The 1200 ha seeded yielded 900 000 kg of seed and generated higher dollar returns than cereal crops. Canada was in the rapeseed business.

The 1943 rapeseed crop was processed into marine lubricating oil in Hamilton, Ontario. J. Gordon Ross formed Prairie Vegetable Oils Ltd., in Moose Jaw, built a processing plant, and handled the 1944 crop. Later he arranged for seed to be processed by the Saskatchewan Wheat Pool plant in Saskatoon. Wilson (85) recounts the rapid development in rapeseed production from 0.5 million kg in 1943–1944 to 28 million kg in 1948–1949, a 50-fold increase in 5 years! After the war, however, production waned to a low of 160 ha in 1950.

At the same time W.J. White of the Dominion Forage Crops Laboratory, Saskatoon, started a variety improvement program through breeding. He was assisted with the necessary chemical analysis by H.R. Sallans and B.M. Craig of the National Research Council's Prairie Regional Laboratory (PRL) in Saskatoon. Cooperation on rapeseed research among these federal agencies and the University of Manitoba, where B.R. Stefansson started a breeding program in the early 1950s, has been close ever since.

Canadian plant breeders were well aware that rapeseed was used to produce an edible oil in other parts of the world. Could Canada use rapeseed for a food, thereby reducing imports of vegetable oils, such as soybean and palm oil? The key was the suitability of rapeseed oil as a base for the production of margarine. At the same time, Ross found a market in the paint and plastic industries where, because of its high erucic acid content, rapeseed oil speeded the hardening process. Growers sold their crops but at prices considerably below those realized in 1944 and 1945 when war created a greater demand. By 1948 Grace (22) and coworkers at the National Research Council in Ottawa succeeded in homogenizing rapeseed oil and with careful refining, bleaching, hydrogenating, and deodorizing concluded that it could be substituted for soybean oil in edible products. This revelation led to the construction of other extraction plants in Manitoba and Alberta as well as in Saskatchewan. By 1950 experimental lots of margarine and salad oils were being prepared, with commercial quantities appearing in 1955.

In 1954 White released Golden, the first licensed rapeseed variety in Canada. Golden matured uniformly and earlier, yielding more seed with a higher percentage of oil than those kinds previously used. In 1957 R.K. Downey, an alfalfa breeder at the Lethbridge Experimental Station, transferred to Saskatoon. Downey inherited the breeding program started by White in 1943 because, as a student, he had had experience with rapeseed.

At that time questions were raised concerning the nutritional value of a main component of rapeseed oil, erucic acid. Downey, in cooperation with Craig at PRL, surveyed the available world rapeseed germ plasm and found the European forage rape variety Liho to have about half the normal level of erucic acid. Selection within Liho resulted in the isolation of the first Low Erucic Acid Rapeseed (LEAR) plants. However, the process of selecting LEAR varieties in the *Brassica campestris* species was slow and required a significant amount of seed. This problem was overcome by splitting seeds into two parts, one with half the cotyledons and the root, the other with the remaining cotyledons. The oil from the latter portion was analyzed for erucic acid. When a half seed showed a low level of erucic acid, its other half was germinated carefully and grown to maturity. This half-seed technique resulted in rapid development of adapted LEAR varieties in both rapeseed species, *B. campestris* and *B. napus.*

225

The first commercial production of a LEAR variety occurred in 1964 and a special market for this new natural oil was developed. In September 1970, scientists from Holland, France, and Canada who attended the International Conference on the Science, Technology, and Marketing of Rapeseed and Rapeseed Products at Sainte-Adele, Quebec, reported that possibly there was a link between feeding rapeseed oil high in erucic acid to young laboratory animals and a fat accumulation around their hearts. The Minister of National Health and Welfare interpreted this possibility as posing a health risk to humans and asked for a switch to LEAR varieties in Canada as soon as practical.

The Research Branch was ready. About 2300 kg of seed of Span, a new LEAR variety, with less than 1 percent of erucic acid, were available at Saskatoon from Downey's breeding. The best way to quickly change to LEAR varieties was to grow a crop during winter. E.D. Mallough, an agronomist from the Research Station, Regina, had had considerable experience growing crops in southern California to accelerate Canadian plant breeding programs by producing two generations each year. Never before, however, had he been asked to multiply 1100 kg into 1 000 000 kg. (Because of the remote possibility of a total crop failure in California, only half the Saskatoon seed was to be used.) It was a test of prairie ingenuity; a daring venture. By September of 1970 Mallough and A.B. Masson, of the department's Production and Marketing Branch, had contracted 770 ha from farmers in the lush Imperial Valley of California to grow Span. None had grown rapeseed before. D.A. Cooke, a plant scientist from the Research Station, Melfort, Saskatchewan, moved to California for the entire winter to supervise the operation. With help from Mallough and J. Capcara, Downey's senior technician, Cooke supervised the seeding rate of 1400 g/ha. By Christmas the crop was growing vigorously. It looked as though it might double the original yield objective, but a severe frost during the last week of January dashed any hope of producing 2 million kg of seed. Then a March frost, just after blooming, caused additional problems. However, only the earliest fields were damaged and the original objective of 1 million kg was achieved. The race against time was won when a fleet of trucks rushed seed to Canadian farmers in time for a June 1971 sowing. By 1972 all 2 million ha of rapeseed grown in Canada for crushing were low in erucic acid and met the Department of National Health and Welfare's standard. The changeover was accomplished within 2 years without the use of legislation or regulation.

The first LEAR varieties, Span and Oro, did not yield as well as the standard high erucic acid varieties; hence by 1973, Midas and Torch, with seed and oil yields superior to all previous varieties, were released.

Downey continued with his intensive breeding. It is not profitable to grow and to crush rapeseed only for oil—meal is an important by-product. Meal of rapeseed contained small amounts of glucosinolates, which are sulfur-based and inhibit growth when fed to some types of livestock. Would it be possible to eliminate these compounds in a manner similar to the way erucic acid was handled? Downey thought so, and with the help of Youngs and Wetter from PRL succeeded in developing a system to analyze rapidly small samples for glucosinolates. In 1968 Jan Krzymanski, a visiting postdoctoral fellow from Poland, found that the Bronowski variety, *B. napus*, which he had brought from his native land, had low glucosinolate meal. Immediately, the research station air-freighted 80 kg of seed from Poland and multiplied it for breeding and feeding purposes. Downey gave Stefansson (University of Manitoba) seeds of the new germ plasm and by 1973 both had bred cultivars low in erucic acid and low in glucosinolates. The cultivar Tower from Manitoba was the best agronomically and was introduced. But the job was only half done. No double-low cultivars of the second species, *B. campestris*, which was the predominant kind of rapeseed grown in Alberta and parts of Saskatchewan, were available. Again, the Saskatoon group was able to meet the challenge. Downey and S.H. Pawlowski crossed the two species and produced Candle in 1977. Now all domestic processors had access to cultivars with seed low in erucic acid and glucosinolates. An expanded poultry and livestock meal market resulted. The term "canola" was adopted by the industry to designate cultivars low in both erucic acid and glucosinolates, and to identify the oil and meal derived from them.

Another problem with rapeseed meal was its high fiber content in relation to its main competitor, soybean meal. Although most livestock need fiber, the most economical source is hay, not concentrates. Again, in close cooperation with other breeders, Downey produced varieties whose meal was low in fiber. Golden yellow seeds were found to produce proportionately more oil and less fiber than the normal brown or black seeds. As with low erucic acid and low glucosinolates, a yellow seed coat and low fiber content were genetically controlled and inherited from generation to generation. The seed coats of Saskatoon's latest *B. campestris* cultivars, Candle and Tobin, are partially yellow.

Meanwhile, Canada was exporting large quantities of seed and oil to many countries, except the United States where vegetable oils "generally recognized as safe" (GRAS) excluded canola. In 1971, scientists in France suspected a link between rapeseed oil consumption and lesions in the hearts of rats. To determine if such a correlation existed, B.B. Migicovsky, Director General of the Research Branch, organized a team of biochemists, nutritionists, pathologists, and toxicologists. Kramer, Mahadevan, Hunt, Sauer, Corner, and Charlton (49) discovered that microscopic lesions in the muscle of the heart occur in male rats regardless of their ingestion of this dietary oil, and that feeding low erucic acid rapeseed oils to male rats tends to increase the incidence of lesions. These scientists also showed that feeding the same oils to hogs and primates produced no heart lesions. Further research by Kramer, Hulan, Mahadevan, Sauer, and Corner (48) demonstrated that some strains of rats were particularly sensitive when the level of fat in their diet exceeded 5 percent. The causal agent was identified as triglyceride and its fatty acids.

By 1977 these data had convinced the Canada Department of Health and Welfare that canola oil was safe for human consumption. European countries agreed in 1979. The United States was still doubtful until Kramer, Farnworth, Thompson, Corner, and Trenholm (47) conclusively demonstrated that canola oil is risk-free and nutritious, and that the problem of heart lesions in rats has no relevance to humans. The GRAS submission, a book edited by Kramer, Sauer, and Pigden (50), led the United States health authorities, in January 1985, to approve the use of canola oil in their country. Fifteen years of cooperation among scientists of many disciplines was required to achieve this goal. Kramer and Sauer were the recipients of a Public Service Merit Award in 1983 for their leadership and research in resolving the canola oil conundrum.

There is much more to the canola story than the few details given here. There are the parts played by commercial crushers, wheat pools, farmers, and several universities. There are the industrial varieties high in erucic acid (50 percent) that are being developed by a team of scientists brought together by Downey. There are the varieties now resistant to diseases. There are measures for controlling flea beetle and bertha armyworm. There is the change in color pattern of the Canadian prairies in June during flowering of canola, its bright yellow flower readily seen and admired by passengers on transcontinental flights. By 1981, less that 40 years from its introduction, canola was more than a billion dollar industry, outproduced in Canada only by the wheat crop.

Charles Saunders bred and introduced Marquis wheat in 1905. He was knighted in 1934 for his outstanding achievement. Keith Downey bred and introduced low erucic acid rapeseed (LEAR) in 1964 and, along with many other honors, was invested as an Officer of the Order of Canada in 1976 by the governor-general—suitable recognition for contributing such a remarkable chapter in our history.

SOYBEAN

William Saunders first planted soja [sic] bean on the Central Experimental Farm in 1897. Because all available varieties were long-seasoned and would not mature sufficiently to produce ripe seed, the soybean was harvested as hay when pods were about half filled. As recently as the late 1930s no variety was available that would reliably mature seed when grown in Canada. Soybean, however, is valuable feed supplement for livestock because of its high protein and fat content. Soybean protein, unlike many other plant proteins, is comparable in quality to that in milk, meat, and eggs.

The first effort to improve soybean for Canadian conditions was in 1923 when F. Dimmock organized extensive variety trials at Harrow. He transferred to the Forage Crops Division, Ottawa, in 1927 but continued to manage the Harrow soybean trials until C.W. Owen was appointed in 1929. Dimmock inaugurated a selection program within the Manchu variety to find earlier maturing varieties for southwestern Ontario. The first selection, A.K. (Harrow), was released in 1931. It was not until 1943, however, that Harosoy was released from the crossbreeding program started in 1936. The introduction of Harosoy from the Experimental Station, Harrow, marked the beginning of the commercial soybean industry in Canada. By 1959 Harosoy was the most important variety in Canada, occupying about 70 800 ha (75 percent of the soybean) in Ontario and about 1 620 000 ha (15 percent) in the United States.

Dimmock at Ottawa and Owen at Harrow used germ plasm obtained from Harbin, China, to develop early varieties. They also freely exchanged parental material with the University of Minnesota; consequently the three programs produced similar varieties.

The objectives of the Harrow and Ottawa programs were to develop varieties that would mature sufficiently early for all seed to ripen before harvest (125–130 days) and to have strong upright branches that held seed pods well above the ground for ease in harvesting. Emphasis toward breeding for resistance to *Phytophthora*, a serious root and stem rot fungus disease, started about 1955 at Harrow when the disease became serious in southwestern Ontario, particularly on poorly drained soils. Breeders at Urbana, Illinois, had similar objectives and cooperated with their Canadian counterparts. By 1963 a new variety identical to Harosoy but resistant to root rot was introduced as Harosoy 63. It was useful for approximately 10 years, after which time races of *Phytophthora* capable of bypassing the defense mechanism of Harosoy 63 developed. To overcome this problem Haas and Buzzell (25) developed a technique to identify soybean varieties tolerant of, rather than resistant to, *Phytophthora* rot. By 1975 the variety Harcor, which has good field tolerance and a high yield potential, was released. Although severe losses in yield have been prevented, complete protection from *Phytophthora* has yet to be provided. Breeders, geneticists, and pathologists such as Buzzell and Anderson (10) continue to seek race-specific resistance.

In 1965 B.R. Buttery at Harrow studied soybean plants in even greater detail and found that their carbon dioxide consumption rate in the photosynthetic process was only about 55 percent that of corn plants. This, reasoned Buttery (9), was why corn grew faster than soybean and accounted for the higher yield of corn. Subsequent research was done on varietal differences in photosynthetic rates of soybean varieties.

The Ottawa program emphasized production of early varieties because the growing season in the northern part of Ontario, Quebec, and Manitoba was about 10 days shorter than at Harrow. In 1961 L.S. Donovan assumed responsibility for the Ottawa soybean (and corn) breeding programs. His objective was to develop varieties of soybean that would mature in the Ottawa River valley of Quebec and Ontario and in southern Manitoba. To achieve this goal he turned to Sweden, which had obtained early maturing, day-neutral varieties from the Sakalin Islands of northern Japan. By using this new germ plasm in combination with material from Germany, Donovan widened the genetic base of his breeding program and made outstanding progress. For instance, older varieties of soybean germinated poorly in cool soil and set seed inadequately if temperatures were low at the time of flower initiation. The Swedish material, in particular, was tolerant of low air temperatures; thus, Donovan was able to improve the percentage of seed setting in his varieties. From this program came Maple Arrow (1976), Maple Amber (1981), and Maple Presto (1982) (82). At the time of introduction, Maple Presto was the earliest maturing soybean licensed in Canada. However, this characteristic resulted in low yields. Plant breeders actually overstepped their

229

goal! Maple Arrow is now the standard short-season (120–123 day) variety for eastern Ontario. Maple Amber, which germinates well under cold conditions, is high in oil (over 20 percent), high in protein, and is the standard for the Great Plains.

Recently H.D. Voldeng of the Ottawa Research Station developed two edible varieties grown specifically for the Japanese market. To obtain the needed small seeds Voldeng turned to wild soybean from China, which has black seeds in small pods. The parents he used had seeds high in protein and were therefore useful for other breeding programs. Voldeng replaced the black seed characteristic with a light-golden seed and, after including low-temperature germination and seed-setting characteristics, he introduced the two varieties Canatto and Nattawa. Because neither variety yields more than 65 percent of the standard varieties of soybean, Canatto and Nattawa are only grown under contract for direct sale to Japan.

Two other recent Ottawa varieties have been introduced for specific reasons. Maple Ridge (83), so named because of a ridge on each seed, yields better than Maple Presto, and is particularly suited to Manitoba conditions. Maple Isle (84), replaces Maple Amber in Ontario and the Maritime Provinces because, unlike Maple Amber, it is tolerant of some widely used herbicides.

In 1978, in cooperation with the Alberta Department of Agriculture, a soybean breeding program was initiated at Lethbridge. H.-H. Mündel was appointed to develop varieties suitable for irrigated lands. One of the first things he needed to know was the water requirements of soybean under Lethbridge conditions. With the aid of E.H. Hobbs, an irrigation engineer, Mündel determined (35) that water use with soybean peaked in late July and early August. The future looks promising, because the water use of soybean will integrate with the early season use of cereal crops. As with other legumes, soybean forms a nitrogen-fixing symbiosis with a group of bacteria called rhizobia; in the case of soybean, it is the bacterium *Rhizobium japonicum*. This particular *Rhizobium* does not exist naturally in western Canadian soils and must be added at the time of planting. Rennie and Dubetz (66) of Lethbridge realized the unique opportunity they had in selecting the appropriate strain of *R. japonicum* for each new variety of soybean in order to maximize the nitrogen-fixing capabilities of soybean under irrigation. The specific strains of rhizobia with which Mündel's new varieties should be grown will be identified.

CORN

The corn that Saunders and Robertson grew was used for ensilage except in the southernmost part of Ontario, where it was shelled. To make ensilage, the whole corn plant (known as maize in Europe) is harvested while still green, chopped, sometimes mixed with alfalfa or clover, frequently supplemented with molasses, and then either blown into an upright silo or spread in a horizontal, or pit, silo. Next, it is compacted, covered, and fermented without air, making a nutritious sauerkraut-like cattle feed for the winter months. After shelled corn (mature seed) is removed from the cob it is ground for cattle or pig feed, milled for corn flour, crushed for its oil and meal, or fermented for alcohol. The North American Indian used corn to make porridge, soup, bread, and alcohol as do people around the world today.

Corn is a grass, known botanically as *Zea mays*. It is a tropical species indigenous to Central America and was cultivated by the native populations there and as far north as the Great Lakes, and in the west as far north as Mandan, North Dakota. Under natural conditions corn cross-pollinates, the male or polliniferous flowers being on the tassel at the top of the plant, and the female flowers with their pollen-receiving stigmas (silks) partway down a strong, central stalk where a cob bears seed. For centuries American corn was an open-pollinated crop, varieties being produced in isolation to prevent wind-borne pollen from mixing germ plasms. In the 1880s new varieties were developed by sprinkling pollen from males of one variety onto the silks of females of another variety. Following several generations of careful selection, a reasonably uniform, new variety would result. This was the method used at Harrow and Ottawa in 1923 by A.E. Mathews and F. Dimmock when they started the corn breeding program of Experimental Farms.

Unlike other cereals such as wheat, oats, and barley, which produce flowers during lengthening days, corn starts to flower as the days shorten. Corn therefore tends to mature late and is subject to damage from frost when grown outside the tropics. Because corn has a more efficient respiratory system than most cereals and thus has a high yield potential, Canadians were interested in adapting it to their shorter growing season. The first five Experimental Farms experimented with corn, and techniques were developed at Agassiz and Ottawa for the production of silage corn. When machinery for harvesting, chopping, and blowing corn and other ensilage crops became available, interest on the part of farmers increased. According to R.I. Hamilton (29) the large amount of land used for growing corn in Canada today is the result of development of appropriate fertilizers, cultivation methods, disease, insect, and weed control, and harvesting methods as well as the breeding of hybrids to conform to Canadian climatic limitations.

The concept of self-pollinating corn families in order to select homozygous (true-breeding) lines with desirable characteristics and then crossing two inbred lines to produce an F_1 was simultaneously conceived by G.H. Shull and by E.M. East in the United States during 1906 following the rediscovery in 1900 of Mendel's Laws of Inheritance (34). Each one proposed single crosses between two unrelated inbred lines. When D.F. Jones of the Connecticut Experiment Station found, in 1918, that two unrelated F_1 hybrids could produce a double-cross of approximately equal value to a single-cross hybrid, hybrid corn became a commercial reality, because seed production costs were drastically reduced. Both Mathews at Harrow and Dimmock at Ottawa tested one of Jones' hybrids as early as 1923 and of 17 varieties grown, the hybrid outyielded all others. With this encouragement, Mathews and Dimmock started Canadian inbreeding programs, using the best varieties from both Canada and the United States as source material. Between 12 000 and 15 000 pollinations were made each year. The race was on to develop commercial hybrids for Canadian weather conditions, because much of the material imported from the United States was too late maturing. In 1937 the first provincial corn committee, consisting of Ontario and federal scientists and representatives from the seed trade, met to arrange for the licensing of hybrid varieties. The Canadian seed corn industry was born in 1940. Using early maturing Wisconsin inbreds the Ontario Ministry of Agriculture Experimental Station at Ridgetown produced the hybrids. The following year O.J. Wilcox, of Woodslee, Ontario, produced the first crop of hybrid seed corn in Canada from early maturing Wisconsin single crosses. Acceptance of hybrid seed by Canadian growers was rapid. In 1939 about 10 percent of land used for growing corn in Canada was planted to hybrids. A year later 50 percent of the corn in southwestern Ontario was hybrid and by 1944 the conversion to the high-yielding varieties was almost completed.

The first hybrid variety from Canadian breeding was introduced by G.F.H. Buckley of Harrow in 1946 as Harvic 300, which remained in commercial production for 17 years, even though five other varieties were licensed shortly after it. Buckley continued to develop hybrid varieties until 1953 when C.G. Mortimore changed the direction of his breeding program to the more pressing problems of inbred lines that would resist diseases such as root rot and stem rot, and attack from the European corn borer. Cooperative research with plant pathologists showed that root rot and stem rot are caused by soil-borne fungi attacking plants as they mature, that plants with high sugar content in their lower nodes were resistant, whereas plants with low sugar content were susceptible, and finally that resistant plants had greater leaf areas and more tillers (suckers). By 1959 two single-cross hybrids produced at Harrow were used in the production of a commercial hybrid resistant to both root rot and stem rot. At the same time, Mortimore selected inbred lines for resistance to corn borer, root rot, and stem rot. The first line with triple resistance was released in 1961, followed by 17 others.

In Canada the European corn borer (*Ostrinia nubilalis*) was first found in southwestern Ontario in 1920 (14), even though there had been an embargo on the importation of corn from the United States for a year. It was reported in Quebec in 1926, was widely spread throughout that province by 1935 and is now in all corn-growing areas of Canada. Corn borer larvae penetrate maize stalks from the end of June until harvest, resulting in breakage and subsequent loss of crop. Cultural methods aimed at destroying corn borer larvae by plowing or burning corn stubble, insecticide sprays, and the release of parasites were all ineffective. Breeding for resistance or tolerance was the best solution but required time. Fortunately, some corn plant families from Iowa consistently escaped severe damage and served as a source of resistance.

Natural borer populations are rarely sufficient to uniformly infest a planting and therefore in 1969 when the research station at Saint-Jean, Quebec, started its program to produce early inbred lines and synthetic varieties resistant to borers, M. Hudon, an entomologist, applied a system from Iowa State to mass-produce egg clusters of corn borers, making it possible for breeders to infest their test plants with artificially reared insects. He now produces more than 250 thousand egg masses each year for breeders throughout Canada. Hudon received a Public Service Merit Award in 1980 for his accomplishment.

Dimmock at Ottawa had different problems to solve. Quebec, the Ottawa Valley, and Manitoba were too far north to grow and mature the varieties produced farther south at Harrow. Therefore, when he moved to Ottawa from Harrow in 1929 to assume the responsibility for the corn, soybean, and sugar beet research, he started an intensive program to develop hybrids that would mature in less than 110 days and, more importantly, would do so with only 2700 heat units rather than the 3400 available at Harrow. He obtained early maturing corn from North American sources and during his 32 years at Ottawa released two outstanding inbred lines, CO109 and CO106, as well as 11 hybrids, all of which produced well commercially. In 1961 Dimmock retired and his assistant since 1958, L.S. Donovan, assumed the responsibility for both the corn and the soybean breeding programs. He increased cooperation with colleagues in the short-season areas and tested their material extensively. His primary intent was to expand Canada's corn belt to include all provinces. He had excellent colleagues in C.G. Mortimore who took over corn breeding from Owen at Harrow in 1943, J. Giesbrecht who joined Morden in 1957, M. Hudon and M.S. Chiang who started their corn breeding at Saint-Jean in 1969, R.I. Hamilton who arrived at Brandon in 1969, M.D. MacDonald at Lethbridge who shifted his genetics program from wheat to corn in 1971, and I.S. Ogilvie who started the L'Assomption corn program in 1982. From these programs have come many outstanding Canadian inbred lines, particularly from Morden and Ottawa. Indeed, Morden CM105 was a parent in one of the seven most-widely sold hybrids in the United States in 1982 and the Ottawa inbred CO255 is expected to rank among the world leaders, as did CO109 and others. For Donovan's work on the corn program he received a Public Service Merit Award in 1972.

On the prairies the first corn breeding program was started in 1939 by S.B. Helgason at the Morden station, where he worked until 1947. His major contribution consisted of the development of inbreds used as parents of commercial hybrids. One was early enough to push corn from the southern counties of Ontario to the area centered around Guelph and also made corn production feasible in many areas of Quebec. W.A. Russell succeeded Helgason in 1947 for 5 years, when J. Giesbrecht replaced him. The first inbred released by Giesbrecht was CM7. Its breeding was started by Helgason, continued under Russell, and finally evaluated, selected, and released by Giesbrecht. It has been a parent of many of the early maturing, fast-drying hybrids grown on the northern edge of the corn belt and has been widely used in Europe. Other inbreds from the Morden program have been used extensively because of their adaptation to the northern edge of the corn belt.

Breeders and seed companies have worked closely together in Canada through several extraordinary organizations. The time required from the initial

233

selection of the first segregating generation until the licensing of a new hybrid is 8–10 years, with extensive testing. In 1937 the Ontario Corn Committee was started to ensure that the Ontario farmer was protected from the marketing of hybrids unsuitable for Ontario conditions. The Ontario Corn Committee in 1937, the Manitoba Corn Committee in 1957, and, more recently, corn committees in nearly all provinces supervise the testing, recommend the licensing of varieties, and are financially independent.

The phenomenal increase in corn production since 1940 has resulted largely from the efforts of breeders but also from the work of soil chemists, plant pathologists, entomologists, agronomists, weed scientists, and economists who have assembled strong corn production systems for each major corn-growing area of Canada. Scientific teams have returned millions of dollars to the economy by producing high-yielding, short-season, cool-tolerant corn hybrids and precisely describing how best to grow them. Shelled corn for human, animal, and industrial use is now produced commercially in all Canadian provinces except Newfoundland.

FORAGE CROPS

In Canada, cultivated perennial forages are seeded on 9 million ha. About one-half of this area is harvested and the forages are stored as hay, silage, or pellets, whereas the rest remains in pasture as feed for grazing ruminant animals. Cultivated forages are distributed fairly uniformly across Canada, with one-half in the Prairie Provinces and one-half in Eastern Canada and British Columbia. In addition to cultivated forages, the Prairie Provinces use 17 million ha of native grasslands for pasture.

In 1886 a few species of grasses and legumes were cultivated in Eastern Canada to improve the soil and to feed cattle and sheep. Director Wm. Saunders realized that eastern farmers required adapted varieties of a wider range of grasses and legumes to support an expanding animal industry. The ranching industry was well established on the prairies. Saunders believed the rapidly developing west needed more than cereals and native range to establish a stable agriculture—it demanded cold-tolerant, drought-resistant, high-yielding forages to maintain soil fertility and to stabilize and expand the forage base.

Saunders wrote to his counterparts throughout the northern hemisphere who responded with hundreds of seed samples for testing at the five original experimental farms. By 1900 he had identified which species might be useful. During the next 20 years he and his colleagues in all parts of Canada learned the best ways to grow the most promising species. In the 1920s Experimental Farm scientists made selections from within these variable populations, seeking improved varieties. Ten years later they started crossbreeding to provide a wider genetic base from which to make selections. Agronomic research kept pace, for, as varieties changed, so did their culture.

Legumes

Alfalfa, the most important legume grown in Canada, is thought to have originated in Iran and Turkey. It was brought to the United States in 1736, and introduced into Canada from France in 1871 but was not winter-hardy in many parts of this country. In 1857, a German immigrant, Wendelin Grimm, imported a sample of the hardy *Medicago media* to Minnesota, where it survived the winters. Seed from it entered Canada as Grimm alfalfa, where it formed the basis for our alfalfa production.

Experimental farms started to evaluate alfalfa before 1900. By 1904 F.S. Grisdale, Chief, Agriculture Division, reported successful overwintering in all parts of Canada with yields varying from 1.75 to 3.50 tons per acre (4–8 t/ha). F.T. Shutt, Chief, Chemistry Division, noted that as a forage alfalfa was rich in flesh-forming nutrients and as a fertilizer it increased soil nitrogen and humus.

The Dominion Agrostologist (see Appendix II) personally conducted research and breeding experiments in the Forage Plants Division (later called the Forage Crops Division) from 1912 until the mid-1930s when specialists were appointed. H.A. McLennan, the first specialist appointed, was responsible for the breeding of both alfalfa and clover. Others such as J.M. Armstrong, W.R. Childers, H. Baenziger, L. Dessureaux, M.A. Faris, and R. Michaud followed. Childers and Baenziger introduced Algonquin and Angus in 1972 and 1973, both of which are cold tolerant and resistant to bacterial wilt.

Michaud at Sainte-Foy, Quebec, in 1982, selected Apica from a long-established field of the variety Saranac. Apica is particularly suitable for Quebec and the Atlantic Provinces.

The Forage Crops Laboratory, Saskatoon, was organized in 1931 as an extension of the Forage Crops Division. L.E. Kirk, Professor of Field Husbandry, University of Saskatchewan, had earned an international reputation for his breeding and genetic studies with grasses and legumes. He was chosen by E.S. Archibald to replace G.P. McRostie who had retired as the Dominion Agrostologist. The agreement between the Honourable Robert Weir, Minister of Agriculture, and W.A. Murray, President of the University of Saskatchewan, was to establish a Forage Crops Laboratory on the university campus to do research on forage crops and to teach forage crop subjects. T.M. Stevenson, who had been one of Kirk's assistants, was appointed as the first officer in charge of the laboratory and given the responsibility for legume breeding. An improvement in the palatability of sweetclover and the production of alfalfa seed were two of his immediate concerns.

Sweetclover is a biennial legume native to temperate Europe and Asia. There are two species, *Melilotus alba*, white flowered, and *M. officinalis*, yellow flowered. Sweetclover's rapid growth and nitrogen-fixing capabilities make it an excellent green-manuring crop. Its high content of coumarin, an anticoalgulant, is a serious disadvantage to the use of sweetclover as a feed. At Brandon, Manitoba, G.F.H. Buckley selected the yellow-blossomed Erector variety from a mixed sample of sweetclover. In 1937, it became the first variety of this forage to

235

be licensed in Canada. Two years later, Buckley introduced Brandon Dwarf, a mutant he had found in a sample of common white sweetclover.

Coumarin develops in poorly cured sweetclover hay, which, if fed in large quantities to livestock, causes ruminant animals to hemorrhage. It was natural, then, that W.J. White, who had been employed by the Saskatchewan Department of Agriculture until 1934, should test seed of many sweetclover plants for their coumarin content. He found one from which, by 1938, he developed a homozygous strain thought to be low in coumarin. Unfortunately, in 1940, a newly developed test revealed the coumarin content of White's selection to be quite high. A new approach was needed.

Plant breeders knew that *M. dentata* was coumarin-free. With considerable difficulty, W.K. Smith, a Canadian working at the University of Wisconsin, crossed *M. dentata* with *M. alba* and obtained a few viable seeds. These he shared with White at Saskatoon. From these seeds White, followed by R.G. Savage, and subsequently J.E.R. Greenshields, developed a low-coumarin sweetclover named Cumino, which was licensed in 1957. An interesting sidelight is that jack rabbits and caragana blister beetles selected the low-coumarin plants as a preferred diet. Rabbits had to be excluded from the test site by closely woven fencing, but breeders used the caragana blister beetles to ferret the low-coumarin plants in segregating populations.

The yellow-flowered sweetclover is preferred by farmers, because it is leafier, has thinner stems, and matures 10–12 days earlier than Cumino of the white-flowered species. Consequently, in 1959, B.P. Goplen started a program to develop a coumarin-free, yellow-flowered variety. The two species would not cross. Goplen therefore transferred the low-coumarin gene from the white sweetclover into the yellow-flowered species by embryo culture. The resulting plants were subsequently crossbred with Yukon, the most widely grown, yellow-flowered sweetclover in Western Canada, and, after 12 generations of breeding requiring 22 years, Goplen produced Norgold. It is the world's first low-coumarin, yellow-flowered sweetclover. Seed was available to farmers in the fall of 1984.

The alfalfa-breeding program at Swift Current was initiated in 1935 by S.E. Clarke. Clarke promoted the culture of alfalfa mixed with crested wheatgrass for the dry ranges of southern Saskatchewan and Alberta. He realized that the standard variety of alfalfa, Ladak, from northern India, did not have the capacity to persist as long as crested wheatgrass, with which it was frequently planted. Clarke therefore began developing drought and cold-tolerant varieties from lines already evaluated by H.J. Kemp. Both Clarke and J.L. Bolton, who joined Clarke in 1936, left Swift Current before a variety was produced. It remained for D.H. Heinrichs, who assumed the direction of the alfalfa program upon Clarke's retirement in 1946, to capitalize on the earlier work. He retained Clarke's objectives to which he added a third—a creeping-rooted habit to provide for spreading under dry conditions. Heinrichs had the support of B.E. Murray (61), a cytologist, J.E. Troelson, an animal nutritionist, and F.G. Warder, a chemist, all of Swift Current.

The following winter-hardy, drought-resistant, creeping-rooted varieties bred to grow with grass in long-term forage stands were released from the Swift Current alfalfa program:

- Rambler 1955 — the world's first creeping-rooted alfalfa, tolerant to bacterial wilt, with good first-cut yield but poor recovery;
- Roamer 1966 — improved resistance to bacterial wilt;
- Drylander 1971 — improved drought resistance and hardiness;
- Rangelander 1978 — improved drought resistance and hardiness;
- Heinrichs 1981 — yield improved over Rambler by 10 percent (bred and released by Irvine and Lawrence (40) following the retirement of D.H. Heinrichs).

On the irrigated lands of southern Alberta and Saskatchewan, qualities other than drought-resistance are required. Recognizing this, M.R. Hanna at Lethbridge, Alberta, started in 1958 to breed a high-yielding, winter-hardy, bacterial-wilt-resistant variety. He had support from plant pathologists M.W. Cormack, J.B. Lebeau, and E.J. Hawn of the same research station. His first variety, Beaver, was bred in cooperation with Saskatoon and made available in 1961. Kane, which has the creeping-rooted characteristic of the Swift Current varieties, followed in 1971. Hanna's 1975 variety, Trek, is the first nematode-resistant variety with sufficient winterhardiness to produce satisfactorily under southern Alberta's irrigation agriculture.

Other legume forages include red and alsike clovers, sainfoin, bird's-foot trefoil, and cicer milkvetch. Varieties from each have been selected by scientists at one or more research stations. At Charlottetown since 1970, T.-M. Choo has identified several superior varieties of red clover, which are now being used in the Atlantic Provinces. H. Baenziger and W.B. Berkenkamp at Lacombe, Alberta, developed the red clover variety Norlac in 1973. It is earlier than the commonly grown variety, Altaswede. Following Baenziger's move to Ottawa he introduced in 1979, a winter-hardy, powdery-mildew-tolerant variety named Bytown. It is especially adapted to northern Ontario conditions.

At Beaverlodge, Alberta, C.R. Elliott, in 1961, selected Aurora from common landrace alsike clover that had been grown in the Peace River District for many years. Aurora was hardier and produced greater yields than the landrace sort. Five years later he had an improved selection, Dawn, that gave faster regrowth after cutting than Aurora.

Hanna, Cooke, and Goplen (30) of Lethbridge, Melfort, and Saskatoon, respectively, in 1969, selected Melrose sainfoin. Hanna, in 1980, released Nova. Each variety resulted from seed originating in Russia. Both were selected as possible replacements for alfalfa because sainfoin does not cause bloat.

The first Canadian cultivar of cicer milkvetch, Oxley, bred and tested at Lethbridge by Johnston, Smoliak, Hironaka, and Hanna (41) became available in 1971. Like sainfoin, cicer milkvetch is a legume not known to induce bloating in livestock. Oxley was selected from Russian seed introduced to the Range Station, Manyberries, Alberta, by S.E. Clarke in 1931. The selected variety produces more forage than the Russian material, although not as much as alfalfa, is winter-hardy, and is broadcast into stands of timothy in the eastern foothills of the Rocky Mountains to improve the quality of these pastures.

Alfalfa seed production

To produce adequate amounts of seed per hectare, alfalfa flowers must be visited by insects, usually bees. Although searching for nectar, bees pick up pollen from the flowers of one plant and deposit some on the stigma of others, thus causing cross-pollination. Honey bees normally approach the nectaries of alfalfa flowers from the side between the wing and the standard petals, failing to trip and pollinate them. The two genera of native bees that successfully pollinate alfalfa are the *Bombus* (bumble) and *Megachile* (alfalfa leafcutter) species. Bumble bees are social and form colonies, whereas alfalfa leafcutters nest alone.

The first Dominion Apiarist, F.W.L. Sladen, was an early student of pollinators for alfalfa (73). In 1916 he observed that alfalfa leafcutter bees visited alfalfa flowers near Medicine Hat and Lethbridge, Alberta, at the rate of 17 flowers per minute. He made similar observations the following year in the Okanagan Valley, British Columbia. He noted that alfalfa leafcutter bees were much more active than honey bees or bumble bees in pollinating alfalfa. It was not until 20 years later in 1940 that Salt (70) strongly recommended the use of wild bees for alfalfa seed production and the preservation of nesting habits.

Two comprehensive studies of alfalfa seed production were made by Knowles (45) in 1943 and Peck and Bolton (63) in 1946 at Saskatoon. They noted that when alfalfa seed crops were first grown in northern Saskatchewan about 1930, yields generally ranged between 220 and 550 kg/ha with some going as high as 1100 kg/ha. Hectarages were rapidly increased by clearing and cultivating wooded land. This, however, destroyed the nesting habitat of native bees, and by 1945, seed yields had dropped to an average of 80 kg/ha, with those of 300 kg being the exception. The reduction was entirely due to lack of pollinators. Alfalfa seed production became uneconomical and Canada imported seed, whereas before it had exported seed. Peck and Bolton again recommended preserving existing nesting sites and establishing new sites to increase the numbers of bumble and alfalfa leafcutter bees. They suggested boring holes in logs in which alfalfa leafcutter bees nested.

G.A. Hobbs and C.E. Lilly started their research on alfalfa pollinators at Lethbridge in the late 1940s. By 1954 they had learned much about the nesting habitats and biology of native alfalfa leafcutter species (37). Hobbs later studied both bumble and alfalfa leafcutter bees as pollinators for red clover and alfalfa. By 1957 he (36) concluded that in southern Alberta bumble bees were good pollinators of red clover but neither bumble nor alfalfa leafcutter bees were completely satisfactory pollinators for alfalfa because competing native food plants bloomed profusely in their nesting habitats, bee populations fluctuated widely from year to year, and alfalfa did not provide sufficient food at the right time. Hobbs concentrated on domesticating bumble bees. He placed 15-cm wooden cube hives underground, on the surface, or aboveground in or near aspen groves on the eastern slopes of the Rocky Mountains. He obtained from 0 to 60 percent occupancy over a 5-year period and learned a great deal about their selection of nesting sites, brood biology, and the factors limiting their populations.

While this work was going on in Canada a fortuitous accident occurred in the United States. During World War II a species commonly known as the European alfalfa leafcutter bee (*Megachile rotundata*) was inadvertently introduced into North Carolina. Somehow the bees were transported to Idaho in the mid-1950s. *M. rotundata*, although solitary, is also gregarious, as it has the inclination to live close to neighbors. Hobbs went to Idaho in 1962 and obtained a few live specimens. Along with his work on bumble bees he studied ways of domesticating these new alfalfa leafcutters. He developed the loose-cell system of alfalfa leafcutter bee management, which makes possible the large production of bees needed to pollinate alfalfa. The system enables easy removal of bee cells from laminated grooved nesting materials for storage over the winter without destroying the nesting material. It enables control of parasites and predators through various management procedures, including hive construction, incubation, and removal and tumbling of cells from the hives. The loose-cell system also makes efficient use of cold-storage and incubation facilities to synchronize bee emergence with the beginning of flower bloom. By so doing, it places the optimum number of bees onto the crop at the appropriate time to obtain a high seed set and an adequate return of viable bees for the following year.

The alfalfa leafcutter bee is an excellent pollinator of alfalfa. Without alfalfa leafcutter bees, seed yields are about 50 kg/ha. With alfalfa leafcutters, seed yields average 340 kg, and experienced managers can obtain up to 850 kg/ha. P. Pankiw at Beaverlodge in the Peace River and D.A. Cooke at Melfort in central Saskatchewan adapted the techniques to their cooler weather conditions. A whole new industry of raising alfalfa leafcutter bees was started and alfalfa seed production in Canada rose from 590 ha in 1955 to 14 240 ha in 1984. Hobbs received a merit award in 1969 for his outstanding accomplishments.

K.W. Richards replaced Hobbs in 1976 upon the latter's retirement and has advanced the work to the point where today alfalfa seed supplies are the highest in 30 years and almost meet the total Canadian market requirement. In 1981, 750 000 kg of seed were exported from Western Canada. This was the largest export of alfalfa seed in 20 years. Due to superior alfalfa leafcutter bee management, the quality of bees produced by Canadian beekeepers has made Canada the world's leading exporter of this valuable pollinator. In the past few years 150 million surplus alfalfa leafcutter bees have been exported annually to the United States, Argentine, the USSR, and several European countries. Richards' publication on the management of alfalfa leafcutter bees (67) is in its sixth revision.

This is an example of the development of a Canadian industry, brought about solely through the dedicated efforts of a few agricultural research scientists and the diligent work of alfalfa seed and leafcutter bee producers.

239

Grasses

Grown alone or with legumes, grass is used for hay, pasture, protection as a cover crop, turfs, and lawns. Nine genera, some represented by several species, are used in Canada to serve one or more of these purposes. Species from some genera are native to North America. Most domesticated species, however, have been imported.

Reference has already been made in Chapter 2 to the value of bromegrass, *Bromus inermis*, introduced into Canada from Russia by Wm. Saunders during the inaugural year of the Experimental Farm System. In 1949, Knowles and White (46) considered bromegrass to be the most important cultivated grass in Western Canada. Over the years, Canada developed bromegrass seed as an export commodity. In 1947, for instance, 3.4 million kg were produced. Various experiment stations in the United States had bred their own "southern" strains of bromegrass, some claiming them to be superior to the northern strains imported from Canada. Using data from nine Canadian experimental stations, Knowles and White demonstrated that there was little difference in yield between southern and northern strains. They did show, however, that southern strains were inferior in seed production and 2–4 days later in maturing than northern strains. In all other respects there were no differences. L.E. Kirk, initially with the University of Saskatchewan, and later with the Dominion Forage Crops Laboratory, Saskatoon, selected within the original population for a reduced creeping habit. In 1936 he introduced Parkland from the Forage Crops Laboratory, already having selected the variety Superior in 1920 while at the university. Both varieties had a reduced creeping habit. As a result, both yielded less forage than the common strain.

Following Kirk, a number of scientists including T.M. Stevenson, W.J. White, H.H. Horner, J.D. Smith, and finally R.P. Knowles have worked at Saskatoon either alone or in cooperation with others on bromegrass. Knowles produced three additional varieties, the most recent of which, Signal, has superior forage and seed yield.

At the Forage Crops Division, Ottawa, others, including R.M. MacVicar, W.R. Childers, and W.R. McElroy, have bred grasses, including bromegrass. Childers introduced the disease-resistant variety, Redpatch, and the high-yielding variety, Tempo, in 1964 and 1965.

Wheatgrass (*Agropyron* sp.) has seven forms—slender wheatgrass, first seeded in Canada at Virden, Manitoba, in 1885; crested wheatgrass, introduced by the University of Saskatchewan from Siberia in 1911; intermediate wheatgrass, introduced to Canada in the mid-1930s; tall wheatgrass, brought from Russia by the University of Saskatchewan; pubescent wheatgrass; and the two native forms, western and northern wheatgrass from Alberta and Saskatchewan.

The first variety of crested wheatgrass, Fairway, was developed by L.E. Kirk while he was at the University of Saskatchewan but released in 1932 when he became Dominion Agrostologist. Kirk had distributed seed of crested wheatgrass to experimental farms and stations on the prairies, where it proved to be superior to all other grasses under harsh, dry conditions, particularly at Manyberries in the southeastern corner of Alberta. Crested wheatgrass was one of the keys to establishing community pastures (see Chapter 13). Parkway, with a yield improvement over Fairway, and less subject to lodging, was released jointly by R.P. Knowles of Saskatoon and D.A Cooke of Melfort in 1969.

Intermediate wheatgrass, studied at Saskatoon and at Swift Current, has two improved varieties, both superior, under Canadian conditions, to those coming from the United States. The first, Chief, in 1961, was again contributed by Knowles at Saskatoon. At Swift Current, T. Lawrence had improved upon both winterhardiness and yield when he released Clarke (after S.E. Clarke) in 1980.

A number of varieties from the other species were produced at both Lethbridge and Swift Current. Each has its own niche in the ecology of the great Canadian rangelands. All play a part in feeding the large cow–calf herds.

Orchardgrass, *Dactylis glomerata*, received attention from R.M. MacVicar when he joined the Forage Crops Division in 1930. Seed had earlier been obtained from Manchuria. By mass selection, MacVicar in 1938 developed Hercules, a productive, early maturing variety that adapted generally to Eastern Canada. The next variety, with excellent winterhardiness, was introduced by W.R. Childers in 1963 and named Rideau. Kay and Juno came later. Juno is particularly useful on heavily grazed pasture because of its earliness, vigor, and leafy growth.

In addition to the work done at the Forage Crops Division, scientists at Lethbridge, Alberta, and Agassiz, British Columbia, bred orchardgrass varieties useful for the local soil and climatic conditions. R.W. Peake developed an early, vigorous, hardy breed for irrigated pastures in southern Alberta that he released as Chinook in 1959. D.K. Taylor and M.F. Clarke sought and found a high-yielding variety adapted to the coastal region of British Columbia. Sumas, made available in 1974, was the result.

The two major grasses in Canada used for turfs and lawns are bluegrass and fescue. In addition, both are found in seed mixtures for pastures. Kentucky bluegrass, *Poa pratensis*, is the main bluegrass species grown in Canada and, as its name suggests, is native to the United States. It was brought to Canada before 1700. Kentucky bluegrass is important for pasture in northern Manitoba, Saskatchewan, and Alberta. MacVicar, of the Forage Crops Division, selected Delta in 1938. It was a variety for pasture, combining well with white clover. In 1974, Lebeau and Hanna (54) at Lethbridge, introduced Banff, an outstandingly uniform, dwarf, persistently green-leafed variety selected from the Banff Springs Golf Course. Banff is particularly useful for golf courses because it is tolerant to close clipping. Dormie, a variety resistant to powdery mildew and snow mold, was introduced by Smith (74) at Saskatoon in 1978.

Creeping red fescue, *Festuca rubra*, is also grown in the northern parts of the four western provinces for pasture and seed. It produces a firm sod and is used in grass mixtures for finished lawns. Creeping red fescue first came into prominence in Canada because of its use for grassing airports during World

241

War II. The following varieties have been introduced from Experimental Farms and Research Branch:

- Duraturf 1943 — R.M. MacVicar, Ottawa, uniform, dense bottom growth;
- Boreal 1966 — C.R. Elliott, Beaverlodge, vigorous, uniform, creeping, high forage and seed yield;
- Duralawn 1971 — R.M. MacVicar, Ottawa, wide, deep-green leaves, resistant to leaf spot, strongly creeping.

Timothy, *Phleum pratense*, was among the first of the grasses evaluated by experimental farms. The species is used extensively as a hay for horses throughout Canada and in mixtures with alfalfa for hay and silage for cattle in Ontario, Quebec, and the Atlantic Provinces. The Forage Crops Division (later, part of the Ottawa Research Station) produced five varieties during the period 1947–1980. The most recent variety, Salvo, released by W.R. Childers, is early, with fast regrowth. The current array of varieties meets most of the needs of timothy growers.

Production systems

Today forage plants are generally considered to be grasses and legumes. When the Division of Forage Plants (later called the Division of Forage Crops) was organized under M.O. Malte in 1912, field roots such as turnips, mangels, carrots, and sugarbeets were widely used as fodder for cattle. These crops, which, with considerable labor, could be stored for the winter, have been replaced with corn and mixed forage that can be stored more conveniently in silos.

Much of the research on managing forage crops has been discussed in Chapter 13, because the culture of legumes and grasses is crucial to good soil husbandry. The Division of Forage Crops and most experimental farms and stations paid attention to their farm rotations, noting which species of legume or grass proved most effective for succeeding crops, which crops were most successful, and which seeding and cultural practices proved satisfactory. A body of knowledge for Canadian conditions and for various environments was therefore collected and made available to farmers through departmental bulletins.

A bulletin (23) by Grisdale, Shutt, and Fletcher, published in 1904, dealt with methods of producing alfalfa, curing it for hay, the feeding value of alfalfa, and many other management considerations. It was updated in 1942 by J.M. Armstrong, F.S. Nowosad, and P.O. Ripley and rewritten in 1982 by a group of scientists representing research stations from Sainte-Foy, Quebec, to Lethbridge, Alberta. In addition, at least 38 other bulletins have been prepared for farmers and describe various aspects of forage management for specific situations. Numerous technical papers describe the research upon which these bulletins are based.

In 1922 the Forage Plants Division started a series of tests to determine which species of grasses and legumes were most promising either singularly or in various combinations for hay in their first 2 years and for pasture in subsequent years. G.P. McRostie reported on 35 mixtures. He concluded that the addition of alfalfa to any mixture results in a marked increased yield of the seeding as a whole and, that within reasonable limits, the amount of seed from various species has little effect upon subsequent yields. He also noted that meadow fescue, redtop, and alsike clover are reasonably moisture tolerant, their use recommended in wet areas.

Native ranges, found in the four western provinces, have always been important to the beef industry. As with any renewable natural resource, they may be harvested, but with care. To determine the best methods, the Dominion Range Experimental Station, Manyberries, Alberta; the Range Substation, Stavely, Alberta; and the Range Experimental Station, Kamloops, British Columbia, were established from 1927 to 1935. Each studied the distribution of native grasses, the effect of different grazing practices on the vegetative cover, the improvement of ranges by reseeding, surface cultivating, fertilizing, and the growing of cultivated forage crops on the better soils within a range.

Annual Italian ryegrass, *Lolium multiflorum*, was identified in 1976 by H.T. Kunelius at Charlottetown, Prince Edward Island, as being useful in extending the grazing season. In a cooperative project, Calder (51) at Nappan, Nova Scotia, and Kunelius identified nitrogen fertilizer requirements appropriate for the species and also learned that harvesting at 4-week intervals (three or four harvests per season) produced a greater yield of dry matter than did longer intervals. This discovery represented a major breakthrough for grassland and potato farming. Now, maritime potato growers, who normally seed a cereal crop the year following potatoes, underseed with annual Italian ryegrass. Following harvest of the cereal, pasture for livestock during cool autumns and much needed additional organic matter for soil fertility are provided.

POTATOES

In Canada, between 100 000 and 125 000 ha of agricultural land have been cultivated for potato production since World War II. When the Experimental Farm Station Act was passed in 1886, 180 000 ha were cultivated for potatoes. Production reached a high of 330 000 ha in 1919, then constantly fell until 1955. The major drop in production occurred in Quebec and Ontario, with smaller but significant reductions in Nova Scotia and British Columbia. Contrarily, areas of production in New Brunswick and Prince Edward Island experienced increases.

Canada enjoys an active export market for seed potatoes because the produce is free from disease and because Agriculture Canada provides a well-respected seed potato certification program. The history of seed potato certification is detailed in Chapter 5.

J. Fletcher, Entomologist and Botanist, in 1887 reported an infestation of potatoes by Colorado potato beetles in Nova Scotia and Manitoba, flea beetles in British Columbia, blister beetles in the North-West Territories (Saskatchewan and Alberta), and wireworms in British Columbia. The first experiments with potatoes were in 1888 when Wm. Saunders tested 251 named varieties and grew

243

237 plants from hybridized true seed. By 1894 Saunders compared rotted manure, fresh manure, and various kinds of mineral and organic fertilizers with potatoes planted on land previously used to grow crops of wheat or barley. In both instances the manured plots yielded nearly double the unmanured check treatment and about 50 percent more than the fertilized plots. Rotted manure was best following wheat, whereas fresh manure produced the largest potato crop following barley.

By 1898 varieties of potatoes had been tested for a sufficient number of years at each of the five experimental farms for W.T. Macoun, Horticulturist, to recommend seven of them. They ranged in season from extra early (Burpee's Extra Early) to late (Late Puritan, American Wonder, and Rural Blush). The list also included recommended varieties for 28 other kinds of vegetables. Few variety names are common knowledge today, although Early Jersey Wakefield cabbage, Chantenay carrot, New York lettuce, Yellow Globe Danvers onion, Improved Stratagem pease [sic], French Breakfast radish, Victoria rhubarb, and Hubbard squash may be recognized by some readers.

In 1895 experiments were started to determine the best planting distance, planting depth, and planting date. Without the help of replication or statistical analyses, Macoun concluded that 30 cm within rows 76 cm apart was the best planting distance; potatoes set 2.5 cm deep, provided weed control was possible without harrowing, was the best planting depth; and the best planting date depended in large measure on the first fall frost, which was unpredictable. With only a 3-year sample, he tentatively concluded 24 June to be the latest date to plant in Ottawa, although in one year a 7 July planting produced a marketable crop. By 1910 Macoun demonstrated the importance of selecting seed potatoes from crops lifted before they matured to produce optimum returns the following year. He also selected within varieties seed potatoes from strong, vigorous plants that were only mildly infected with late blight. In some varieties these seed potatoes were shown to produce up to 50 percent more crop when used as seed the following year. Similar experiments were conducted until 1921. Of 39 varieties tested, the yields of 27 benefited from using immature seed, some as much as doubling the yield derived from mature seed.

At Nappan, Nova Scotia, W.W. Baird compared the use of sprouted seed with dormant seed. Six years' data from 1924 to 1930 showed dormant seed to yield 20 percent more crop than sprouted seed. During the same period he found early planting (22 May) to outyield late planting (6 June) by 15 percent. At the same time C.F. Bailey at Fredericton, New Brunswick, in cooperation with the Fredericton Laboratory of Plant Pathology, tuber-indexed seed of the four main potato varieties in a greenhouse, then planted those that were disease-free in tuber units with the objective of providing foundation seed of each variety to seed growers.

Pascal Fortier, Superintendent, La Ferme, Quebec, in 1930, confirmed Macoun's findings that 30 cm between sets in 76-cm row spacing yielded more with Green Mountain potatoes than wider spacings. With Irish Cobbler, however, 35 cm between sets in 90-cm row spacing was the best. Fortier, too, determined that unsprouted seed was better than sprouted seed.

Many experimental stations compared rates and kinds of fertilizers. H.F. Murwin, Superintendent, Harrow, Ontario, demonstrated that 27 kg of ammonium sulfate produced slightly more potatoes than 27 kg of dried-blood fertilizer. Both yielded 30 percent more crop than with no nitrogen. Phosphoric acid in the form of superphosphate at a ratio of only one-half the amount of nitrogen was needed, whereas potash at about 66 percent of nitrogen was required. These conclusions were based upon 4 years' data (1927–1930) from single, one-tenth acre (400 m^2) plots.

The number of potato varieties marketed in Canada, often by traveling sales personnel representing United States seed firms, were in the hundreds. As a result, vegetable wholesalers, when buying potatoes at harvest, were frequently unable to assemble uniform carlots for shipment. To resolve this caotic situation, licensing of potato varieties began in 1923 when a number of varieties were included on a list of vegetables and herbs. Only in 1937 were potatoes officially registered in the varieties order of the Seeds Act.

The potato breeding program at the Experimental Station, Fredericton, was initiated in 1933. Because potatoes are particularly susceptible to attack from diseases and insects (see Chapter 18) a cooperative project was established with L.C. Young and H.T. Davies of the experimental station responsible for the horticultural aspects of the program, including making the crosses, multiplying seedlings, and assessing the commercial value of potential varieties; D.J. Mac-Leod and J.L. Howatt of the Laboratory of Plant Pathology responsible for disease control; and R.P. Gorham of the Entomological Branch responsible for providing expertise.

Breeding any crop is complicated. Potatoes, in particular, require adequate greenhouse space, suitable field plots well isolated from all other potatoes (distances of kilometres), and properly controlled cold-storage facilities. For these reasons Fredericton was designated as the site for the national potato breeding program. During the first few years the two main objectives were to breed varieties resistant to mild mosaic and late blight and to meet the needs of various potato industries across Canada. Other specific characteristics have been added since. Because of the difficulty of keeping seedlings free from aphid-borne virus diseases, an isolation station well separated from other potato fields and on the Atlantic coast where aphid populations are usually low was soon established at Alma, New Brunswick.

During the first 15 years of the program nearly 150 000 potato seedlings were produced. More than one-half were bred for resistance to attack by the late blight-causing fungus, *Phytophthora infestans*. A wild species of potato, *Solanum demissum*, immune to late blight, generally was used as one parent. *S. demissum* possesses many undesirable horticultural characteristics, including long stolons (underground stems on which are borne the potato tubers), and tubers about the size of hazel nuts. Following the original cross between cultivated potatoes, *S. tuberosum* and *S. demissum*, a series of backcrosses was needed to recover the desirable horticultural characteristics. Each progeny plant used in the program had to be tested for immunity to attack from late blight in order to assure the correct genes were present. Nonetheless, good progress was made and varieties resistant to late blight were released. Similar methods were used to meet the other objectives of the program.

In March 1943, a potato policy committee of the Canada Department of Agriculture was instituted by H.T. Güssow, Dominion Plant Pathologist, and included C.F. Bailey, Superintendent, Experimental Station, Fredericton, and M.B. Davis, Dominion Horticulturist. National potato seedling and variety tests under the supervision of N.M. Parks were started in 1945 as a result of deliberations by the potato policy committee. These trials, at many experimental stations and universities, received from Fredericton seedlings that had been screened for resistance to one or more diseases. Horticulturists at each station selected those seedlings adapted to their regional conditions and, in cooperation with Fredericton, named and released new varieties. In the first 10 years, 25 000 seedlings and 100 named varieties were tested, from which two varieties, Canso and Keswick of Fredericton breeding were licensed for sale. Both are resistant to some strains of late blight. Six other named varieties were licensed from the same program.

The program had some difficulties. A few plant breeders, particularly at western stations, thought they should be given the opportunity of making their own crosses, screening the seedlings for diseases peculiar to their conditions, and releasing appropriate varieties. M.B. Davis, the Dominion Horticulturist, believed that it would be too expensive to duplicate the Fredericton facilities. He felt that since D.L. Young at Fredericton was prepared to make any crosses requested by cooperators, their needs should be met. To partially correct the problem, regional trials, differing in entries, were started by Parks in 1955. A prairie regional committee, followed by a maritime committee in 1956, and Ontario and British Columbia committees in 1957 helped speed the evaluation of seedlings and varieties for entry into the advanced trials. Recent research by Lynch, Tai, Young, and Schaalje (55) has conclusively demonstrated that pre-selection in the first generation at Fredericton improves the maturity, yield, and tuber traits of the same seedlings when grown in Alberta, without causing a reduction in their variability. Thus, Davis was correct in his decision to have all crosses and initial screening done at Fredericton.

Potato chips and french fries gained popularity in mid-century with the advent of fast foods. By 1960 about one-fifth of potatoes used for food were processed, many for chips and fries. Both products depend largely on their color for sales. Because color develops as a result of variety and storage conditions, Townsend and Hope (81) at Kentville, and Hyde and Shewfelt (39) at Morden studied the relationship to browning of reducing sugars, amino acids, and varieties. They found that the initial presence of reducing sugars was not a requirement for the browning reaction, provided sucrose was present and conditions were favorable for its hydrolysis. Actual frying of chips was needed to determine precise acceptability until 1968 when Chubey and Walkof (12) at Morden found they could duplicate the color of fried chips without sacrificing a potato tuber, although it did require cutting each tuber once. By pressing a filter paper disc between two cut surfaces of a tuber and then frying the moistened disc in fat produced an almost identical color ($R = +0.929$) to that obtained when an actual potato chip was fried.

W.A. Russell, in 1970, moved from Scott, Saskatchewan, to Morden, Manitoba, where he was in closer contact with other horticulturists. The following year he crossed the Netted Gem variety with a number of others. Netted Gem is notoriously difficult to use as a female parent because of poor seed development. To date, no varieties have been introduced from these crosses. Russell (69) did select an early, white-skinned, shallow-eyed seedling from Fredericton and introduced it as Carlton in 1982. Carlton is expected to replace the early, deep-eyed Warba.

The Research Station, Lethbridge, Alberta, in 1977, appointed D.R. Lynch. Until then, although always part of the national potato seedling and variety testing program, Lethbridge did not have a staff scientist whose prime responsibility was potatoes.

There were 59 potato varieties licensed for sale in Canada in 1982; 11 were from the Fredericton program. Only Belleisle, however, was in the top 10. Shepody, with good french fry qualities, was 11th and showing a rapid increase in popularity. It was bred by Young, Tarn, and Davies (87) of Fredericton.

FRUIT PRODUCTION

One of the first experimental programs initiated by William Saunders when he became director in 1886 was the development of new varieties of fruit. Indeed, he had started fruit experimentation as a horticultural hobby in London, Ontario, and, as recorded in Part I, he moved much of his collection of raspberries and gooseberries to Ottawa. Only a few of the contributions made by the Research Branch in fruit breeding and production are mentioned here, but they illustrate how research is conducted and the time needed to complete it.

Even before 1900, the year in which Mendel's Laws of Inheritance were rediscovered, Saunders recognized that new varieties resulted from crossing two plants within the same species, the progeny of which might combine some of the desirable characteristics of each parent. By 1894 he was the first in the world to make an interspecific cross of apple by hybridizing *Malus baccata* with the common apple, *M. pumila*, to develop a hardier sort. He again crossed (backcrossed) these hybrids, including Osman and Columbia, with the common apple and produced Piotosh, Rosilda, Trail, and other acceptable varieties, which he tested at both Brandon and Indian Head.

Saunders published his results in 1911 (72), 24 years after receiving his first seeds from the Royal Botanic Gardens, St. Petersburg, Russia, in 1887. He concluded that with persistent efforts a number of apple varieties would be available to settlers in the north where ordinary apples could not be grown. He was correct. Starting from this base, plant breeders at Ottawa, Morden, and elsewhere have developed many good-quality, hardy apples and apple crabs. Following Saunders' retirement in 1911 W.T. Macoun, Dominion Horticulturist, continued the apple-breeding work and by 1926 had named 174 varieties, the

247

most noteworthy of which are Lobo, Melba, and Joyce. All were open-pollinated
seedlings of McIntosh. D.S. Blair at Ottawa continued to breed and select early,
high-quality apples from seedlings of controlled crosses. He used Melba as one
parent and frequently found hardy trees that bore fruit earlier than Melba.
Spangelo (75) describes Quinte, Ranger, and Caravel released between 1942
and 1954 as being winter-hardy with early, high-quality fruit.

The Horticulture Division studied the hardiness of rootstocks to which named
varieties of apples, pears, and plums could be budded. By 1921 Macoun recommended hardy rootstock from seedlings of Russian crab apple, Chinese pear, and
native plum for the three major fruit types. As fruit culture became more intensified,
variation in tree response to untested seedling rootstocks became costly. The East
Malling Research Station, England, had developed a series of vegetatively propagated rootstocks, the use of which gave predictable tree sizes. Most, however, were
tender under Canadian conditions. In 1961 L.P.S. Spangelo started to develop a
series of hardy, hybrid, seedling apple rootstocks suitable for Canadian orchards. By
1971 he had six (76) proven hardy stocks that were also tolerant of latent viruses
commonly found in commercial apple varieties.

As other experimental stations became established, particularly ones specializing in horticultural crops such as Kentville, Nova Scotia; Morden, Manitoba;
and Summerland, British Columbia; additional fruit-breeding programs were
commenced. At Kentville, in 1928, W.H. Brittain started a series of pollination
studies on apples to determine which varieties should be interplanted. He made
hundreds of pollinations between known varieties and, as a measure of their
effectiveness, he counted the number of seeds in the resulting fruit. By 1934,
through planting the resulting seed, he had established an orchard of 31 168
trees, one of the largest plantings of its kind in the world. Brittain and his
successors R.P. Longley and C.J. Bishop were looking for a late, high-quality, red
apple that would store well. Later they also selected for resistance to scab with
the cooperation of J.F. Hockey, officer in charge of the Plant Pathology Laboratory, Kentville. From this program came two varieties: Nova Easy Grow in 1971,
and Novamac in 1978.

The driving force at Morden, the principal horticultural experimental station
on the Great Plains, was its second superintendent, W.R. Leslie. E.M. Straight,
the first superintendent, moved to Saanichton 3 years after his appointment at
Morden, leaving the development of the station to his successor. Following
7 years in forestry and horticulture at various Canadian stations, Leslie was
appointed in 1921 to Morden where he remained until his retirement 35 years
later. He was an extroverted, knowledgeable, enthusiastic horticulturist who,
together with his staff, introduced in 1929 the Mantet summer apple, an open-
pollinated seedling of Tetofsky. Mantet gained wide popularity throughout the
prairies and the Maritime Provinces as well as in the northern United States.
Eighteen additional apple varieties as well as other fruits, vegetables, and
ornamentals were introduced before Leslie retired.

Leslie was in large measure responsible for developing the Prairie Cooperative Fruit Breeding Project, with Morden as the lead station, the universities of

Saskatchewan and Alberta as the main test orchards, and with other orchards at Beaverlodge, Lacombe, Brooks (Alberta Department of Agriculture), Scott, and Melfort. Since 1949, 130 000 apple and crab apple seedlings, along with 13 000 plum and 5000 cherry seedlings have been distributed to cooperating orchards. From this population 53 apple seedlings were chosen for a second test between 1968 and 1974. The varieties Noret, Norcue, and Norhey (32) were released in 1976 and Norland, Parkland, and Westland in 1979. The Prairie Cooperative Fruit Breeding Project is still active. Final decisions regarding the release of a few more varieties are pending. When these decisions are made the project will close, having achieved its goals.

R.C. Palmer and A.J. Mann joined the staff of the Experimental Station, Summerland, British Columbia, in 1921, 7 years after the station was opened by R.H. Helmer. Palmer, the pomologist, bred apples, apricots, peaches, and cherries. Mann studied the possibility of growing tobacco in the Okanagan. Such a venture proved uneconomical, so upon Palmer's appointment as superintendent, Mann became the fruit breeder. Palmer crossed McIntosh with Newtown in 1926 and planted the resulting seedlings in 1928. By 1932 these seedlings had matured sufficiently to bear fruit. In 1934 Mann (56) selected one seedling because of the robust nature of the tree and the outstanding color and flavor of the fruit. He named the variety Spartan in 1936. Since then, Spartan has been planted widely throughout the north temperate zone. It has received worldwide acceptance, and several times has been voted the best-tasting apple at English fruit shows.

In 1948 Bishop (4) at Kentville introduced an innovative technique to improve fruit yield—the use of X rays to produce fruit tree mutations. Mutations are sudden genetic changes starting in a single cell of a growing shoot tip. Natural mutations are rare, and when they do occur, very few result in an improvement over the original variety. Until 1948 only a Swedish research station had successfully used X rays. Bishop found the resulting fruit to be of giant size, irregular shape, and with more or less intensity of color than the original variety. He concluded that irradiation had potential in fruit-breeding programs. After further research, he and Aalders (5) determined that neutron irradiation was even more effective than X rays. Their studies have led to many superior mutations in the variety Cortland. These are being cleared of virus diseases in preparation for release.

Later, at Summerland, K.O. Lapins (who replaced Mann) had remarkable success in producing new commercial varieties of several kinds of tree fruits, using gamma radiation from a cobalt 60 source. He started his work in 1956 and by 1963 had produced (52, 53) a compact Lambert cherry, which was released commercially in 1969. Dwarf, or compact, fruit trees are advantageous to orchardists because of their ease of pruning, spraying, and harvesting. They can be planted more densely than standard-sized trees. Although apple trees had been dwarfed, using selected rootstock and spur-type natural mutants, no dwarf cherry varieties had previously been found. Since cherry is the largest of the deciduous fruit trees, Lapins' discovery was particularly significant.

Another sweet cherry, Stella, was released by Lapins at the same time as his compact Lambert. Stella is the first sweet cherry to have the remarkable characteristic of being self-fertile. For Stella to set fruit, there is no need for pollinator trees, although Stella itself is a universal pollinator. Stella is the result of a cross between Lambert and John Innes Seedling 2420. The self-fertile parent seedling was produced at the John Innes Institute in England by irradiating developing pollen with X rays.

Many other kinds of fruits such as strawberries and raspberries have been produced by scientists of the Research Branch. Strawberries have received the greatest attention because of their commercial value and their popularity with home gardeners. Initially, breeders sought varieties with improved yields and fruit characteristics. Following Saunders, A.W.S. Hunter and L.P.S. Spangelo continued and expanded the research. From 1949 to 1953, approximately 300 selections were propagated and tested under commercial conditions. From this population they selected nine for commercial test in 1954, following which Cavalier, Grenadier, Guardsman, and Redcoat were named in 1957. Redcoat was still considered a valuable variety 25 years later.

Immediately following World War II a soil-borne root rot disease called red stele caused damage to west coast strawberry plants. Horticulturists at the Agassiz Experimental Farm collected varieties from around the world that were resistant to the disease and used them as parents in their breeding program. In 1956 J.A. Freeman (20) released the variety Agassiz, which was partially resistant to the fungus. Agassiz could be grown in the Fraser Valley and in colder climates because of its hardiness. The variety was replaced by other more productive ones from the State of Washington until 1971 when H.A. Daubeny, also of the Agassiz Experimental Farm, released Totem (17). Within 5 years, Totem had gained the dominant position in British Columbia with its excellent fruit, plant vigor, winterhardiness, and resistance to red stele, powdery mildew, and viruses. Totem received the Outstanding Cultivar Award from the Canadian Society for Horticultural Science in 1984. Daubeny (15), released a second, improved variety, Tyee, in 1980 from the Vancouver Research Station. In addition to breeding for improved varieties, Daubeny worked with virologists in determining ways in which strawberries inherit a tolerance to complex virus diseases.

During this period R.E. Harris at Beaverlodge was searching for a hardy, large-fruited strawberry to replace Senator Dunlap, a variety that was only moderately hardy and had soft fruit. In 1957 he crossed Glenheart with Cheyenne and 7 years later named the resulting variety Protem (31); it was hardier than Dunlap and produced firm, high-quality fruit. Protem has become the standard variety in the Peace River District and is grown throughout the Prairie Provinces.

In the meantime, D.L. Craig at Kentville introduced a number of strawberry varieties suitable for the Maritime Provinces. Probably his most successful introduction was his last variety, Kent, which has yielded a record 40 t/ha. Craig was instrumental in developing a certification program to provide virus-free strawberry plants for commercial growers in the Maritime Provinces as had been done in Ontario and British Columbia.

Since Saunders' work, the major advancements to the breeding of red raspberries have been achieved by A.W.S. Hunter and L.P.S. Spangelo of Ottawa, and H.A. Daubeny of Agassiz and Vancouver. Hunter, followed by Spangelo, made particular efforts to breed varieties resistant to virus and fungus diseases. Some of the varieties they released from Ottawa, such as Madawaska, seemed to be resistant to mosaic, whereas all but the variety Ottawa were resistant to powdery mildew. As with his strawberry work, Daubeny has contributed to the improvement of varieties as well as to the knowledge of ways in which raspberries resist pests. He (16) is now using disease- and insect-resistant characteristics from the indigenous North American species *Rubus strigosus* and incorporating these into the cultivated raspberry, *R. idaeus*. He has also selected plants that shortened the ripening season of their fruit and produced berries which easily shake from their receptacles, as he searched for machine-harvestable varieties. In 1973 Daubeny released Haida and in 1978 he released Chilcotin, Skeena, and Nootka—all of which have excellent fruit characteristics, fewer thorns, resistance to cane and fruit diseases, and abundant crops. Daubeny's program has increased the efficiency of raspberry production in British Columbia. Additional varieties that will be available shortly are expected to diversify the industry.

Two rarer kinds of fruit deserve note. Inhabitants of the Canadian prairies look forward to summer harvests of saskatoon berries. They are not true berries but are pomes, as are apples. They make delicious fresh desserts and tasty pies. For many years R.E. Harris at Beaverlodge collected seeds from the largest fruit he could find, planted them, and again selected those bushes that yielded large fruits. By this process he developed varieties that are now parent stock in commercial plantations. Harris tested the fruit in cooperation with the Summerland fruit-processing laboratory and found it made an appealing canned or frozen product. He also made attractively colored wine and juice. The possibility of saskatoon berries becoming commercially viable and a part of the Canadian diet as a result of Harris' work is real.

A totally new Canadian fruit is the kiwi. In 1976 Henri Bailley, a gardener on the Saanich Peninsula, noted that the climate and soil in that area of British Columbia were similar to those of his native France, where kiwi fruit grows successfully. Bailley was encouraged by the knowledge that some gardeners in Victoria had grown kiwi plants as ornamentals. He grew some from seed and gave them to the Saanichton Research Station.

Kiwi fruit was introduced (57) into New Zealand from China in 1906 and into California in 1935. It was first named after the Chinese goose and known as Chinese gooseberry, but later it was named after another bird, the kiwi, native to New Zealand. The fruit resembles an overgrown gooseberry with a fuzzy, brown skin and soft, completely edible, attractive, light-green flesh, which tastes like a cross between a strawberry and a gooseberry. Botanically, the kiwi plant is *Actinidia chinensis*, which falls between the cola and the tea families. Kiwi fruit is usually eaten fresh, but it also makes delicious jam and preserves. Its vitamin C content is several times that of an orange; thus, with its slight acidity, it makes an

251

excellent breakfast juice. If cooled immediately after harvest, it will store up to 6 months without loss of quality.

The plants that grew at Saanichton in 1976 resembled grapevines, so they were pruned and trained over supports like grapes. The fruit of the grape is borne on the current year's wood; therefore, after each vine has attained its desirable size, all but a few buds of the previous year's wood are removed. When this method of pruning was used on the kiwi plant, no flowers or fruit were produced. Then, after several winters when the plants survived temperatures as low as −12°C, it was learned that kiwi plants flower and fruit on the previous year's wood, and that the plants are dioecious (separate male and female plants). Most of the plants grown from seed germinated by Bailley were male and thus produced no fruit. J.M. Molnar, director of the research station, imported named varieties from California and New Zealand. The imports started to bear fruit in 1984 but were no better than one of those already in fruit at the station.

Meanwhile, Bailley moved to Halifax, Nova Scotia. Still interested in kiwi fruit, he persuaded the director of the Kentville Research Station, G.M. Weaver, to grow the fruit. Plantings have been made on the station and also by a cooperating grower who has good winter protection through the use of a plastic greenhouse. It is too early to make predictions for the success of kiwi fruit in the colder conditions of the Annapolis Valley, but there is promise for this new crop.

Saanichton has demonstrated that kiwi fruit can be grown on southern Vancouver Island. The next step is to learn if it can become economically important. Molnar foresees the day when the soils, from which potatoes and tomatoes are excluded because of infestation by golden nematode, will be used to grow kiwi plants.

ORNAMENTALS

Beautifying Canadian gardens has always been an objective of Experimental Farms. William Saunders, while still a pharmacist in London, Ontario, collected named varieties and grew seedlings of ornamental plants and fruit such as gooseberries, raspberries, and tree fruits. He brought many of his varieties and seedlings to Ottawa shortly after 1886 where he planted, propagated, and distributed them to the other four experimental farms for testing under Maritime, Great Plains, and British Columbia conditions.

The five experimental farms took pride in their hedge trials (some still stand) and introductions of flowers and shrubs from around the world. Their lovely grounds became favorite places of relaxation and celebration for local citizens and tourists. This tradition of hospitality is perpetuated at the Central Experimental Farm and at many research stations, as they have become oases in the midst of a bustling city or a prairie landscape.

Lilies

The first person at Experimental Farms to concentrate on breeding ornamental plants was Isabella Preston. She was born in Lancaster, England, in 1881, and emigrated to Canada in 1912 after graduating from a horticultural college in Kent. She was employed to make a special study of plant breeding at the Ontario Agricultural College, Guelph. Between 1912 and 1920 Preston bred various vegetables, fruits, and flowering plants, including garden lilies, and gained international recognition by introducing the variety George C. Creelman, *Lilium princeps*.

In 1920, Preston joined the staff of the Division of Horticulture, Central Experimental Farm, where she continued her studies with lilies under the direction of the Dominion Horticulturist, W.T. Macoun. Among many hardy varieties resulting from her crosses was a group designated the Stenographer series because each variety was named for one of seven stenographers at the Central Experimental Farm. The varieties in the series were particularly hardy (11) and their dark red or orange flowers that faced outward and upward, were a new characteristic. The Stenographer series resulted from a cross made in 1929 between *Lilium davidii* var. *willmottiae* and a seedling of *L. dauricum*. Each of the seven varieties was easily propagated from stem bulblets and bulb scales. Five won Awards of Merit from the Royal Horticultural Society, London, and/or the Massachusetts Horticultural Society, Boston, and were widely distributed commercially.

Others in experimental farms who bred lilies were R.C. Palmer, superintendent at Summerland, and W.R. Leslie, superintendent at Morden. Both achieved success in developing varieties suitable for their own regions.

Roses

Roses have been cultivated in Canada for nearly 300 years (78), the first varieties having been imported from France. Only in the past century, however, has anyone in Canada developed varieties hardy enough to withstand our extreme climatic conditions. Most world introductions were either hybrid tea or floribunda roses, both requiring winter protection in all parts of Canada except in the more temperate climates of coastal British Columbia and southwestern Ontario. More than 150 varieties have been introduced by Canadian breeders since the 1890s; only one-quarter are of the tender sort, the remainder are hardy.

Wm. Saunders was the first rose breeder at Experimental Farms. In 1900 he bred Agnes, a hybrid of *Rosa rugosa* × *R. foetida* 'Persiana', which was introduced into the trade in 1922. It is still available today. Agnes is very hardy and has double, open, fragrant, pale amber flowers and nonrecurrent bloom. The following year Saunders released Grace, a cross between *R. rugosa* × *R. harisonii*. Grace is no longer available.

Several varieties of roses were also introduced by Isabella Preston of the Division of Horticulture. These included Cree, a hybrid of *R. rugosa* and *R. spinosissima hispida* in 1931; Conestoga, a variety of the native *R. blanda* in 1946, and Erie, from a complicated cross, in the same year. Later, H.F. Harp, at Morden, Manitoba, released Prairie Sailor, Prairie Wren, and Prairie Maid.

Prairie Youth was developed by W. Godfrey during the period 1946 to 1959. The four Morden varieties were from complex crosses involving *R. rugosa*, *R. altaica*, and a hybrid tea. In 1953 the Melfort, Saskatchewan, Experimental Station produced Cumberland, a hardy variety, whose parentage is unrecorded. H. Marshall at the Experimental Farm, Brandon, and later at the Research Station, Morden, introduced three varieties of floribunda roses that were judged by Svejda (78) to be quite hardy, namely, Assiniboine (1962), Cuthbert Grant (1967), and Adelaide Hoodless (1975).

In 1961 Felicitas Svejda at the Ottawa Research Station reactivated the Ottawa program of breeding completely winter-hardy, recurrent-flowering rose varieties. She also sought varieties resistant to black spot and mildew, and ones with good ornamental features of flower and shrub (79). Svejda was apprehensive regarding the possibility of combining the recurrent flowering habit with hardiness. Hardiness depends upon the maturation of rose wood early in the summer, whereas recurrent flowering depends upon the production of new wood on which the flowers are borne. However, a check of available varieties revealed that some did combine these two features. Thus encouraged, she proceeded.

For 8 years she used as parents diploid (two sets of individual chromosomes) varieties and species. As a result, in 1968 she introduced a vigorous, fragrant, pink-flowered, hardy, disease-resistant, free-flowering variety, Martin Frobisher, the first in her series of Canadian Explorer Roses. Four other diploid Canadian Explorers were released between 1976 and 1982. They were all hardy bush roses, they bloomed over a 9–14-week period, and were disease-resistant.

Svejda had found hardy offspring from tetraploid (plants with four rather than two chromosomes of the same kind) hybrids. A few of these were sterile, but not all; so she abandoned further breeding with diploids and focused on tetraploid varieties, and several species (autotetraploid offspring from the diploid variety Max Graf, which is a hybrid of *R. rugosa* × *R. wichuraiana* and *R. kordesii*) as parents. A new hardy, disease-resistant hybrid from *R. kordesii*, John Cabot, was registered in 1977. It is a climber with vigorous arching branches bearing a profusion of medium red, slightly fragrant clusters of 7–10 flowers. William Baffin, released in 1983, also a hybrid of *R. kordesii*, has more flowers per cluster but only a slight fragrance. Champlain, also from the tetraploid program, is a bush rose, and flowers throughout the summer and fall.

The persistance in Svejda's search for hardy, recurrent-flowering, disease-resistant, climbing and bush roses has provided Canadian gardens and parks with great beauty throughout the summer and fall months. She hopes to add attractive fall foliage colors and hips (seed pods) for beauty and bird feed during the winter months.

Lilacs

The common lilac, *Syringa vulgaris*, is not a reliable plant in many parts of Canada, where flower buds or blooms are frequently nipped by late spring frosts. In an effort to develop lilac varieties for problem areas, Preston, in 1920, became the first plant breeder to cross two hardy, late-blooming species *S. villosa* and *S. reflexa*. Numerous hardy seedlings that attracted worldwide attention resulted. In her honor, the new species was named *Syringa prestoniae*. *S. prestoniae* and other crosses made by Preston produced varieties that received awards from the Royal Horticultural Society. In all, Preston named 52 varieties of lilacs well adapted to Canadian climatic conditions. Many were released to the trade and marketed internationally. Donald Wyman, horticulturist, Arnold Arboretum, Boston, grew over 40 of the Preston varieties in the 1950s. Six years' comparative testing with varieties from other breeders, indicated 11 were strongly recommended for the Boston area. During the more than 10 years following formal testing (86), their performance has substantiated Wyman's ratings. In 1972 the President's Award of the International Lilac Society was presented to the Ottawa Research Station for developing *Syringa* cultivars and for educating the public on the beauty and use of lilacs. The society also presented its award of Outstanding Merit to A.R. Buckley, curator emeritus, Dominion Arboretum, Central Experimental Farm, Ottawa, for his 35 years in promoting the use of lilacs.

Rhododendrons

Evangeline, a Kentville rhododendron, was chosen as the dedication plant for the opening of the new Rhododendron Garden in Montreal in 1976. This honor was the culmination of a 25-year Kentville Research Station breeding program. The first rhododendron hybrids were planted in 1919 at Kentville by W.S. Blair to search for varieties suitable for maritime gardens. D.S. Craig expanded upon the variety trials in 1952 by gathering a collection of 80 species and 150 varieties from around the world. Twenty-seven of these, ranging from dwarf species for rock gardens to brilliantly colored, standard rhododendrons and azaleas, have proven themselves both useful and splendid in Canada's Atlantic region. He also inaugurated a breeding program of crossing several *Rhododendron* species to obtain compact, hardy, prolific-blooming varieties of which Evangeline is one. Some selections from the cross *R. catawbiense compactum* × *R. williamsianum* grew less than 0.76 m in both height and diameter in 9 years, their bell-shaped flowers suspended in loose clusters above the foliage.

At Saanichton, British Columbia, J.H. Crossley introduced dwarfed potted rhododendrons by treating the soil with growth retardants. The plants, when forced in a cool greenhouse, bloomed 10 days later and bore flowers 0.5 cm less in diameter than untreated plants. During the subsequent 3 years the treated plants were only half their normal height, even though repotted in untreated soil.

Rosybloom crab apples

A limited number of flowering trees and shrubs are hardy in the colder areas of Canada. Those with purplish-colored foliage such as some *Prunus* (plum) and

Fagus (beech) species are not hardy in Ottawa nor in the colder parts of Canada. Preston recognized the need in Canada for hardy trees with colored foliage and attractive flowers. In 1920 she crossed (64) *Malus pumila* 'Niedzwetzkyana' with *M. baccata*. The first parent has purplish foliage, deep-rose flowers, and medium-sized, deep-maroon or purplish-colored apples. The tree is not completely hardy and has an unattractive shape. The second parent grows into a large, round-headed tree with green foliage, masses of white flowers, and is extremely hardy.

Preston made many pollinations between the two parent species. Unfortunately, both parent plants were growing on lawns in public view. The large maroon apples on the *M. pumila* were too attractive to resist and all but four of the hand-pollinated apples at the treetop were pilfered. Preston carefully harvested the remaining fruit, including some that were naturally pollinated on the same tree. Judging from the trees, flowers, and fruit produced from the naturally pollinated seed, they were crosses with an *M. baccata* hybrid, which was growing nearby.

More than 100 seedlings were grown, those with green foliage being discarded after 5 years. By 1932 Preston had selected and named 15 of them after Canadian lakes. All the varieties bear colored flowers; many have colored leaves and fruit. W.T. Macoun, the Dominion Horticulturist, named the series rosybloom crab apples.

Chrysanthemums

The chrysanthemum is one of the most popular perennial fall flowers in Canadian gardens. Several varieties are hardy, easy to propagate, and to care for. Horticulturists at Lethbridge, Beaverlodge, Morden, Brandon, and Ottawa have developed dozens of varieties suitable for local use. The largest chrysanthemum display in the Research Branch has been exhibited each November since 1912 in the main greenhouses on the Central Experimental Farm. This artful showing of 2000 plants representing more than 100 varieties is open to the public for 2 weeks. Besides its display value, it also forms part of a program seeking greenhouse varieties that grow well in low temperatures (below 18°C), with the objective of conserving heat.

MINOR CROPS

Some minor crops, grown on relatively small areas in Canada, are valuable to farmers whose enterprises have benefited from crop research.

Fiber flax and hemp

Once the Division of Economic Fibre Production was established in 1918, with R.J. Hitchinson as its chief (G.G. Bramhill remained for only the 1st year), variety tests were started with flax at Charlottetown, Nappan, and Kapuskasing and with fibre hemp in the Prairie Provinces. The production of fiber flax decreased to about 400 ha following World War I, slowly increasing to 4000 ha by 1939. During World War II production peaked at 20 000 ha in 1942, then decreased rapidly to 3500 ha by 1947. It did not recover to compete with sythetic fibers.

Some hectarages of hemp for fiber were grown in Canada until its culture was prohibited in 1938 under the Narcotics Drug Act. Experimental work was resumed during World War II with variety, rates and dates of seeding, fertilizer, and retting trials until 1944 when the supply of cordage fiber improved. In 1977, a small plot of *Cannabis* varieties was grown on the Central Experimental Farm under Royal Canadian Mounted Police supervision for the use of scientists in their research at the Department of National Health and Welfare.

The Cereal Division bred flax prior to the formation of the Division of Economic Fibre Production. C.E. Saunders named two fiber varieties, Blanc and Damask, and one linseed variety, Diadem. Breeding continued and expanded in the Cereal Division, particularly when diseases such as wilt, rust, and pasmo (caused by a seed and soil-borne fungus) became epidemic about 1945. Most of the research relating to diseases was done by W.E. Sackston at the Laboratory of Plant Pathology, Winnipeg, and T.C. Vanterpool of the University of Saskatchewan, Saskatoon, both of whom made outstanding contributions toward the production of disease-resistant varieties.

Breeding of fiber flax by the Fiber Division started in 1930. Hutchinson made hundreds of selections from commercial fields in the De Beaujeu area of Quebec. Individual plants with strong, straight, but not unduly long straw and a good seed yield were brought to Ottawa for microscopic examination of their fiber count in each vascular bundle of the straw and number and weight of seed per bole. Seed from the best was kept for propagation and reselection. In cooperation with the Cereal Division, two new varieties, Lira Prince and Stormont Cirrus, were multiplied for foundation seed status in 1947 and released to seed growers.

Other research with fiber flax showed little economic advantage to applying fertilizer at Ottawa, Sainte-Anne-de-la-Pocatière, or at a number of illustration stations in flax-growing areas. Such was not the case with herbicides to control weed growth. In 1947 selective herbicides were new and some organic compounds proved to selectively control weeds in flax. In 1945 a pilot flax mill was completed at Portage la Prairie, Manitoba. E.M. MacKey, the assistant to Hutchinson in Ottawa, moved to Portage la Prairie as its officer in charge. Its principal research was intended to promote the production of fiber flax in Manitoba and the industrial use of linseed flax straw, primarily for making paper and wallboard. J.C. Woodward was its first resident chemist. The flax industry was superseded by chemical fibers, however, and the Portage la Prairie station changed its program to the study of vegetable crop production.

Linseed

Flax seed is a traditional crop of the Prairie Provinces. Because of its high quality for industrial purposes in paints, soaps, imitation leather, and salt-resistant coatings for cement surfaces (43), production in Canada has been relatively stable. Although production fell from a 1910 high of 1.4 million ha to a 1930–1950 low of 0.2 million ha, it recovered in 1970 to its 1910 level. Since then, 0.6 million ha of flax seed has been grown.

Since 1959, linseed breeding has been the responsibility of the Research Station, Morden, with related management studies being conducted at Brandon and Melfort. The principal characteristic sought in breeding flax for seed is high yield; oil content and quality are of secondary importance. Disease resistance and early maturity are part of the yield component. The first variety to come from the Morden program, Dufferin, was released by Kenaschuk (44) in 1975. It is resistant to three major races of rust and has a high yield and oil content, an improvement in both characteristics over the previous standard, Redwood 65 variety. The first cross in the program was made in 1966. By 1980 Dufferin occupied more than 40 percent of the flax area and thus had a considerable impact on the industry. By 1982 two other improved varieties, McGregor and NorLin, were released and, in 1984, NorMan was released; these varieties are expected to replace Dufferin.

Plant pathologists J.A. Hoes and E.E. Zimmer at Morden, working closely with plant breeders, have been instrumental in producing varieties resistant to flax rust. Resistance in flax is dominant and dependent upon six genes at different loci. To eliminate the disease, Hoes and Zimmer (38) believe that the complete replacement of susceptible varieties by resistant ones without any admixture of susceptible plants is needed. As long as there is any susceptible flax, new races of rust will continue to evolve by hybridization among the races present.

Breeders at Morden were careful not to release varieties with low oil content. Crushing plants encountered considerable variability in the pressure needed to extract oil from seed originating in different places. Therefore, Dedio and Dorrell (18) in their research at Morden found that the moisture content of pressed seed had a dramatic effect on the percentage of oil recovered. In dried seed with a moisture content of 3 percent, or less, more than 50 percent of the oil could be recovered in 2 minutes, whereas with moist seed of 8 percent moisture, only 35 percent of its oil could be recovered with a 2-minute press and no more than 45 percent after 5 minutes.

Gubbles (24) at Morden demonstrated the remarkable capacity of flax plants to compensate for low plant densities in the field by their correspondingly higher rates of basal branching. In addition, he found that loss from lodging following strong wind or heavy rain was only half that in the branched plants; thus high rates of seeding proved to be a disadvantage.

Tobacco

The first tobacco grown on experimental farms was at the request of the Honourable Auguste Real Angers, Minister of Agriculture, in 1893. From these experiments, J. Craig, Chief, Horticulture Division, was able to recommend suitable varieties and improved cultural methods for eastern Ontario and western Quebec areas. By 1906 the Minister of Agriculture, the Honourable S.A. Fisher, and his deputy, G.F. O'Halloran, recognized the economic possibilities of growing tobacco based on the results at Ottawa, and established a Tobacco Branch separate from the Experimental Farms Branch. The new branch, with its three tobacco experimental stations, returned to Experimental Farms as a division in 1912.

The early work by F. Charlan, the first tobacco specialist in the department, and later divisional chief, dealt with selecting suitable American varieties of burley and cigar tobacco and comparing their yields and quality, using different fertilizers. The same type of investigations were conducted at Harrow, Ontario, and at Saint-Césaire and Saint-Jacques, Quebec. In 1935 L.W. Koch of the St. Catharines plant pathology laboratory found that black root rot caused by the fungus *Thielaviopsis basicola* resulted in extensive damage to tobacco crops. The following year, Koch, with R.J. Haslam, also of the St. Catharines laboratory, discovered other diseases that caused damage to tobacco crops. As a result of these investigations, Koch moved to Harrow with Haslam to establish a plant pathology laboratory nearer the problem site. None of the varieties was resistant or tolerant to black root rot and therefore Haslam started a breeding program to produce one. By 1948 he was successful and released Delcrest, the first Canadian flue-cured tobacco variety. By 1951 Delcrest was grown on 70 percent of flue-cured tobacco hectarage, although American varieties were dominant in 65 percent of the industry as a whole until 1969. Further Canadian varieties followed and today they supply 95 percent of the flue-cured tobacco needs.

At the L'Assomption Experimental Station, cigar tobacco breeding started in 1964 with two objectives: to satisfy the growers by developing a high-yield, good-curing, disease-resistant variety, and to provide the manufacturers with a quality product that handles and sells well. To determine consumer reaction to a new variety of cigar tobacco, a volunteer smoker evaluation program was established in 1968. Every 2 weeks 250 smokers in Canada and Europe received two cigars that they were asked to evaluate. In 1982, the breeding program was phased out because of the considerable drop in production resulting from a low cost–benefit ratio at the growers' level. Lines from crosses made in 1964 are now undergoing field trials and may be licensed commercially in 1986.

The feasibility of flue-cured tobacco production on Prince Edward Island was first examined in 1959. Results from the initial 2 years were encouraging and to maximize the crop's potential, K.E. LeLacheur at the Research Station, Charlottetown, began studies on tobacco production. His early research included the identification of fertilizer requirements; evaluation of new cultivars from the breeding program at the Research Station, Delhi; control of nematodes with chemical nematicides; and the effect of different rates of zinc and boron on yield and quality.

In 1980, Islangold was selected from the variety evaluation trials and recommended as a superior cultivar for the Maritime region.

Recent studies by LeLacheur and W.J. Arsenault show that yield and quality are increased when tobacco is topped at the floral development stage as opposed to first, or full, flower. From studies resulting in a change in nitrogen fertility the use of more ammonia and less nitrate is recommended to increase yields.

The four major tobacco research stations have contributed much to the economic viability of the tobacco industry. Scientists have provided critical information on fertilizer use, the need for adequate crop rotations, the control of nematodes, fungi, viruses, and weeds, and the chemical content of tobacco leaves. Although other crops such as rye, corn, and beans are grown in rotation with tobacco on light, sandy soils, tobacco has been the main cash crop and source of income. As the health hazards of smoking became apparent and the yields of tobacco increased, less land was needed to satisfy Canadian demands. Other profitable crops for sandy soils were sought to replace tobacco. In 1970 Delhi tested the possibility of growing peanuts. This proved so promising that in 1974 the University of Guelph carried on the work for another 11 years at the Delhi Research Station; this research has led to the establishment of a new industry.

As a result of research findings of the branch, land used by the Canadian tobacco industry has expanded from just over 2000 ha in 1913 to 64 000 ha in 1960, with a high of 264 000 ha in 1982. Canada grows nearly all the tobacco it uses and exports 20 percent of its production. The total farm value of the crop is $250 million.

NORTHERN AGRICULTURE

"The north" means various things depending upon one's viewpoint. To produce food north of the 55th parallel in Eastern Canada is much more difficult than doing so in the west, where temperatures are moderated by the Japanese current of the Pacific Ocean. Northern agriculture, then, means producing food in subarctic or polar regions. Because of the climatic conditions of these regions, soils are cold and the growing season is short. The one redeeming feature, however, is the farther north, the longer the summer days.

As church missionaries moved north, particularly along the Mackenzie River valley, they reared livestock and gardened to provide their own food. With improved transportation throughout northern Canada, food from "the outside" became available; hence, many northern farmers and gardeners diverted their energies to other activities. Once Saunders had the first five experimental farms organized he sent seeds and plants to settlers across Canada, including those in the north. In return, they provided him with information concerning plant adaptability. The first tests started in 1905 in cooperation with the Royal North-West Mounted Police to determine the agricultural possibilities in the Yukon. Several grasses such as timothy, brome, and western ryegrass grew well, as did barley, oats, and potatoes.

The first permanent northern station was established by F.S. Lawrence at Fort Vermilion in 1907 (see Chapter 4), followed by W.D. Albright opening the Beaverlodge station in 1915. Also by 1915, experimental plots were extended to Fort Smith and Fort Resolution in the Northwest Territories and Grouard near Lesser Slave Lake in Alberta. Many crops matured, even though some seeding was not completed until 30 May. No additional stations were established until the end of World War II when Mile 1019 in the Yukon, Fort Simpson in MacKenzie District, and Fort Chimo in northern Quebec were opened. Each operated summer plots, the farthest north being at Inuvik, at the mouth of the Mackenzie River on the Arctic Ocean. During the nearly 25 years scientists at these stations studied food production under polar climates (the stations were closed about 1970), much was added to our knowledge of growing crops and vegetables in subarctic or polar regions. R.E. Harris and coworkers (33) identified the growing requirements of cool-season crops as being a minimum of 80 frost-free days, a growing season of 110 days when daily mean temperatures are at least 5.6°C, an accumulation of 1000, or more, degree-day heat units, and adequate precipitation. These requirements may be reduced slightly in the most northerly locations where there are more daylight hours or in areas where shelters and plastic covers can be used. Nitrogen and phosphorus fertilizers are usually needed.

All cool-season plants such as strawberries, potatoes, cabbages, carrots, and salad crops can be grown successfully. Early maturing cereals such as barley and, at some locations, winter rye and wheat survive. Current varieties of spring wheat do not mature before first fall frosts. Many of the cultivated grasses produce satisfactorily in the north, but since the pasture season is only from early June to late September adequate hay must be produced for the 8 months of winter feeding. Legumes such as alfalfa and sweetclover were found to increase hay yields at those locations where they survived. At Mile 1019, beef cattle could remain outside protected by wind shelters over a 17-year period, but they required feeding from October through May. Although one privately owned dairy farm was maintained in the Yukon for 50 years, competition from canned, powdered, and sterile milk now make dairy production uneconomical. Swine and poultry require additional heat and the economics of producing these meats under polar conditions are marginal.

In 1909 Saunders arranged with F. Moberley to grow grains and vegetables on his property at Whitefish River, near Lake Abitibi, Quebec. Moberley was disappointed with his results during the 1st year because the growing season was late and cold. However, subsequent years were more encouraging. In 1916 experimental stations were opened at Kapuskasing, Ontario, by J.P.S. Ballantyne and at La Ferme, Quebec, by P. Fortier to serve the clay belt of 20 million ha in northeastern Ontario and northwestern Quebec. These stations opened in response to the increased growth of agriculture resulting from the completion of the National Transcontinental Railway in 1913. Research on soils, fertilizers, all kinds of crops, and livestock gave struggling pioneer farmers definitive answers to their many questions. Farm populations increased until the early 1930s; they have since decreased through farm abandonment.

In 1956 the last northern station to open was located at Fort Chimo, Quebec, near the coast of Ungava Bay. Ten years of research by agronomists demonstrated that the growing seasons were too short for commercial farming to be viable. R.I. Hamilton and H. Gasser demonstrated that cool-season, garden vegetables could be grown when plastic tunnels and mulches were used to capture and retain heat and when adequate plant nutrients were applied. The practicality of growing vegetables at Fort Chimo, however, was restricted and the station was closed in 1965.

Food production in the Canadian arctic and subarctic regions is limited. Early settlers and agricultural scientists identified the kinds of plants and livestock that could be grown, and they learned many techniques to enhance food production in these areas. Perhaps eventually there will be a need to produce food under such difficult climatic conditions, but for now, more southerly parts of Canada offer brighter prospects.

262

References

1. Anderson, J.A.; Morrison, J.W. 1967. Canadian wheat research. *In* Proc. Can. Cent. Wheat Symp. 31–43. Modern Press, Saskatoon, Sask.
2. Andrews, J.E.; Grant, M.N. 1962. Note on Winalta winter wheat. Can. J. Plant Sci. 42:206–207.
3. Archibald, E.S. 1949. The story of Canadian wheat. Hilgendorf Memorial Address. 23 p.
4. Bishop, C.J. 1954. Mutations in apples induced by X-radiation. J. Hered. 45:99–104.
5. Bishop, C.J.; Aalders, L.E. 1955. A comparison of the morphological effects of thermal neutron and X irradiation of apple scions. Am. J. Bot. 42:618–623.
6. Boulter, G.S. 1983. The history and marketing of rapeseed oil in Canada. *In* Kramer, John K.G.; Sauer, Frank D.; Pigden, Wallace J. High and low erucic acid rapeseed oils. Academic Press, Canada. 582 p.
7. Bourdon, Mary B. 1984. Charlottetown Research Station 1909–1984. Agric. Can. Hist. Ser. 19. 82 p.
8. Buller, A.H. Reginald. 1919. Essays on wheat. Macmillan Co., N.Y. 339 p.
9. Buttery, B.R. 1970. Effects of variation in leaf area index on growth of maize and soybeans. Crop Sci. 10:9–13.
10. Buzzell, R.I.; Anderson, T.R. 1982. Plant loss of soybean cultivars to *Phytophthora megasperma* f.sp. *glycinea*. Plant Dis. Rep. 66:1146–1148.
11. Cameron, D.F. 1949. Ornamental plant breeding. Pages 139–153 *in* Div. Hort. Prog. Rep. 1934–1948.
12. Chubey, B.B.; Walkof, C. 1968. A rapid, non-destructive method for estimating chipping quality of potatoes. Can. J. Plant Sci. 48:636–637.
13. Cowan, P.R. 1952. The development of barley varieties in Canada. Agric. Inst. Rev. 7(Mar.)23–24.
14. Crawford, H.G.; Spencer, J.G. (undated). The control of the European corn borer. Crop Prot. Leafl. 16. Can. Dep. Agric., Ottawa. 3 p.
15. Daubeny, Hugh A. 1980. Tyee strawberry. Can. J. Plant Sci. 60:743–746.
16. Daubeny, Hugh A. 1980. Red raspberry cultivar development in British Columbia with special reference to pest response and germplasm exploitation. Acta Hortic. 112:59–67.
17. Daubeny, Hugh A. 1979. The strawberry cultivars of the Pacific Northwest. Fruit Var. J. 33:44–45.
18. Dedio, W.; Dorrell, D.G. 1977. Factors affecting the pressure extraction of oil from flaxseed. J. Am. Oil Chem. Soc. 54:313–315.
19. Downey, R.K. 1983. The origin and description of *Brassica* oilseed crops. *In* Kramer, John K.G.; Sauer, Frank D.; Pigden, Wallace J. High and low erucic acid rapeseed oils. Academic Press, Canada. 582 p.
20. Freeman, J.A. 1957. The Agassiz strawberry. Fruit Var. Hortic. Dig., Am. Pom. Soc. 11:45.
21. Fulcher, R.G.; Wood, P.J.; Yiu, S.H. 1984. Insights into food carbohydrates through fluorescence microscopy. Food Technol. 38(1):101–106.
22. Gridgeman, N.T. 1979. Biological sciences at the National Research Council of Canada: The early years to 1952. Wilfrid Laurier University Press, Canada. 153 p.
23. Grisdale, J.H.; Shutt, F.T.; Fletcher, J. 1904. Alfalfa or lucern: Its culture, use and value. Can. Dep. Agric. Bull. No. 46. 19 p.

263

24. Gubbles, G.H. 1978. Interaction of cultivar and seeding rate on various agronomic characteristics of flax. Can. J. Plant Sci. 58:303–309.

25. Haas, J.H.; Buzzell, R.I. 1976. New races 5 and 6 of *Phytophthora magasperma* var. *sojae* and differential reactions of soybean cultivars for races 1 to 6. Phytopathology 66:1361–1362.

26. Hamilton, D.G. 1951. Culm, crown and root development in oats as related to lodging. Sci. Agric. 31:286–315.

27. Hamilton, D.G.; Clark, R.V.; Hannah, A.E.; Loiselle, R. 1960. Reaction of barley varieties and selections to root rot and seedling blight incited by *Helminthosporium sativum* P.K. and B. Can. J. Plant Sci. 40:713–720.

28. Hamilton, D.G.; Keegan, R.; Lods, W.A. 1955. Barley in eastern Canada. Barley Improvement Institute, Winnipeg. Bull. 4. 30 p.

29. Hamilton, R.I. 1980. Corn and sorghum in Canada, 1985 and 1990. *In* Prairie production symposium, Advisory Committee to the Canadian Wheat Board, University of Saskatchewan, Saskatoon.

30. Hanna, M.R.; Cooke, D.A.; Goplen, B.P. 1970. Melrose sainfoin. Can. J. Plant Sci. 50:750–751.

31. Harris, R.E. 1965. Protem strawberry. Can. J. Plant Sci. 45:299.

32. Harris, R.E.; Knowles, R.H. 1976. Three new early apples for northwestern Canada: Cultivars Noret, Norcue, and Norhey. Can. J. Plant Sci. 56:997–998.

33. Harris, Robert E.; Carder, Alfred C.; Pringle, William L.; Hoyt, Paul B.; Faris, Donald G.; Pankiw, Peter. 1972. Farming potential of the Canadian northwest. Can. Dep. Agric. Publ. 1466. 26 p.

34. Hayes, Herbert Kendall. 1963. A professor's story of hybrid corn. Burgess Publishing Co., Minneapolis. 237 p.

35. Hobbs, E.H.; Mündel, H.-H. 1983. Water requirements of irrigated soybeans in southern Alberta. Can. J. Plant Sci. 63:855–860.

36. Hobbs, G.A. 1957. Alfalfa and red clover as sources of nectar and pollen for honey, bumble, and leaf-cutter bees (Hymenoptera: Apoidea). Can. Entomol. 89:230–235.

37. Hobbs, G.A.; Lilly, C.E. 1954. Ecology of species of *Megachile* Latreille in the mixed prairie region of southern Alberta with special reference to pollination of alfalfa. Ecology 35:453–462.

38. Hoes, J.A.; Zimmer, D.E. 1976. Ne North American races of flax rust. Probable products of natural hybridization. Plant Dis. Rep. 60:1010–1013.

39. Hyde, R.B.; Shewfelt, A.L. 1960. Measurement of chipping qualities in Manitoba-grown potatoes. Can. J. Plant Sci. 40:607–610.

40. Irvine, R.B.; Lawrence, T. 1982. Heinrichs alfalfa. Can. J. Plant Sci. 62:805–808.

41. Johnston, A.; Smoliak, S.; Hironaka, R.; Hanna, M.R. 1971. Oxley cicer milkvetch. Can. J. Plant Sci. 51:428–429.

42. Kaufmann, M.L. 1971. The random method of oat breeding for productivity. Can. J. Plant Sci. 51:13–16.

43. Kenaschuk, E.O. 1975. Flax breeding and genetics. *In* Oilseed and pulse crops in western Canada. Western Cooperative Fertilizers, Calgary. 703 p.

44. Kenaschuk, E.O. 1977. Dufferin flax. Can. J. Plant Sci. 57:977–978.

45. Knowles, R.P. 1943. The role of insects, weather conditions, and plant character in seed setting of alfalfa. Sci. Agric. 24:29–50.

46. Knowles, R.P.; White, W.J. 1949. The performance of southern strains of bromegrass in western Canada. Sci. Agric. 29:437–450.

47. Kramer, J.K.G.; Farnworth, E.R.; Thompson, B.K.; Corner, A.H.; Trenholm, H.L. 1982. Reduction of myocardial necrosis in male albino rats by manipulation of dietary fatty acid levels. Lipids 17:372–382.

48. Kramer, J.K.G.; Hulan, H.W.; Mahadevan, S.; Sauer, F.D.; Corner, A.H. 1975. *Brassica campestris* var. Span: II. Cardiopathogenicity of fractions isolated from Span rapeseed oil when fed to male rats. Lipids 10:511–516.

49. Kramer, J.K.G.; Mahadevan, S.; Hunt, J.R.; Sauer, F.D.; Corner, A.H.; Charlton, K.M. 1973. Growth rate, lipid composition, metabolism and myocardial lesions of rats fed rapeseed oils (*Brassica campestris* var. Arlo, Echo and Span, and *B. napus* var. Oro). J. Nutrition 103:1696–1708.

50. Kramer, K.G.; Sauer, Frank D.; Pigden, Wallace J. 1983. High and low erucic acid rapeseed oils. Academic Press, Canada. 582 p.

51. Kunelius, H.T.; Calder, F.W. 1978. Effects of rates of N and regrowth intervals on yields and quality of Italian ryegrass grown as a summer annual. Can. J. Plant Sci. 58:691–697.

52. Lapins, K. 1963. Note on compact mutants of Lambert cherry produced by ionizing radiation. Can. J. Plant Sci. 43:424–425.

53. Lapins, K.O.; Schmid, Hans. 1976. New fruits from Summerland British Columbia 1956–1974. Can. Dep. Agric. Publ. 1471. 19 p.

54. Lebeau, J.B.; Hanna, M.R. 1975. Banff Kentucky bluegrass. Can. J. Plant Sci. 55:1073–1074.

55. Lynch, D.R.; Tai, G.C.C.; Young, D.A.; Schaalje, G.B. 1983. Effect of selecting potato seedlings under maritime conditions on the performance of the first clonal generation under prairie conditions. Am. Potato J. 60:607–615.

56. Mann, A.J.; Keane, F.W.L. 1956. Tree fruits of the Okanagan Valley. Exp. Stn Summerland, Misc. Publ. 46 p.

57. McKeever, Harry P. 1983. Kiwifruit "find their feet" in western Canada climate. Western People, March issue, p.6.

58. Metcalfe, D.R.; Buchannon, K.W.; McDonald, W.C.; Reinbergs, E. 1970. Relationships between the "Jet" and "Milton" genes for resistance to loose smut and genes for resistance to other barley diseases. Can. J. Plant Sci. 50:423–427.

59. Metcalfe, D.R.; Johnston, W.H. 1963. Inheritance of loose smut resistance II. Inheritance of resistance in three barley varieties to races 1, 2, and 3 of *Ustilago nuda* (Jens.) Rostr. Can. J. Plant Sci. 42:390–396.

60. Morrison, J.W. 1960. Marquis wheat—a triumph of scientific endeavor. Agric. Hist. 34(4):183–188.

61. Murray, Beatrice M. 1957. The ontogeny of adventitious roots of creeping-rooted alfalfa. Can. J. Bot. 35:463–475.

62. Newman, L.H. 1939. New wheat creations and their significance. Can. Geogr. 18:208–216.

63. Peck, O.; Bolton, J.L. 1946. Alfalfa seed production in northern Saskatchewan as affected by bees, with a report on means of increasing the populations of native bees. Sci. Agric. 26:388–418.

64. Preston, Isabella. 1944. Rosybloom crabapples for northern gardens. J. N.Y. Bot. Gard. 45:169–174.

65. Rajhathy, Tibor; Thomas, Hugh. 1974. Cytogenetics of oats (*Avena* L.) Genet. Soc. Can., Misc. Publ. 2. 90 p.

66. Rennie, Robert J.; Dubetz, Steve. 1983. Soybeans as a N_2-fixing crop. Pages 76–80 *in* Research highlights, Agric. Can., Res. Stn, Lethbridge.

265

67. Richards, K.W. 1984. Alfalfa leafcutter bee management in Western Canada. Agric. Can. Publ. 1495. 53 p.

68. Roberts, D.W.A. 1984. Winterkill of fall-seeded winter wheat. Agric. Can. Tech. Bull. 1984-10E. 13 p.

69. Russell, W.A.; Dabbs, D.H.; Young, L.C.; Molnar, S.A. 1983. Carlton: A new early tablestock potato variety. Am. Potato J. 60:599–605.

70. Salt, R.W. 1940. Utilize wild bees in alfalfa seed production. Can. Dep. Agric., Field Crops Insect Invest. Mim. Circ. 266.

71. Sampson, D.R. 1972. Evaluation of nine oat varieties as parents in breeding for short stout straw with high grain yield using F_1, F_2, and F_3 bulked progenies. Can. J. Plant Sci. 52:21–28.

72. Saunders, Wm. 1911. Progress in the breeding of hardy apples for the Canadian northwest. Dom. Can., Dep. Agric. Bull. 68. 14 p.

73. Sladen, F.W.L. 1918. Pollination of alfalfa by bees of the genus *Megachile*. Table of Canadian species of the *Latimanus* group. Can. Entomol. 50:301–304.

74. Smith, J.D. 1978. Dormie Kentucky bluegrass. Can. J. Plant Sci. 58:291–292.

75. Spangelo, Lloyd P.S. 1964. New early apple varieties. Can. Dep. Agric. Publ. 1241. 4 p.

76. Spangelo, Lloyd P.S. 1971. Hybrid seedling rootstocks for apple. Can. Dep. Agric. Publ. 1431. 4 p.

77. Sterling, J.D.E.; MacLaren, R.B. 1967. Cabot oats. Can.J. Plant Sci. 47:715–716.

78. Svejda, Felicitas. 1974. Canadian hybridized roses. Can. Rose Annu. 1974:29–34. Toronto.

79. Svejda, Felicitas. 1984. Canadian explorer roses. Am. Rose Annu. 69:70–82.

80. Taylor, D.K. 1968. Fraser oats. Can. J. Plant Sci. 48:107.

81. Townsend, L.R.; Hope, G.W. 1960. Factors influencing the colour of potato chips. Can. J. Plant Sci. 40:58–64.

82. Voldeng, H.D.; Seitzer, J.F.; Donovan, L.S. 1982. Maple Presto soybeans. Can. J. Plant Sci. 62:501–503.

83. Voldeng, H.D.; Seitzer, J.F.; Hamilton, R.I. 1985. Maple Ridge soybean. Can. J. Plant Sci. 65: in press.

84. Voldeng, H.D.; Seitzer, J.F.; Hamilton, R.I. 1985. Maple Isle soybean. Can. J. Plant Sci. 65: in press.

85. Wilson, C.F. 1978. A century of Canadian grain. Western Producer Prairie Books, Saskatoon, Sask. 1137 p.

86. Wyman, Donald. 1970. The Preston lilacs. Am. Nurseryman, 132(Dec.):11–13.

87. Young, D.A.; Tarn, T.R.; Davies, H.T. 1983. Shepody: A long, smooth, white-skinned potato of medium maturity with excellent french fry quality. Am. Potato J. 60:109–113.

Chapter 18
Production Support

The discussions in this chapter involve entomology, both crop and livestock; plant pathology; weed agronomy; plant physiology; chemistry; and physics. Insects, arachnids, fungi, bacteria, viruses, and weeds have a marked influence on the production of food. Those researchers who devote their lives to understanding the biology of pests, and from this understanding learn how to control or contain them, are an integral part of our food system as necessary to it as are those who produce the final product.

GRASSHOPPER AND LOCUST CONTROL

The orthopteroid insects, such as grasshoppers and locusts, are the most destructive to agricultural crops. In Canada, as early as 1800, Alexander Henry reported dead grasshoppers piled on the shore of Lake Winnipeg to a depth of 20 cm. That plague continued until at least 1808. Norman Criddle, as quoted by Gibson (53), believed that locusts have always existed in the Prairie Provinces, posing severe problems one year in four. During the current century, general outbreaks have occurred with similar regularity.

Grasshoppers are voracious eaters. A grasshopper plague will completely devour a standing crop, binder twine, clothing, and even wooden handles on plows, shovels, and other tools. The ravenous appetite of the grasshopper extends to anything that contains organic matter, including human perspiration.

Outbreaks were reduced for two reasons. First, buffalo herds were eliminated. In years of high temperatures and low rainfall, buffalo overgrazed the plains and, with their weight and sharp hooves, loosened large areas of soil. This created an excellent breeding ground for grasshoppers. Second, methods to control grasshoppers were developed.

As more of the plains were cultivated for wheat production, grasshopper control became essential. In 1890 the most popular control was a mechanical hopperdozer. Hopperdozers were horse-drawn machines that swept grasshoppers into pans of tar which were then burned. This was the standard method of control until about 1910 when Criddle of Treesbank, Manitoba, demonstrated that poison bait was more effective than any mechanical means. Mixtures of bran, sawdust or horse manure, molasses or oranges, Paris green (an insecticide), and water were used in various combinations as the bait. During the 10-year period of 1913 to 1923 Gibson (53) estimated that 72 000 tonnes of bait costing $1.8 million were used, resulting in a saving of crops worth at least $80 million—a return of 44 to 1! Baits continued to be used until the mid-1940s, when they almost disappeared following the introduction of organochlorine insecticide sprays.

N. Criddle of Treesbank, Manitoba, was the pioneer Canadian entomologist who studied the taxonomy of nymphs (immature grasshoppers). In
order to develop an effective pest control for each species he would first have to
identify which of the 21 prairie species of *Melanoplus* individual grasshoppers
belonged. Between 1924 and 1933 he published six technical papers (see R.H.
Handford); he died before completing his investigations. Thirteen years later
Handford (59) of the Entomological Laboratory, Brandon, Manitoba, had
advanced Criddle's study. He found, as expected, that early instars were difficult,
almost impossible, to identify but devised a key by which he was able to separate
the various species.

To effect control with poison bait in the 1933–1935 outbreak, farmers,
municipal councils, provincial extension officers, and Canada Department of
Agriculture scientists cooperated, as described in Chapter 6. Many meetings
were held to explain to farmers how and when to use the bait. An extensive
system of observers was established to detect the start of an outbreak. Supplies
of bait were rushed to such sites for immediate distribution.

In 1931 Criddle started surveying egg and adult grasshopper populations in
the Prairie Provinces to predict where and to what extent grasshopper outbreaks
might be expected the following year. These surveys have been conducted
continuously since then, and are done now with the help of provincial departments of agriculture. The techniques for estimating potential grasshopper populations have improved as a result of studies such as those of Randell and Mukerji
(116). They used the number of eggs and the air temperatures to devise a
mathematical model that accounted for 96 percent of the variation in predicting
hatching dates. These dates were shown by Pickford (105), also of Saskatoon, to
be important because females hatching early in the season produce 20 percent
more eggs than do those hatching later. Smith (142) at Lethbridge, in 1972,
demonstrated that grasshopper females from parents raised in uncrowded cages
(10 nymphs per 35-L cage) were 30 percent heavier but produced only one-half
the number of eggs as did those females from crowded cages (150 nymphs per
35-L cage). This explained how, when crowded, grasshopper populations can
increase so quickly. Mukerji, Gage, and Randell (97), using 16 years of data,
revealed that a general increase in density should occur when accumulated heat
units reached 1600 degree-days above 10°C provided soil moisture was at or
above 13.5 percent. This led Mukerji and Gage (96) to refine their grasshopper
prediction model to include estimates of hatch and mortality rates of eggs.

When a moderate-to-severe infestation is identified new methods involving
biodegradable insecticides are employed. If applied immediately to the hatching
area, only small amounts are required to control large numbers of grasshoppers,
thereby possibly reducing crop loss. For instance, Gage and Mukerji (51) related
estimates of grasshopper populations to estimates of crop yields. They found
economic losses caused by grasshoppers tended to occur during periods of poor
crop yields due to drought. In 1961, despite extensive measures to protect the
crop, economic loss of wheat from grasshoppers was estimated at $40 million
annually, or 17 percent of the crops' total value.

Some species of grasshoppers are damaging to native ranges. Hardman and Smoliak (63), using 1928–1944 data collected by R.M. White at Manyberries, Alberta, supplemented with information collected between 1970 and 1975 by D.S. Smith, showed that three of the 26 species collected accounted for 80 percent of the grasshopper population. Only two of the 17 most abundant species were regarded as innocuous. All others have the potential to cause economic damage should they become numerous. Indeed, all four species that develop in early spring are grass feeders and do some damage to rangeland.

Recently, Mukerji and colleagues (95) have reintroduced the use of poisoned bran as a bait for grasshoppers. By applying the insecticide dimethoate to bran and spreading it at key points around fields, the amount of insecticide needed is reduced by 65 percent compared with spray applications. Nontarget insects such as bees remain clear of the insecticide, and residues of the insecticide on the crop itself are negligible. A further refinement by Ewen and Mukerji (33) is a bran bait onto which is sprayed a suspension containing the parasitic protozoan *Nosema locustae*. Target grasshoppers ingest the parasite that debilitates its host by competing for food reserves in blood and fat. *N. locustae* is specific to grasshoppers and locusts, and therefore is safe to use. It is slow to act but greatly reduces the grasshopper population. The protozoan is at this time registered in Canada only for experimental use.

Much of the fear of grasshoppers blocking the sun, eating clothes, and decimating crops has been dispelled. Gone are the days when accidental deaths of farm and wild animals, and even people, occurred as a result of ingesting arsenic-poisoned grasshopper baits. Serious localized infestations have plagued the southern prairies as recently as 1984—a reminder that the threat of a massive grasshopper infestation remains. Indeed, it is aggravated by land conservation practices in which soil is left undisturbed, maximizing the hatching of grasshopper eggs. Scientists continue to search for a solution to this problem.

WHEAT STEM SAWFLY

A small bee known as wheat stem sawfly, *Cephus cinctus*, lays eggs through minute holes cut by the saw-like ovipositor of the female in the hollow stem of wheat plants just below the head. After hatching, the young larvae grow rapidly and eat their way down the stem to the lowest joint, where they gnaw the inside of the stem, cutting a ring almost through to the outside. They then spin thin, delicate cocoons and remain torpid until spring, when they turn into pupae and shortly leave the wheat stems as adult sawflies and repeat the cycle.

James Fletcher first reported (36) wheat stem sawfly damage in wheat near Souris, Manitoba, in 1896. Five years later similar damage was reported at Bozeman, Montana. The insect is native to North America and lives in grasses, chiefly of the *Agropyron* species. Further severe damage occurred in 1931 and 1943 when 4 million ha were infested, causing a loss of 1.3 million t of wheat. Something needed to be done quickly to prevent further loss to farmers and the

Canadian economy. Similar devastation of wheat crops in North Dakota and Montana came a few years later.

Agricultural experts in Canada and the United States sought control measures. H.J. (Shorty) Kemp at Swift Current noticed that, although some cultivation practices reduced sawfly damage, the strip and trash farming methods used to control wind erosion actually helped to increase sawfly populations. Strip farming, for instance, extended the borders of a wheat field. It was in plants along these borders that female sawflies laid their eggs.

Before 1929, Kemp observed that the solid-stem Golden Ball durum wheat was less seriously damaged by sawflies than the bread wheats, Marquis and Reward. He then obtained seeds from the U.S. Department of Agriculture of a supposedly solid-stemmed wheat strain of Egyptian origin. Some of the plants from these seeds did have solid stems and were undamaged by sawfly, whereas others were hollow-stemmed and became infested and cut. Next, he obtained from O. Frankel, 38 varieties of New Zealand wheat originally collected in Spain, Portugal, and Morocco. Two of the Portugal strains, S-615 and S-633, having some solid-stemmed plants, were grown and selected until the solid stem characteristic became stable. Between 1930 and 1946 Farstad (34) studied the biology of the insect; Kemp (73), the relationship between insect and solid-stemmed wheat; Platt(106), the influence of environment on the solid stem characteristic of wheat varieties; and Platt and Larson (108), the transfer of the solid stem characteristic from *Triticum durum* to *T. vulgare*. By 1946 Platt and Farstad (107) had developed uniform tests of wheat varieties for resistance to sawfly. Such tests were essential for the development of resistant varieties.

A.W. Platt of Swift Current, while on a work transfer to the Cereal Division, Ottawa, in 1937, crossed a selection from S-615 with both Apex and Thatcher. After 9 years of selection and cooperation from cerealists and entomologists at many other experimental stations and universities, Platt and Farstad introduced Rescue from the Apex cross and in 1952, Chinook from the Thatcher cross. By 1949, Rescue was the second most widely grown wheat in Canada, exceeded only by Thatcher. When Chinook became available both varieties were widely grown in the sawfly areas of the United States as well as Canada.

Platt, R.I. Larson, and S.A. Wells moved to Lethbridge in 1948 to join entomologists led by C.W. Farstad at the new Science Service Laboratory. Four other sawfly resistant varieties have come from the Swift Current and Lethbridge programs. One of the most interesting is Leader, released in 1981 by De Pauw and colleagues (30). This variety, besides its solid stem characteristic, has a long seed dormancy, which means it will not sprout in the fall before threshing, even though it remains wet in the swath.

All solid-stemmed bread wheat varieties developed in Canada and the United States trace their ancestry of the solid stem characteristic to the S-615 strain introduced and developed by Kemp in 1931. The magnitude of the wheat stem sawfly problem has been reduced to a manageable level through the close cooperation of plant breeders, entomologists, and many others in the Research

Branch in what has proved to be a classic example of using genetic resistance to control an insect pest.

INTEGRATED CONTROL OF INSECT PESTS

At least 10 years before Rachel Carson published *Silent spring* (20), pointing out the dangers of excessive use of chemical insecticides, entomologists of Science Service were concerned with that very subject in the control of insects damaging orchard crops. In 1957 E.J. LeRoux (86) found that the codling moth on apples survived despite spray applications. First-generation larvae, and possibly even second-generation larvae, emerged after the cover spray.

The codling moth has been a problem for a long time. It was first observed in Quebec in 1872 by Leon Provancher who found considerable economic damage to fruit in apple orchards. He recommended banding the trunks of trees with a sticky material to trap larvae, and burning wormy fruit to destroy the larvae. Other growers introduced Paris green and lead arsenate as sprays. V.A. Huard discovered the importance of woodpeckers as predators of overwintering larvae. When C.E. Petch was appointed officer in charge at Covey Hill, he made intensive studies of the insect and recommended up to five sprays of lead arsenate. Starting in 1939 A.A. Beaulieu of Saint-Jean assumed responsibility for the research project. He found there was a second brood of codling moth and in 1946 introduced dichlorodiphenyltrichloroethane (DDT). Five sprays were still needed, however. This insecticide had an immediate but short-term success.

Similar work by A.D. Pickett in Nova Scotia, W.L. Putman in Ontario, and J. Marshall in British Columbia, yielded parallel results under different growing conditions. They were all concerned about the excessive use of chemical insecticides and the ultimate development of strains of insects resistant to particular chemicals. By 1956 Pickett, Putman, and LeRoux (103) found that the use of pesticides often interfered with natural control by other species because the numbers of natural enemies were sometimes reduced, the reproductive or survival potential of a target species was sometimes increased, and the target species often developed resistance to the pesticide. Their research was based on studies of the ecological impact of spray applications rather than on the economic effect of such sprays.

In the mid-1970s Pickett (104) reported that the use of integrated control methods (also called integrated pest management) by Nova Scotian orchardists, was generally accepted. He noted that oystershell scale was virtually eliminated by removing sulfur from the plant disease spray program. Sulfur had killed the predatory fungus of the scale.

The control of codling moth was more challenging. A truly satisfactory control method became available only when D.W. Clancy of West Virginia produced ryania, an insecticide made by grinding stems of a tropical shrub into fine powder. This insecticide was effective in controlling codling moth, yet harmless to the beneficial insects. It remains in use today.

In the intensive fruit-growing area of the Okanagan Valley, British Columbia, orchardists have supported integrated control. Entomolgists at Summerland found that one particular predaceous mite fed on another mite that harmed fruit trees. The spray program was changed so that the predaceous mites were unharmed, their population increased such that they kept destructive mites under control and so prevented damage to fruit and trees. To foster confidence in the system, a mite-counting service was provided, first by the Entomological Laboratory, then commercially, to monitor insect populations. On rare occasions, when harmful mite populations pose a threat, sprays are applied for their control. The monitoring service has proved so effective that commercial packing houses throughout the Okanagan Valley now provide it.

Another example of individual orchard sampling again involves the Okanagan Valley codling moth control program. Recently, a sex phermone (smell) of female codling moths was synthesized by scientists at Summerland. The phermone is placed in a trap, male codling moths are attracted to it, and their numbers are estimated. Spraying is based on the calculated codling moth population within each orchard. By so doing, pest control costs and the level of chemicals in the biosphere have been reduced, and parasite and predator survival has increased.

If onion maggots in southwestern Ontario were uncontrolled, reliable estimates indicate that the loss to farmers of marketable onions would be 78 percent of the crop, or about $8 million. At first, DDT was an effective insecticide, but onion maggots mutated and populations developed resistance to the insecticide. Later, when organophosphates were used, resistance to them was also developed by the pest. As many as 20 sprays per year had to be applied to regain control. The sprays were applied on a timed schedule, regardless of need. When integrated control methods were developed by Harris and coworkers (64) at London, fields were monitored for onion maggots and sprays applied only when needed—as few as four sprays per year.

Integrated control has reduced farmers' production costs and reassured consumers that fruit and vegetables, which previously could have been contaminated with pesticides, are now safe. The research funds expended in developing these methods have been nominal in relation to the benefits derived. Much information has been imported from other countries and successfully applied by Canadian scientists to assist Canadian orchardists in providing better-quality produce.

BIOLOGICAL CONTROL

The ideal system of controlling a disease, an insect, or a weed is to balance nature in such a way that the control mechanism and the pest are self-propagating, the pest being kept at a level such that it causes no economic damage to crops or animals. To achieve this situation either plants or animals that are resistant to attack from pests are required or growing conditions need to be adjusted to enable natural or introduced parasites or predators to keep the pest in check. In Canada, as in other countries, scientists have successfully developed such mechanisms. A review in 1961 by Turnbull and Chant (146) notes a Canadian success rate of 40 percent in 31 attempts prior to 1958. From 1959 to 1968 the success rate, as estimated by Munroe (98), is 15 percent.

The first person to suggest that parasites be imported into Canada for biological control purposes was the Reverend C.T.S. Bethune (87) in 1864. Later, in 1882, Saunders imported *Trichogramma* species (calcids) from the United States for the control of the imported currant worm. A calcid is a very small insect that lays its eggs in many target species, thus parasitizing them. Between 1893 and 1910 both Fletcher and Hewitt made a number of favorable imports and the biological control of insects in Canada was well established. Generally, predators or parasites are most effective in the control of pests that have been imported and have not adapted to the Canadian environment. A few are described here.

The oystershell scale was accidentally introduced into Canada from Europe as early as 1869 and eventually infested all apple- and pear-growing areas in Canada. It is particularly debilitating to young trees, because it extracts the sap from the cambium (growing) layer just under the bark. In Eastern Canada, a native predaceous mite usually gives effective control; only occasionally is a spray needed. In British Columbia, the mite was introduced to Vancouver Island, Mission City, and the Okanagan Valley in 1917 by Tothill of the Fredericton Entomology Laboratory. The predator became well established by 1925 and since then has been an important factor in controlling oystershell scale.

Another successful introduction of a native parasite was against the woolly apple aphid (plant lice) in the Okanagan Valley. As with the oystershell scale, this aphid is thought to have been introduced from Europe as early as 1819. The parasite *Aphelinus mali*, another calcid native to Eastern Canada, kept the aphid under reasonable control, but when the aphid was accidentally transported to Vancouver Island in 1892, the parasite did not accompany it. The aphid then moved to the Okanagan in 1912; serious outbreaks occurred in the 1920s. The control organism was shipped to British Columbia and within a few years the populations of woolly apple aphids were reduced and they were no longer regarded as a problem. Some believe that DDT, introduced in 1950, killed the parasite, causing the aphid population to increase to epidemic proportions by 1952. It decreased sharply in 1953 and British Columbia has been substantially free from this pest since then.

Biological control measures to regulate insects may take forms other than predator insects. The green apple bug (*Lygus communis*) bites into young fruit, causing the fruit to be seriously deformed at maturity. The bug, first reported in Nova Scotia by Brittain in 1914, was lethally attacked by a fungus. In 1920, therefore, entomologists reared the fungus artificially and disseminated it throughout orchards in Nova Scotia. Within 5 years the *Lygus* bug was essentially eliminated. It became a pest again in the mid-1940s with the introduction of modern fungicides, which "controlled" the pathogenic fungus. Fortunately, this control was only partially successful because the fungus continues to keep the *Lygus* bug within tolerable bounds.

Some attempts to control orchard insects by introducing parasitic or predaceous insects or parasitic fungi or bacteria have been disappointing. To biologically regulate codling moths, which severely damage apples by puncturing the fruit, four different insects and one bacterium have been introduced from Europe. Each was widely distributed throughout apple-growing areas in Canada, but none was successful in affording any control. The biological solution came from an entirely different source.

In 1931 E.F. Knipling of the U.S. Department of Agriculture devised a method of controlling the screwworm, a devastating livestock pest, by releasing large numbers of sterilized male screwworms. The principle behind this biological control method is to inundate the target insect population with sterile males, thus reducing the number of offspring. For this method to work, the following conditions must be met: the target population has to be small and be contained within a geographically isolated area, the females must mate only once, and a technique to rear and sterilize males, yet sustain their viability must be available. The system was successfully tested in 1954 on the 440-km^2 island of Curacao in the Caribbean.

Proverbs, Newton, and Logan (111) of the Summerland Entomology Laboratory started eradicating codling moths in 1956, using the same principle. Two of the conditions were readily met, but rearing and sterilizing male codling moths, and ensuring that the females mated only once were not. The first problem involved learning how to rear and to sterilize males. To begin, Proverbs reared codling moths on immature apples but this was too time-consuming. With the help of food processors he developed an artificial diet, the management of which could be completely automated and upon which immature moths thrived. Having raised codling moths, he then sterilized them with gamma rays from a cobalt 60 source and released them in an isolated orchard at the rate of 40 sterilized males to one native male. The population of native codling moths was greatly reduced, and by 1966 fewer apples (0.003 percent) were damaged in the isolated orchard by moths than in orchards where chemicals were used. Without broad-spectrum persistent insecticides, pests such as mites and aphids and other pests were controlled by their predators and parasites, resulting in little or no need for spraying.

The unqualified success of this research project attracted worldwide attention. M.D. Proverbs was the recipient of a Public Service Merit Award in 1978 for his progressive research. The program was expanded to include the Similkameen Valley, located just west of the Okanagan Valley. From the biological point of view, the system again worked as expected—after 2 years, codling moths were virtually eradicated from the valley. However, the cost of producing and distributing thousands of sterile moths was greater than the annual cost of spraying. Even though the effects lasted 4–5 years, when it came time to again treat the orchards, growers returned to sprays. Realizing the overall benefits of the biological method of controlling codling moths, the British Columbia Fruit Growers' Association, in 1984, planned its resumption.

FOUR FOREST INSECTS

Forests can be devastated by insects and sometimes by diseases. Until 1960 research scientists from Science Service and Research Branch monitored the many insects capable of damaging forests. They also cooperated with provincial departments of forestry on control measures. Then, in 1960, when the Forest Biology Division of Research Branch became part of the newly created Department of Forestry, most of the forest entomologists and pathologists were transferred. Later, two taxonomists and support staff were reassigned to the Biosystematics Research Institute to classify those insects affecting forests. Four of the several dozen insects that cause economic damage to Canadian forests are reported here. Their control provides examples of the type of research conducted by the former Forest Biology Division.

The European spruce sawfly

Entomologists regard the near eradication of the European spruce sawfly, *Gilpinia hercyniae*, as a model for the biological control of an insect species. This insect was first recorded in North America near Ottawa in 1922 (12). It was then discovered in the United States in 1929 and in the Gaspe Peninsula in 1930. By 1932 it had caused serious damage over a 5000-km^2 area of the Gaspe. The damage was compounded by the spruce beetle that vigorously worked on the weakened trees. The sawfly infestation soon reached epidemic proportions in the spruce forests of Quebec, the Maritime Provinces, and the New England States. Exactly when the species was imported remains a mystery; Balch, Reeks, and Smith (14) hypothesize that biologically it could have been sometime since 1880.

Adults of the sawfly emerge in May or early June, lay their eggs in slits they have cut in spruce needles of the previous year's growth, and in a few days these eggs hatch. Damage to spruce trees, particularly white spruce, a favorite host, occurs when the larvae feed on small pieces of their needles and later eat entire needles. This debilitates the trees until, after several years, they become weakened and eventually die. When larvae are fully grown they drop to the ground, burrow into the litter, and spin a tough parchment-like cocoon. Strangely, male spruce sawflies are rare in the Canadian populations. In Europe some populations have about one-half males, whereas others are nearly all females. Neilson and Morris (99) of the Forest Entomology and Plant Pathology Laboratory, Fredericton, say that mating has not been reported in the Canadian populations and probably does not occur.

Once it was found that no native parasite existed, forest entomologists quickly introduced parasites of the species from eastern Europe, through the Commonwealth Institute of Biological Control. They also learned that sawfly succumbed readily to calcium arsenate dust, and affected forests would have been sprayed with this dust by aircraft had finances during the depression permitted. Instead, intensive programs of parasite introduction were conducted from 1933 to 1945 in both Canada and the United States. At the same time, investigations by Neilson and Morris measured the effect of the parasite on sawfly populations.

During 1934 and 1935 R.E. Balch and F.T. Bird, also of the Fredericton Laboratory, reared 25 generations of sawfly that were apparently free from disease. Early in 1936, however, small numbers of larvae began to die, and the mortality rate increased to the point where, by 1939, the scientists found it impossible to rear a single larva in the laboratory. Balch and Bird (13) observed that numbers of larvae in their sampling squares fell from nine in 1939, to two in 1940, and to less than one-half in 1941. Furthermore, well over 90 percent of the few larvae that were recorded in 1940 were dead from a disease. Only the early stages of larval development were affected, causing lesions on the gut. Adult sawflies were resistant but some became vectors of the disease and transmitted it to their offspring.

De Gryse reported (29) diseased larvae in the wild in 1937 and correctly suspected the disease was caused by a virus. By 1943 the sawfly population had decreased in many Canadian and American forests to the point where defoliation was no longer a concern. The origin of the virus is unknown, but McGugan and Coppel (90) consider it to have been introduced with parasites from Europe. The first of 10 species of European parasites were released in New Brunswick in 1933. Releases were repeated annually until 1947. Two species were parasites of cocoons, five others were parasites of the larval stage of the European spruce sawfly. Over 890 million parasites were reared and released during these 15 years.

W.A. Reeks (118) concluded that despite low densities of sawfly, some parasites would become permanent additions to the biological control program. This has occurred, even though an unidentified virus caused the rapid demise of the sawfly during the 1940s and saved our spruce forests.

Ambrosia beetles

Attacks on softwood logs on the Pacific coast by ambrosia beetles destroy commercial timber. According to Prebble and Graham (109) there are at least five species that overwinter as adults. In the spring they bore galleries into felled timber within which the females lay their eggs. The galleries are lined with ambrosia fungus on which both young and adults feed. During the summer the young beetles fly to nearby standing timber and hibernate in dark sites. The galleries and stain from the ambrosia fungus damage wood, lowering its commercial value. In Douglas-fir, damage is restricted to sapwood, but in other firs and in hemlock, galleries may penetrate heartwood to depths of 100 mm (32).

Entomologists could find no parasite to control these beetles as they had with the sawfly, and tried other methods. One was to remove logs from their cutting areas. Log booms stored in water and sprayed with insecticide during the spring was another. This method, however, had detrimental side effects. Spray-

276

ing of logs in the woods was not effective because of the difficulty of reaching their undersides. A third method was to remove slash and other debris in which beetles breed. Since stumps were left standing, any significant reduction of populations by such sanitary methods was next to impossible.

Douglas-fir beetle

Other beetles attack standing timber. One such insect is the Douglas-fir beetle, *Dendroctonus pseudotsugae*, which is highly destructive of sawtimber as well as standing trees. Unlike the ambrosia beetles, the Douglas-fir beetle has a varied life cycle, all stages occurring throughout the year. Adult females attack mature, weakened, or fallen trees by excavating galleries between the bark and wood. Beetles mate within the galleries where the females lay their eggs, which hatch within 2–3 weeks. After hatching, larvae continue to mine the wood, pupate, and remain there as adults until the following spring. The original adults may produce two or three broods during the year. These beetles not only damage wood identified for lumber, but they also weaken, and often kill, other trees.

Because young vigorous trees are rarely attacked, Wright and Lejeune (161) found that the proper management of forests is the best control method. This includes harvesting trees before they mature, cutting stumps close to the ground, and cleaning up logging slash. Direct control methods have not been effective except when Douglas-fir trees reach moderate size. Then, insecticides sprayed from aircraft have proved useful.

Spruce budworm

The eggs of the spruce budworm, *Choristoneura fumiferana*, hatch on the underside of spruce needles in 5–7 days. The larvae then hibernate at the base of a needle, spinning a tent-like covering. Here they remain until the following spring when they feed on new leaf buds, stunting further growth of the tree.

The first mention of this destructive forest insect in a Canada Department of Agriculture report was by James Fletcher, honorary entomologist, in 1885, the year before the Experimental Farm Station Act was passed. Fletcher noted injuries to fir and spruce forests in both Quebec and New Brunswick. The moth was first collected in Canada at London, Ontario, by H.S. Saunders, the second son of William Saunders. In 1907 Fletcher observed larvae damaging white spruce in the Spruce Woods Forest Reserve in southern Manitoba. Native to North America, spruce budworm has damaged commercial forests of spruce, *Abies* fir, Douglas-fir, and jack pine in all provinces. Its populations, and hence its damage to forests, fluctuate because of many factors. A century after Fletcher first reported spruce budworm it still occasionally multiplies beyond control.

Losses of timber from spruce budworm infestation can be severe. In mature stands losses may reach 90–100 percent (53). In young stands, after 3 years of defoliation even the most vigorous trees are often killed. The actual dollar loss per year, conservatively considered to be in the millions, is difficult to estimate. Budworm outbreaks are most devastating in the spruce–fir forests of Eastern Canada. During the past 25 years more than 50 million ha of forests in Quebec and New Brunswick have been sprayed—making this the largest chemical control program in the world. In the long-term, sound forest management practices provide the only solution to the problem. The alternative is close monitoring of budworm populations, followed by aerial spraying of safe chemicals on susceptible stands.

LIVESTOCK INSECTS

Cattle grubs

There are two species of warble flies that lay grub-producing eggs on cattle. Both occur in Canada, do substantial damage, and are controlled by using the same methods. Control is far from simple. It has been the subject of study by veterinary entomologists in the Research Branch for several decades.

The egg-laying activities of warble flies start in May and continue through June. The flies lay their eggs on the hairs of the legs, bellies, and flanks of cattle, which agitates the animals into running and gadding. This interferes with normal grazing, reduces milk flow, and sometimes causes injuries. Upon hatching, the larvae (grubs) burrow through the skin of the animal and migrate within the body for about 9 months until they reach the animal's back where they produce the characteristic lumps, or warbles. The migrating larvae lodge either in the esophagus, making it difficult for the animal to eat, or in the spinal canal, sometimes causing paralysis. Each grub makes a breathing hole in the hide of the animal through which infection can occur. In March or April of the following year the grubs emerge from the warbles, drop to the ground, pupate, and emerge as flies in May or June to start another cycle.

Infested animals have lower market values because the impaired carcasses require trimming and the damaged hides are devalued as sources of leather. K.K. Klein of the Lethbridge Research Station estimates (76) the annual packing plant losses attributed to grubs to be $14 million (1979 dollars). If warble flies could be controlled or eliminated, industry would profit.

John Ware was a rancher in the Brooks, Alberta, district in 1875. He started his own ranch about 1880 and became the first rancher to try to eradicate cattle grubs. At round-up time in the spring and fall he dug a hole large enough to hold a cow, filled it with creosote, soap, and water and drove his cattle through the mixture. His system was adopted by other ranchers and was the only preventative method until rotenone, an insecticide from the root of a South American plant, became available in the 1920s. Hadwin (57) in 1923 refers to the use of creosote and derris (rotenone) as a control measure. By 1938 (70) rotenone was used regularly as a spray or wash on the backs of cattle and still is used on dairy animals because it is completely metabolized.

In 1942, grubs were causing considerable damage to meat and hides, both of which were needed more than ever for the war effort. R.H. Painter, at Lethbridge, was placed in charge of the Warble Control Program for the four western provinces. He introduced high-pressure sprayers to apply rotenone to the backs of livestock in the early spring before the grubs dropped to the ground. The control worked well but required two or three applications at 2-week intervals. Early in 1944 (8) he organized a campaign that he hoped would reduce, and perhaps even eliminate, warbles from the west. Associations of cattlemen, provincial departments of agriculture, Science Service, and Experimental Farms all urged ranchers to use rotenone on their cattle. They did; millions of animals were sprayed. At first there was considerable reduction of grub populations, but the system was too labor-intensive and untimely treatments were counterproductive. The grub infestations gradually returned.

The search for better control methods continued at Kamloops, British Columbia, and at Lethbridge, Alberta. In 1956, one large chemical company released trolene, a systemic insecticide,[1] on a trial basis for the control of cattle grubs. Its application is usually simple, but its side effects can be complicated. Weintraub, Rich, and Thompson (152) tested the systemic trolene on groups of 25 to 30 calves at both Kamloops and Lethbridge, administering the insecticide orally as either a drench or a large pill at various times between November 1956 and April 1957. Results were spectacular! Untreated cattle at Kamloops averaged 30 grubs per animal and at Lethbridge 80 grubs per animal. The best two treatments were one and three grubs per animal. Although the control was good, scientists found both the drench and the large pill awkward to administer. Furthermore, some animals showed minor toxic symptoms. The search for another control method continued.

R.H. Robertson at Lethbridge had been studying the production of antibodies in animals infested with grubs (121). Other workers had found that old, previously infested cattle were less subject to reinfestation. Robertson therefore reasoned that an antiserum could possibly be developed that would enable young animals to produce these antibodies. Khan, Connell, and Darcel (75) put the proposal to the test on more than 100 calves in 1959 and obtained a reduction from 48 to 13 grubs per animal, an improvement, but advances were still sought. If the system was to work, some controlled method of artificially infesting animals was needed to produce antiserum. Weintraub (151) tethered male and female flies to threads with paraffin wax, they mated in flight, and he then caused the mated female to lay eggs either on the hairs of his own arm or on the hairs of a cow. This gave entomologists a known number of eggs to place on an animal for experimental purposes. Weintraub's experiments would have other far-reaching results.

G.B. Rich at Kamloops continued his work on the systemic insecticide trolene (119) to evaluate its use under ranch conditions. He was able to treat

279

---

[1]A systemic insecticide or fungicide is one that, when administered to an animal or a plant, is absorbed into the blood stream or sap stream and provides protection from infestation.

between 800 and 1300 cattle in each of 6 years in an isolated area. After two treatments Rich reduced the average number of grubs per animal from 30 to less than two. They remained at this level during the 6 years of treatment, and for 1 year after treatment was suspended. The following year the number rose to more than 10 grubs per animal. Rich concluded there was not much reason for optimism on an area basis with the insecticides then available.

M.A. Khan, meanwhile, treated over 5000 animals per year from 1959 to 1961 in central Alberta in an attempt to exterminate cattle grubs in that region (74). He used newer sprays of nontoxic systemic insecticides each fall and reduced the grub infestation from 14 per animal to zero in the central zone of his experiment. Even in the peripheral areas where there had been no intensive spraying, the grub population was reduced by about 75 percent. Two years after the experiment was complete and Khan was reasonably confident extermination was possible, the grub population in the central zone was still less than one grub per animal. These studies led to organized control programs of warble flies in Alberta and Saskatchewan. The incidence of infestation was reduced to less than 10 percent by 1978. How could those last few be eliminated?

J. Weintraub had shown warble fly populations could recover rapidly from insecticidal decimation and concluded some integrated control system was needed. Other entomologists had shown that when an insect population was reduced to very low levels the last few survivors might be eliminated by flooding the population with sterile males. When sterilized males breed with females of the native population the females produce either infertile eggs or sterile males. Starting in 1975 Weintraub applied this knowledge on a large southern Alberta ranch, first reducing the grub population by using a systemic insecticide spray, then rearing and collecting grubs with the tethered fly technique and sterilizing them by irradiation with a cobalt 60 source, and finally releasing them to mate with the few native females left on the ranch. He was successful. As a result, United States scientists sought his cooperation in making a joint, large-scale test of 4200 km^2, one-half of it in Alberta and one-half to the south in Montana (79). At the time of writing the results from the 1st year's release show great promise. Eradication seems likely by the integration of systemic insecticide sprays with sterile male releases. Many millions of dollars will be saved by ranchers throughout North America should they choose to apply the technique on a continental scale.

Black flies

Periodically, and often without warning, livestock in Manitoba, Saskatchewan, and Alberta are attacked by black flies, also known as sand flies or buffalo gnats. They attacked buffalo initially but transferred their attention to cattle as the buffalo population declined. Attacks, always during daylight hours, are so severe that hundreds of animals may succumb within 6–24 hours. Black flies inflict painful bites, inject a toxic saliva, and suck blood. The toxin causes symptoms of an anaphylactic-type shock, which is usually fatal in new-born calves as well as in cattle not previously conditioned to such bites.

A.E. Cameron of the Entomology Division identified and reported on black flies in 1913 (19) when approximately 100 animals near Duck Lake, Saskatchewan, died as a result of black fly attacks. F.J.H. Fredeen, J.G. Rempel, and A.P. Arnason, scientists at Saskatoon, made a detailed biological study of this pest between 1947 and 1955 (42). Fredeen (38) lists 27 species, five of which attack cattle, and two of these are particularly vicious. Large swarms of these two species travel 5–225 km from their breeding grounds. Their life cycles are similar and for purposes of this discussion have been treated as one.

The species overwinter as eggs in the sand of riverbeds. They hatch after the breakup of ice in the spring. The larvae are carried downriver and attach themselves to pebbles in shallow rapids. As the larvae grow they continue downriver and amass on larger boulders in the swiftest flowing water. Fredeen counted populations as dense as $70/cm^2$. He calculated there were seven billion on a rock-filled weir across the North Saskatchewan River at Prince Albert. Flies emerge in late May and early June, swarm, and mate up to 65 km from their hatching place. Only females take a blood meal and may do so, again in swarms, at great distances from their home river. It is these swarms of females, blown by winds, that indiscriminately attack livestock. Males feed on plant nectar and sap as do females between blood meals. Having fed, they oviposit in late June and early July, again in swarms, over the surface of quiet rivers. The eggs settle in the mud or sand bottom where they remain until the following spring.

Cameron unsuccessfully tried to control larvae by applying miscible oils on the rivers. However, he did find that an effective repellent was either to graze cattle and horses in the lee of smudge fires or spray them with oil. In reviewing the work done by the Entomology Division during World War II for the Defence Research Board, C.R. Twinn (147) noted that river applications of DDT as an insecticide, and oil, as a larvicide, were effective. The Fredeen–Arnason team (41), from 1949 to 1951, applied DDT to reaches of the North and South Saskatchewan rivers just above rapids where black fly larvae lodged. The resulting concentration of DDT in the river varied from one part per billion to one part in four billion—an exceedingly small amount! Nonetheless, they obtained better than 80 percent control of black fly larvae at distances up to 60 km, and 30 percent control as far as 150 km downstream from the point of application. This was attributed to adsorption of DDT onto silt particles that larvae ingested, accumulating fatal dosages. They also found that insect-feeding bottom fish such as suckers were killed when the insecticide concentrate was heavier than water. By the late 1960s increasing evidence of the persistence of DDT in nature indicated that it was undesirable for use as a black fly control. Fredeen (40) showed that methoxychlor was also an effective but much less persistent larvicide.

In the early 1960s the black fly problem in the Prairie Provinces increased to the point of public outcry. A severe outbreak in 1962 in the Athabaska River area of central Alberta demanded immediate attention. As a result, an extensive

281

cooperative study was initiated by the Alberta Department of Agriculture, the Alberta Fish and Wildlife Service, the University of Alberta, the Fresh Water Institute, the Alberta Research Council, and the Research Station, Lethbridge. W.O. Haufe of the Research Station headed the group, which exhaustively studied the biology, hydrology, and environmental impact that proposed control measures might have upon the Athabaska River (62). They called upon Fredeen to study the biology of the species (39) and concluded that methoxychlor, which Fredeen had already tested, would be safe to use if applied when larval numbers on standard river bottom sampling devices had reached a predetermined level.

The Athabaska River was treated with methoxychlor over a 3-year period at a cost of less than $0.30 per animal and without damage to fish. Other studies showed that stock exposed to mosquitoes and biting flies, though not killed by their attacks, were less thrifty than protected stock. Haufe calculated that the economic returns from the control of black flies was 30 times the cost, even excluding losses from the erratic deaths.

Because all rivers and streams cannot be treated, and because black flies may become resistant to methoxychlor with repeated application, Shemanchuk conducted experiments with repellents (126). He found two sprays, which, when applied to the entire body of animals in the field, repelled attacks from black flies for up to 10 days. Ten applications were needed to protect an animal for an entire season. The method was applicable to small, confined herds but almost useless on large, widely ranging ones. Shemanchuk and W.G. Taylor are seeking better, longer-lasting repellents and improved methods of applying them.

An effective, environmentally safe system of control has taken 60 years to develop. It still can be improved upon.

WART ON POTATOES

H.T. Güssow, Chief, Botany Division, had been employed with the Experimental Farms Service for only a few months in 1909 when wart disease was suspected in Newfoundland. Güssow visited Newfoundland and discovered (55) that many potatoes, especially those grown in home gardens, were infected with this serious disease. The cause is a primitive fungus, *Synchytrium endobioticum*, that can live in the soil for 30 years, or more, particularly under cool, moist conditions. The fungus produces large, wart-like growths on infected potato tubers, and a marked reduction in yield.

The discovery of wart disease and of winter nests of browntail moth, *Nygmia phaeorrhoea*, in shipments of fruit seedlings from France (53) motivated C.G. Hewitt, Chief, Entomology Division, and Güssow to prepare what became the Destructive Insect and Pest Act of 1910. It replaced the San Jose Scale Act passed by Parliament in 1898 and amended in 1900. The updated, broader Act, covering both insects and diseases, has made it possible for mainland Canada to stay free from wart as well as from other diseases and dangerous insects.

In 1949 Labrador and the island of Newfoundland became the 10th province of Canada. Strict adherence to the policy of washing earth from all automobiles, trucks, and railcars before leaving the island of Newfoundland confined the disease. The annual cost of about half a million dollars seems high but is far less costly than permitting the seed potato fields of other provinces in the Atlantic region to become infected.

Potato wart has been a concern at St. John's West ever since the Experimental Station was established there in 1950. The method of control is to breed varieties that are resistant to attack from the disease. The Dutch (Ultimus) and the German (Mira) varieties of potato were known to be resistant to wart disease, as were some wild and cultivated South American species. These, then, were used by K.G. Proudfoot, O.A. Olsen, and M.C. Hampson in their potato breeding programs. Since 1969 these scientists have introduced five excellent varieties that range in maturity and are resistant to nearly all known races of wart. The first two, Pink Pearl and Mirton Pearl (110), have white flesh and their yields are comparable to Kennebec. Pink Pearl is resistant to wart races 1, 2, and 6 but not to race 8. Mirton Pearl is resistant to all races of wart detected in Newfoundland. The most recent variety, Brigus, is blue-skinned, wart-resistant, and substantially free from scab.

The research program at St. John's West makes it possible for home gardeners and commercial vegetable growers to successfully produce potatoes in Newfoundland. The incidence of the disease will decrease when all potatoes grown on the island are varieties resistant to wart. Eventually the fungus may be eliminated from the soil; however, several decades of constant surveillance will be required to accomplish this.

ROOT ROTS

There are many organisms and soil conditions that cause roots to rot. Some can be attributed exclusively to one fungus, bacterium, nematode, or soil condition. More often, however, the cause is complex. To illustrate this fact, two of the several Canadian situations are described here.

Peach replant problem

Peach growers on the Niagara Peninsula and in other areas of southwestern Ontario, particularly Essex County, noted that peach trees, when planted on old peach orchard sites, had their growth retarded, often to the point of stunting, and their leaves showed signs of chlorosis. Koch (78), in 1955, found that roots displayed symptoms of greatest diagnostic value. They showed varying degrees of discoloration and necrosis. Even more startling was the observation that within 24 hours of new lateral roots emerging from parent root tissue, they discolored and developed lesions. A microscopic examination of these lesions revealed that some were caused by nematodes, but that in others dead cells or groups of cells were void of visible organisms. Severely affected trees died.

Although peach and citrus growers in parts of the United States had experienced the problem in their orchards, Ontario peach growers, of which

283

over one-half had noted their trees were affected by the condition, were alone in Canada with their problem. Some scientists in the United States theorized that peach roots produced a toxin, but they lacked sufficient experimental evidence as proof. Patrick (101) of the Harrow team presented the evidence in 1955 by showing that substances leached from peach roots inhibited the respiration of excised peach root tips. The same reaction was obtained when a cyanoglycoside, amygdalin, was substituted for the peach root leachates. Other experiments led Patrick to the conclusion that microbial action on the amygdalin fraction of peach roots resulted in the production of hydrocyanic gas (HCN) and was the main toxic factor encountered in old peach orchard soils. Ward and Durkee (149) chemically confirmed that some peach root tissue, particularly bark, contained more than 5 percent amygdalin on a dry-weight basis. When Wensley (155) fumigated problem soils to inhibit the activities of the fungi found in the rhizosphere, he induced improved peach seedling growth in most soils, further indicating that the problem was one of bacteria reacting upon the amygdalin.

By 1958 Mountain and Boyce (93, 94) had shown that one species of nematode, *Pratylenchus penetrans*, was in greater number in Essex County than it was in the Niagara Peninsula where the peach replant problem was more severe. They also discovered that the replant problem decreased as the interval increased between the removal of old trees and the planting of new ones. They hypothesized that this ameliorating effect may have been related to the decline of the nematode population. One way of reducing nematode populations in peach orchards is to grow tall fescue, *Festuca arundinacea*, or creeping red fescue, *F. rubra*, grasses, which have recently been shown by Townshend, Cline, Dirks, and Marks (144) to suppress nematodes.

Although it has diminished, the peach tree replant problem remains with Ontario peach growers. They now know how to minimize the problem by growing some other crop for 2 years prior to replanting peach trees, by fumigating to destroy nematodes when the population is high, and by using appropriate kinds of grasses as cover crops in orchards.

Cereal root rots

G.B. Sanford, in the first systematic survey of cereal diseases of Alberta (23) in 1927, noted that root rots were severe in wheat grown on stubble but usually insignificant following fallow. W.C. Broadfoot and M.W. Cormack made a second survey the next year, concluding it to be the major disease of wheat and barley.

Cereal root rots are caused by a complex of organisms. *Ophiobolus graminis* causes take-all disease and *Cochliobolus sativus* is primarily responsible for common root rot. W.C. Broadfoot and L.E. Tyner from the Plant Pathology Laboratory, Edmonton, showed (23) these diseases to be most severe when plants were short of potassium, nitrogen, and carbon. Phosphorus levels had no influence in the manifestation of the disease. Broadfoot's study of 1935 (18) showed that there was a marked tendency for one of the two fungi to predominate at different locations. He therefore concluded that environmental factors played an important role in the occurrence of the disease.

Simmonds and Sallans (132) at the Dominion Laboratory of Plant Pathology, Saskatoon, were anxious to learn the effect on crop yields of different levels of root rots. In their 1930 field work at the Experimental Farm, Indian Head, they excised seminal (seed) and crown roots of wheat plants at varying stages of growth. The loss of seminal roots caused a reduction in yield and a tendency toward delayed maturity. The loss of crown roots, however, resulted in little reduction of yield and tended to encourage the maturation process. Simmonds and Ledingham (131), working from the same Indian Head plots, identified 44 kinds of fungi associated with the wheat roots. Of these, 15 were classed as pathogenic. More than 50 percent of the individual isolates were in this latter group and were generally found in the top 30 cm of soil.

The extent of damage from root rots continued to elude plant pathologists and agronomists until 1947 when Sallans (124) analyzed 10 years' data from the Saskatchewan plant disease survey. By using correlation and partial regression on this large body of data (before electronic data processing was available to most scientists), Sallans estimated the annual loss of wheat yields due to common root rot to be 350 kg of grain per hectare. A more recent estimate (83) places the loss at 800 000 t in the Prairie Provinces each year.

Simmonds, Sallans, and Ledingham (133) then found that conidia of *Helminthosporium sativum* (the imperfect stage of *C. sativus*) were abundant on the stubble and in the surface soil of some fields. Their experiments showed these conidia to be important as primary infectors of wheat seedlings. They also demonstrated that the germination of conidia was suppressed by microflora on the stubble and in the soil surface, from which they concluded that some form of antibiosis could be used to reduce root rot infection. Between 1951 and 1958 Ledingham, Sallans, and Wenhardt (84) compared various methods of seedbed preparation for the second crop of wheat in a fallow–wheat–wheat rotation at the Swift Current Experimental Station. Plows rather than cultivators or one-way disks reduced the incidence of root rot in the seedling stage. As the season progressed, however, plants became infected. By maturation, with continued infection, the advantage of plowing had disappeared.

S.H.F. Chinn and R.D. Tinline made extensive studies on the mechanisms involved in the survival of pathogenic fungi in soil. In their 1964 report (22) they showed that white- and tan-spored isolates of *C. sativus* germinated in soil and disappeared, whereas brown-spored isolates, without this characteristic, remained viable in the soil for extended periods. H. Katznelson at the Microbiological Research Institute, Ottawa, reviewed (72) the research on the rhizosphere conducted prior to 1965 and concluded that additional knowledge was needed before the rhizosphere could be altered enough to control soil-borne pathogens.

Scientists continued to seek control measures for root rots. Ledingham (82) showed that straw and stubble incorporated into the seedbed just before seeding reduced the severity of common root rot, whereas supplemental nitrogen intensified the disease. Ledingham's 1970 research also supports recommended soil erosion control measures. Atkinson, Neal, and Larson (9) at Lethbridge, in 1974, demonstrated that a single chromosome in wheat is critical for providing wheat varieties with resistance to common root rot. By 1981 Verma, Chinn, Crowle, Spurr, and Tinline (148) were able to reduce root rot in seedling and mature wheat plants by treating seed with a complex organic systemic-type fungicide called imazalil. Treating wheat seed with imazalil partially controlled root rot but did not guarantee higher crop yields. The problem of root rots in cereals awaits a complete solution.

PLANT VIRUSES

The first indication that diseases could be transmitted by organisms other than fungi or bacteria occurred in 1892. A Russian bacteriologist, Ivanovski, found that sap from a tobacco plant contaminated with mosaic disease could transfer the infection to healthy plants, even though the sap had been moved through a porcelain filter known to prevent the passage of fungi and bacteria. Viruses are self-reproducing agents smaller than bacteria. Bacteria can be seen with the aid of only an optical microscope, whereas an electron microscope is required to see viruses. Viruses multiply within living, susceptible cells, making them elusive and difficult to control. They are responsible for a wide range of infectious diseases. In 1902 foot-and-mouth disease, which affects cattle, was attributed to a virus. In 1915, viruses of bacteria (bacteriophages) were described. The study of virology intensified after 1930 when scientists realized how hazardous viruses can be to the health of humans, animals, and plants.

The running-out, or degeneration, disease of potatoes in the Maritime Provinces in 1914 was the first virus disease of plants to be identified in Canada. Initially, it was combatted by discarding seed lots that failed to pass inspection (23), then Paul A. Murphy, officer in charge of the Plant Pathology Laboratory, Charlottetown, Prince Edward Island, started the Seed Potato Certification Service in 1915 (see Chapter 5). In 1916 C.G. Cunningham, of the Plant Pathology Laboratory, Fredericton, New Brunswick, located some lots of relatively virus-free potatoes that became the nucleus of certified seed production in New Brunswick. The staff at St. Catharines, Ontario, under W.A. McCubbin, extended the research on viruses of potatoes and included raspberries in their investigations. In 1920 H.R. McLarty at Ottawa demonstrated the viral nature of sweetclover mottle disease. The following year he was appointed the first plant pathologist at Summerland, British Columbia.

Rankin and Hockey (117) at St. Catharines, in 1922, made the first major contribution to the control of viruses. They proved that both mosaic and leaf curl viruses in raspberries were transmitted by plant aphids and they developed control methods. In the meantime W. Newton, then with the British Columbia Department of Agriculture, showed that aphid populations in the Pemberton Valley, north of Vancouver, were in fewer number than elsewhere in the province. He concluded, therefore, that this would be a good location to grow certified seed potatoes. In 1928, L.C. Young at the Experimental Station, Fredericton, began breeding potatoes for resistance to virus diseases and later, in 1932, he cooperated with D.J. MacLeod of the Plant Pathology Laboratory on the same subject. Thousands of seedlings from this program have been tested at numerous experimental farms and stations across the country.

New viral information was reported almost monthly during the 1930s. J.W. Eastham, the British Columbia provincial plant pathologist, noted a mottled-leaf condition of sweet cherry at Nelson in the Kootenay Valley in 1932. Little cherry disease, as it is known today (140), soon devastated the sweet cherry industry in that valley. By 1946 virologists were able to transmit the disease by grafting. R.D. McMullen, an entomologist at Summerland, British Columbia, suspects the apple mealy bug as being the vector of the causal virus. By 1955 Wilks and Welsh (157) identified one sweet cherry variety (Star) and a number of seedlings which, when inoculated with the causal virus of the disease developed a reddening of their leaves. This gave the scientists an excellent indicator host. Plant pathologists and horticulturists have prevented the Okanagan Valley from suffering the same fate as the Kootenay Valley through careful monitoring and with considerable expense. Nearly 2000 cherry trees suspected of having the disease were removed during the period 1969–1984 by Okanagan Valley orchardists.

In 1938 T.B. Lott, at Summerland, noticed two cherry trees in a commercial orchard that were much smaller than others of the same age. By 1943 (88) he had found 11 additional stunted trees and showed that this condition could be transmitted to young trees by budding or grafting. He and F.W.L. Keane (89) also identified the host range of twisted leaf in cherry. Lott's lifetime study of virus diseases on stone-fruits contributed much to understanding the relationship between viruses in native and ornamental cherries and those in commercial stone-fruits.

G.H. Berkeley, at the St. Catharines Laboratory, also studied virus diseases of stone-fruits. He collaborated with E.E. Hildebrand of New York State and D. Cation of Michigan in the compilation of an authoritative, widely acclaimed handbook published in 1942.

Newton, by now in the new plant pathology laboratory at Saanichton, British Columbia, was one of the pioneers in plant virus serology. He produced (100) a crude but workable antiserum against the latent potato virus X, using chickens. Davidson and Sanford (28), at Edmonton, Alberta, determined that potatoes infected with the virus early in the season showed primary leaf curl; those infected after mid-season did not, but their progeny developed secondary leaf curl the following year. If, however, infection occurred much later, the virus might not reach the tubers at all. They eventually attributed the tuber condition, known as net necrosis, to infections that occurred during the phase of rapid tuber enlargement. Their discovery led to the development of a system of tuber indexing during the winter months that consists of removing one bud or eye from each tuber and growing it in a greenhouse. If the resulting plant shows no

287

symptom of virus, the remaining portion of the tuber is deemed to be healthy and is planted the following spring.

R.H. Bagnall moved from Charlottetown to Fredericton in 1946 to study the control of potato virus Y by restricting the aphid vector and by encouraging early harvest; neither method proved effective. He then turned his attention to developing resistance in the potato. At the same time, Gilpatrick and Weintraub (54) at St. Catharines found that when some leaves of *Dianthus* spp. (pinks) were inoculated with the causal virus of carnation mosaic, the remaining leaves on the same plant were later immune to infection from that virus. The mechanism was elucidated by M. Weintraub and W.G. Kemp in 1961. This discovery of acquired resistance caused considerable investigation by virologists throughout the world, resulting in the publication of several hundred papers on the subject. Although the protection mechanism ceased abruptly in that clone of *D. barbatus*, the phenomenon later was found to be a universal consequence of localized infections. Attempts to isolate the antiviral factors, which seem to be similar to mammalian interferon, are high priority projects in many laboratories. The new techniques of recombinant deoxyribose nucleic acid (DNA) are being applied to clone the gene responsible for the acquired resistance.

The decline of pear trees on the Pacific coast of North America began in 1948 at which time McLarty, of Summerland, noted in the Canadian Plant Disease Survey that pear trees of several varieties growing in the area from Penticton south to Oliver were dying from some unknown cause. He suspected a virus. In 1970 pear decline was listed (77) as catastrophic in Washington State, Oregon, and California.

In 1952 J.T. Slykhuis at Lethbridge, Alberta, raveled out the problems of stunted growth and low yield of winter wheat that had confounded agronomists for 20 years. He showed (139) that wheat streak mosaic was the viral culprit and that a mite of the genus *Aceria* was the vector. By the following year, he had learned how to control the mites, and hence the virus. The practical application of this one discovery increased yields of winter wheat in Alberta and Saskatchewan by thousands of tonnes.

Virus research accelerated when R.E. Fitzpatrick moved from Summerland to establish a Science Service plant pathology laboratory on the campus of the University of British Columbia in 1946. He decided to concentrate the research effort on virus diseases of strawberries, raspberries, and potatoes. In 1955, Fitzpatrick and his team of virologists, chemists, and entomologists renewed interest in Newton's seriological method of identifying viruses, but they used rabbits instead of chickens. Rabbits injected with a purified plant virus build a blood antiserum, which, when brought in contact with juices of a plant infected with the same virus, separates out a visible precipitate. Should the unknown virus be different, no precipitate forms. The procedure is widely applied, including its use by Hamilton and Nichols (58), to positively identify a particular virus of peas. Previously, viruses had been identified by observing host plant symptoms and by studying insect vectors. These methods were time-consuming and often inconclusive.

288

Fitzpatrick, Stace-Smith, and Mellor (50) were also able to free the British Sovereign strawberry and Cuthbert red raspberry from viruses by growing plants for 9–11 days at between 35° and 41°C. Among the plants that survived such harsh treatment were some apparently freed from all viruses. This was a great achievement, for with normal multiplication practices and care in preventing insect infestation, whole fields of commercial, virus-free strawberries and raspberries could be grown. As might be expected, virus-free plants produce more fruit than infected plants—about 20 percent more. Propagation of the healthy plants was done in screenhouses at the Agassiz Experimental Farm where plants free from red stele root rot were already being produced.

In 1967 Raine (115) discovered leafhopper vectors of potato witches'-broom. Wright, MacCarthy, and Forbes (162), in 1970, associated the potato leaf roll and green peach aphid problems in the lower Fraser Valley with the existence of a sugarbeet seed industry. The eradication of potato viruses X and S from North American varieties and the encouragement of a commercial, virus-free potato seed industry were achievements of considerable significance. In 1972 N.S. Wright, R. Stace-Smith, F.C. Mellor, and E.F. Cole were honored with a Public Service of Canada Merit Award for their background research in clearing potato varieties of virus diseases. Their efforts led to the establishment of isolated Pemberton Meadows as a commercial, virus-free seed potato area.

In 1964, Ragetli and Weintraub (114) introduced immuno-osmophoresis, a method for quantitative virus assay in non-purified preparations, which simplified the positive identification of virotic plants. Using electron microscopy, Jacoli (71) observed the translocation of mycoplasma-like bodies through the sieve tubes of aster yellows-infected tissues, and Weintraub, Ragetli, and Leung (153) detected elongated virus particles in the intercellular passageways. These were indications of a probable means of spread from one cell to another. The physical and chemical properties of viruses aid in their identification and classification.

Major progress has been made in the control of virus diseases of strawberries, raspberries, and potatoes and extensive knowledge of plant viruses has been accumulated from four decades of research at Vancouver. Due to the advances made by its scientists, the Vancouver Research Station has gained recognition as the national center for plant virus study.

In 1959 Bagnall of Fredericton, working at the University of Wisconsin, separated three viruses from a mosaic disease complex in potato (10). Two of the viruses, S and M, though distinct in several ways and readily distinguishable from each other, bore a slight serological relationship. Bagnall continued his investigations with the aid of two German scientists, and developed a high titer antiserum. They demonstrated that the three viruses had essentially identical normal particle lengths—predicting that "distant serological relationships might be more common than had been supposed." Today, with 20 virus groups established, more than a dozen have such serological relationships.

R.H.E. Bradley, also at Fredericton, used several ingenious experiments (16) to show that aphids most readily acquired potato virus Y by inserting their

289

stylets into only the epidermal layer of a leaf. His techniques included measuring exposed stylets after the aphid had been anesthetized with CO_2, inactivating the virus on exposed stylet tips with dilute formalin or ultraviolet irradiation, and permitting aphids to feed through membranes such as parafilm and the stripped epidermis from different plants. This, and other studies, led Bradley, Wade, and Wood (17) to the concept of spraying a dilute oil emulsion onto plants. In practice, this reduced the spread of potato virus Y by as much as 80 percent. It is used extensively.

In Summerland, M.F. Welsh and coworkers had focused upon preventing the introduction of viruses into the tree fruits of the Okanagan and Kootenay valleys. D.V. Fisher, the research station pomologist, in a cooperative move with the British Columbia Department of Agriculture and the British Columbia Nurserymen's Association, established a budwood nursery. In 1962 the British Columbia Fruit Growers' Association assumed responsibility for the nursery's supply of true-to-name, virus-free budwood and its distribution to commercial nurseries.

Following the discovery by Slykhuis of the vector for wheat streak mosaic, P.H. Westdal at Winnipeg, Manitoba, searched for vectors of other cereal virus diseases. In 1957 he (156) found that a high incidence of aster yellows virus (now called mycoplasma) coincided with heavy migrations of sixspotted leafhoppers into Manitoba from the United States. C.C. Gill, in 1964, and A.W. Chiko, in 1970, joined Westdal to study five viruses (one proved to be a mycoplasma) affecting cereals, and they showed that barley yellow dwarf, also transmitted by infective migrant aphids from the United States, caused the most severe economic losses. Control of barley yellow dwarf has been achieved by incorporating one resistant gene into barley cultivars and several into oat cultivars. Seeding needs to be as early in the spring as possible to escape attack by the infective aphids.

Through the 1950s and 1960s scientists at St. Catharines concentrated on the study of tree fruit viruses. George and Davidson (52) discovered that both necrotic ringspot and prune dwarf viruses were spread in infected pollen, which led to a clearer understanding of the epidemiology of viruses. Tremaine, Willison, and Allen (145) were among the first North American researchers to initiate studies on the isolation and characterization of tree fruit viruses. Their efforts resulted in the discovery of the multicomponent nature of this group of viruses. Allen and Davidson, with the goal of upgrading nursery stock, developed the virus-tested fruit stock program for Ontario. In 1971, raspberries and strawberries were added (2) to the virus-tested stock program, which now serves nurseries and research establishments throughout Canada.

H.F. Dias developed a virus-tested stock program for grapes, which was transferred to the Ontario Grape Growers' Marketing Board in 1978 under a federal grant program. He demonstrated (31) that the cucumber necrosis virus was transmitted by a soil-borne fungus. Allen and Dias (3) collaborated on the characterization of Ontario nematode-transmitted fruit and grape viruses. Studies on soil-borne viruses have become a major program at Vineland, Ontario, the station to which St. Catharines staff moved in 1959.

290

On the west coast, J.A. Freeman at Agassiz and F.C. Mellor at Vancouver, illustrated in 1962 (45) that three latent viruses (ones showing no symptoms on the host plant) reduced the vigor, yield, and fruit size of strawberries—yield by 25 percent and fruit size by 20 percent. An attempt at strawberry plant certification in 1955 was unsuccessful, but later Freeman and Mellor were even more certain they had virus-free plants and renewed their efforts to establish a certification scheme. They succeeded and growers are now provided with healthy plants. At the same time, Welsh at Summerland had found a number of damaging apple viruses in Okanagan orchards. He and Nyland (154) from the University of California found they could inactivate viruses by subjecting them to temperatures of 38°C for periods of up to 7 weeks.

In 1965 Freeman and Stace-Smith (46) started a productive study on viruses affecting red raspberries. They demonstrated that there was a gradation in varietal response to the nematode-borne tomato ringspot virus (47) and learned how to reliably detect raspberry plants carrying it (48). By 1975, working with H.A. Daubeny, small-fruit breeder from Vancouver, they showed (26) how tomato ringspot caused crumbly fruit in susceptible cultivars. In 1978 the same three scientists (27) were the first in North America to identify raspberry bushy dwarf virus.

Upon the urging of Welsh, a quarantine station was established in 1965 at Saanichton through the cooperative efforts of the Research Branch and the Plant Protection Division of Agriculture Canada. Here, at what is now regarded as the world's leading tree-fruit virus quarantine station, vegetative plant material destined for import into Canada is grown, tested for viruses, and released when proved healthy.

A new team of scientists was assembled in 1967 at the Cell Biology Research Institute (later called the Chemistry and Biololgy Research Institute), Ottawa. Slykhuis transferred from Lethbridge to join R.C. Sinha, L.N. Chiykowski, and Y.C. Paliwal. They would specialize in virus–vector relationships and complement the virus team at Vancouver. They began by studying yellows-type diseases in cereals, forage legumes, and fruit trees, believing them to be viral. It soon became clear, however, that these diseases were mycoplasmal in nature. Mycoplasmas are larger than viruses but smaller than bacteria. They produce symptoms on plants similar to those caused by viruses and until 1967 were thought to be viruses. However, unlike viruses, some plant mycoplasmas can be cultured in artificial media, the diseases they cause respond to antibiotics, they vary in size and in shape, and are transmitted only by leafhoppers. The team has made a number of important contributions to the knowledge of transmission mechanisms of viruses and mycoplasmas (137) and to serological detection of mycoplasmas (138) in plants.

In 1970 Singh (134) at Fredericton showed that the spindle tuber disease of potato is transmissible through the true seeds of potato—a first for potato diseases. Later, in cooperation with Bagnall (135) and Clark (136), Singh revealed that potato spindle tuber disease is caused by a protein-free infectious nucleic acid, which was later proved to be of very low-molecular weight. This

291

low-molecular ribonucleic acid (RNA) since has been termed a "viroid"—a new class of agent that causes several plant diseases throughout the world.

At Fredericton in the 1930s, Young and MacLeod had cooperated to find potatoes resistant to virus. In 1972 Bagnall and Young (11) showed that field resistance was recessive. This led Young, Davies, and Johnston (163) to breed the variety, Jemseg, which possessed resistance to latent and mosaic viruses.

Until 1977 three methods of obtaining virus-free plants were available—selection, heat treatment, and propagation of meristems followed by micrografting. In 1972 medical researchers discovered that an organic compound, ribavirin, was a virucide. Five years later, plant virologists demonstrated ribavirin to be an effective chemotherapeutant against a potato virus. A.J. Hansen, at Summerland, made similar successful tests with an apple virus (60). He also found that ribavirin was effective (61) with *Prunus* species and less expensive than other methods of obtaining virus-free plant material.

Vigilant plant virologists have kept a careful watch over Canada's wealth of food and ornamental plants. These men and women frequently have been in the forefront of scientific discoveries, without losing sight of how their science might benefit Canadian farmers. It was reassuring to them when two American virologists, Shepard and Claflin (127), pointed out in 1975 that "the only serious attempt being made in North America to produce [virus-]free certified seed [potatoes] at the grower level is in British Columbia, Canada." This is in the same valley that Newton had identified 50 years before as ideal for growing certified seed potatoes because its aphid population was infinitesimal.

WEED MANAGEMENT

Weed management is young compared to many other sciences affecting agriculture. Nonetheless, weeds were of concern to both farmers and the staff of experimental farms a decade before the twentieth century. Everyone realized that there were heavy losses to crops infested by weeds and that crop quality deteriorated when contaminated with weed seeds.

J. Fletcher, entomologist and botanist, in his second report to the director, 1888, noted that millers expressed concern about weed seeds in the wheat. Fletcher identified most of these seeds as being *Polygonum convolvulus*, or black bind-weed (now called wild buckwheat). They had been inadvertently imported from Europe and, by 1888, had spread across Canada. Fletcher suggested that farmers separate the fine weed seeds from the grain by sieving before planting. He also received samples of perennial sow-thistle for him to identify and to recommend controls. In both his writings and his many addresses to western farmers (56), he advised "constant vigilance and summerfallowing."

The first experimental farm bulletin on weeds (35) was written by Fletcher, published in July 1897, and reprinted in February 1907. In it he outlined the various cultural practices needed to destroy annual and perennial weeds. These included frequently cultivating row crops, preventing weeds from seeding, summerfallowing, and seeding down to grass or clover. He recommended building a straw stack or manure pile over particularly troublesome patches of perennial weeds. Fletcher published line drawings of the more common weeds and included a list of 164 Canadian weeds complete with their descriptions and suggested methods of eradication.

Fletcher concerned himself with the control of weeds by mechanical methods and F.T. Shutt, chief chemist, investigated the use of chemicals. Fletcher, in 1899, wrote: "the introduction of weeders into the dry regions of the West, I consider an event of enormous importance to all grain growers." The weeders, which consisted of harrows, often ganged in pairs, were used when weeds were very small and the grain crop ranged from 2.5 to 20 cm high. This method of controlling weeds was used and strongly recommended by A. Mackay, Superintendent, Experimental Farm, Indian Head, Saskatchewan. Shutt experimented with iron and copper sulfate sprays in 1899 following reports of their successful use in England and France. He found that a 2-percent solution of copper sulfate sprayed at the rate of 560 L/ha before the growth reached 15 cm high destroyed all mustard plants. He also noted a species interaction—true Charlock, *Brassica Sinapistrum*, was destroyed, whereas smooth-leaved Charlock, *B. campestris*, was not affected. The use of copper sulfate and later, iron sulfate, to control wild mustard continued through the 1940s at Kentville, Nova Scotia; L'Assomption, Quebec; Ottawa, Ontario; and Regina, Saskatchewan. In 1932 Regina added a 6-percent solution of sulfuric acid at the rate of 1100 L/ha to its treatments. It worked well that year and in 1933, but in 1934 sulfuric acid proved to be harmful because of the severe drought. Sodium chlorate at rates of up to 1100 kg/ha was needed to control Canada thistle, *Cirsium arvense*, but the residual effects prevented the growth of grain for up to 4 years following treatment.

The first full-time weed scientist in Canada was G. Knowles (143). He was appointed to the Fibre Division, Ottawa, in 1923 as assistant to R.J. Hutchinson, the divisional chief, and in 1935 as weed scientist for fiber crops. Weeds tend to choke out flax and some cause trouble in the scutching of fiber and cleaning of seed. Until 1942, mustard was controlled in flax as well as in cereal crops with sprays of 4 percent copper sulfate. Knowles then introduced the sodium salt of dinitro-ortho-cresol (Sinox) as a selective herbicide to control broad-leaved weeds in flax. He found it to be slightly slower than but as effective as copper sulfate with the added advantage that it did not corrode equipment.

Methods of controlling weeds continued to occupy the attention of agronomists at an increasing number of Research Branch establishments. Wild oats were kept in check at Brandon by a 3-year rotation of hay, hay followed by fallow, and grain. Scientists at Lethbridge drew similar conclusions, using alfalfa as the hay crop. They recommended, however, 5–10 years in hay, or 2 years of sugarbeets, either of which would eradicate wild oats. Agronomists at Lethbridge were able to keep Canada thistle under control with 3 years, or more, of irrigated alfalfa.

To control weeds, their precise identification must be known. Many species of plants within a genus are superficially alike but often react differently to treatment. Weed scientists, therefore, rely upon taxonomic botanists to assist in

293

differentiating among the various species. N. Criddle, Treesbank, Manitoba, was one of the first in Canada to make a detailed study of wild oats. While employed by the Seed Branch, he conducted a 5-year study (25) that was completed in 1912. In this study, he compared wild oats, *Avena fatua*, with five types of false wild oats (genetic aberrants of domestic oats, *A. sativa*) to determine whether false wild oats would be as damaging to grain crops as true wild oats. He concluded they would not be, because seeds of false wild oats germinated readily when fresh and so could be destroyed at once, whereas those of true wild oats often remained dormant in the soil for several years before they germinated. The work of Criddle is outstanding for its thoroughness and for the accompanying pen and ink illustrations, which he drew himself.

In 1937, when studying ways in which to maximize crop production under the low moisture conditions that existed on the prairies at the time, W.S. Chepil at Swift Current noted the damaging effect of weeds. However, he also recognized their benefit—dormant seeds, when given moisture, will germinate and protect soil from various types of wind and water erosion. After 6 years of examining seeds of 58 weed species, he (21) concluded that seeds of different species varied widely in their behavior in cultivated soils. He also determined that the period of dormancy, resulting in uniform germination, is one of the greatest factors contributing to the damage caused by any weed.

More than 50 years after Criddle completed his research, dormancy of wild oat seed was still a subject for research. Hay (69) at Ottawa, in his 1961 experiments, found that seed lay dormant when unfavorable germination conditions such as lack of oxygen, restriction of gas movement around the embryo, and temperatures above the optimum existed. Sexsmith (125) at Lethbridge, found that various strains of wild oats reacted differently to the temperature and moisture regimes under which they were grown. Seed produced on plants grown at 15°C in moist soil were 30–100 percent more dormant than seed produced on plants grown at 27°C in dry soil. Sexsmith concluded that hot, dry conditions should cause the production of small quantities of wild oat seed with a limited dormancy that would therefore be readily destroyed. The reverse was true for seed produced under cooler, moist conditions.

Others at the Biosystematics Research Institute such as I.J. Bassett, R.J. Moore, and C. Frankton have made detailed studies of plantains and thistles. Bassett (15) studied seven species of *Plantago* growing in North America, one of which had been introduced. Up until this time, there had been confusion in separating some of the species, but Bassett demonstrated how it could be achieved. Each species was found to have a distinct geographical distribution. Similar research by Moore and Frankton (92) with four North American *Cirsium* (thistle) species clarified confusion among these and other species.

In 1945, the whole concept of weed control changed with the introduction of 2,4-dichlorophenoxyacetic acid, commonly known as 2,4-D, and 4-chloro-2-methylphenoxyacetic acid, MCPA. A.E. Smith of Regina reports (141) that 2,4-D was first tested in Canada during 1945, and within 5 years nearly 6 million ha of western Canadian cereal crops were sprayed annually. It is used almost universally because 2,4-D is a selective herbicide that kills broad-leaved plants but permits cereals to grow relatively unharmed. The introduction of 2,4-D sparked the development of many other selective chemical herbicides in the ensuing years.

Scientists such as W.H. Minshall at the Research Institute, London, Ontario, studied the mode of action of various herbicides. He demonstrated (91) in 1967 that applying nitrogen fertilizer either in the form of urea or as potassium nitrate in combination with the herbicide atrazine, increased the concentration of atrazine in the exudate of a cut tomato stem. On the west coast, a year later, Freeman and Finlayson (43) were among the first to find an interaction between a herbicide and an insecticide when applied together. In some instances, yields of cabbage were seriously reduced. Over a 7-year period they (44) found 29 of 215 herbicide–insecticide combinations were phytotoxic to field-seeded rutabaga. In the Maritime Provinces, Leefe (85) demonstrated that damage to strawberry plants was minimal when the herbicide simazine was applied on acid soils (pH 4.2) but was considerably increased when the acidity was reduced with limestone (pH 6.5). From these and many other such experiments it was learned that herbicides applied with care were useful.

Aquatic weeds

Open irrigation canals, shallow irrigation storage lakes, and open drainage ditches are subject to pollution by aquatic weeds. These weeds are frequently cleaned from canals and ditches with backhoes or draglines—an expensive but effective operation. In lakes, weed pollution is combatted with various underwater weed harvesting machines. These are usually of limited use because only the growing stems and leaves, not the roots, are removed.

R.J. Allan, Research Station, Lethbridge, was among the first in Canada to investigate the efficacy of selective herbicides in combatting aquatic weeds. By 1969 Allan, McDonald, and Hall (1) had determined that a number of herbicides were effective in controlling aquatic weeds provided the herbicides were injected 1 m, or more, below the water surface.

Biological control

One of the program objectives of the Institute for Biological Control, established in Belleville, Ontario, in 1929, was to discover insects and fungi that would attack weeds but leave crop plants undamaged. This was a difficult assignment requiring care in selecting potential parasites that would harm only the target weed and not themselves become a pest. Harris and Zwölfer (68) compiled a list of six major steps to be taken before an introduction should be made.

In an attempt to control tansy ragwort, *Senecio jacobaea*, which is poisonous to cattle, Harris, Wilkinson, Neary, and Thompson (67) successfully released the cinnabar moth, *Tyria jacobaeae*, in British Columbia, Nova Scotia, and Prince Edward Island. The larvae of the cinnabar moth feed exclusively on the leaves of tansy ragwort, stripping the plant of its leaves and keeping it under control but never, of course, eradicating it. In theory, as the moths increase in number the vigor of the ragwort plants is reduced, keeping the system in balance. Progeny of the cinnabar moths introduced and released by Harris and coworkers between 1961 and 1967 were recovered from three of the 15 release sites—one each in Prince Edward Island, Nova Scotia, and British Columbia. Three to four generations were required for these European moths to adapt to their new environment. By 1984 Harris, Wilkinson, and Myers (66) were able to report that "the cinnabar moth is now present throughout the ragwort-infested regions of Canada, including Newfoundland where no releases were made."

Another successful introduction has been a seed-eating weevil, *Rhinocyllus conicus*, from Europe that feeds on the seed heads of nodding thistle, *Carduus nutans*. Nodding thistle forms dense stands on dry, uncultivated grasslands in many parts of North America. The project began in 1962 with Canadian-sponsored surveys for insects on the weed in Europe. Initial releases of *R. conicus* in 1968 were made in Canada by Harris (65). By 1980, all stands of nodding thistle in central Saskatchewan were affected by the insect, reducing the thistle to such an extent that it no longer causes losses in pasture stands in Canada.

Other attempts to control weeds by using biological means have been less successful. Peschken (102) found the control of Canada thistle, *Cirsium arvense*, to be particularly difficult because eggs and larvae of several parasites are destroyed when farmers cultivate. However, Saidak and Marriage (122) at Harrow, Ontario, were able to kill both tops and roots of Canada thistle with amitrole and glyphosate herbicides to provide effective control.

The control of most weeds is now taken as a matter of course by farmers. The sale of wild oat herbicides in Canada total $320 million annually—the largest item in the current $700 million pesticide market. However, with this expenditure cereal farmers produce crops free from wild oats, consequently increasing their yields and net returns. Freyman and colleagues (49) at Lethbridge, in studying 70 years' data of dryland rotations, concluded that chemical weed control was the main contributing factor to the yield increases.

LOW TEMPERATURE RESEARCH

One of the first problems facing Saunders in 1887 was damage to plants caused by low temperatures in winter. Some apple and pear varieties could withstand the severe winters, whereas others had their fruit buds and upper branches damaged or killed. Sometimes trees were killed to their crowns.

His approach was to search for hardy varieties suitable for Canadian culture. He obtained many species and varieties first imported into Canada from Russia by the famous plantsman Charles Gibb of Abbotsford, Quebec. Among this material was *Malus baccata*, a Russian species of apple that was exceptionally hardy under all Canadian conditions. It was widely used as a root stock and intermediate framework for apples and pears. Its use meant that the main tree was protected, but the fruit-bearing branches of commercial varieties remained vulnerable to low temperature damage. Hardier fruiting varieties were needed for grafting.

The breeding of apples, pears, cherries, peaches, and other fruits had begun in Canada even before 1886. Most of the promising varieties were tested at all five Experimental Farms. Saunders started the lengthy procedure of growing seedlings from the hardier varieties and making controlled crosses (see Chapter 17). There was an interminable wait while plants matured enough to bear fruit and then several "test" winters were required to learn whether any selections could withstand severe cold. Were there any tests available, or could some be devised, to hasten the process and make it unnecessary to wait for the one "test" winter in 10?

M. MacArthur of the Horticulture Division, in 1940, soaked frozen plants in water before she applied an electric charge to the resulting solution. She found that the more easily the charge passed through the liquid the more susceptible the plant was to low temperature damage. This simple test worked because cells that were damaged by low temperatures released their contents into the water. Electricity passed through the solution with increasing ease as the cell contents became more concentrated in the solution. Various scientists refined the equipment used to measure the electrical charge but in all instances at least part of the plant was destroyed during the test.

In 1952, J. Wilner and W.A. Russell, scientists at the Forest Nursery Station, Indian Head, Saskatchewan, used MacArthur's methods in judging the winterhardiness of various woody species. By 1955 Wilner (158) had refined the method to where he was confident he could subject potted trees to artificially low temperatures and obtain results similar to those in the field. Andrews (7) at Lethbridge, working with winter wheat between 1953 and 1955, devised a system of germinating seed for 16 hours, subjecting them to freezing temperatures, and then growing the seedlings to determine which ones survived. Results were closer to field observations than were those of controlled freezing tests with growing plants. Shortly after Andrews made his studies, Lapins (80) at Summerland obtained good agreement between artificial freezing tests made in early winter or midwinter (but not late winter) and the known hardiness of 41 apple varieties. Recovery tests made by growing plant material in a greenhouse gave consistent results only when sufficient cold injury was produced during the artificial freezing tests. Electrical conductivity measurements of water extracts were most valuable in differentiating fine degrees of injury.

Wilner, now at Ottawa with the Division of Horticulture, in 1960, worked with W. Kalbfleisch and W.J. Mason of the Engineering Research Service to adapt (160) a new technique of inserting electrodes directly into the plant tissue to be tested. They showed that the electrical resistance between the electrodes was negatively correlated with the conductivity of the exudate, previously shown to be associated with cell injury caused by freezing temperatures. This new method kept plants intact and could be used on trees in orchards (159) and on

297

grasses (24). Weaver, Jackson, and Stroud (150) at the Research Station, Harrow, Ontario, while breeding winter-hardy peach trees, found Wilner's conductivity test valuable as a primary index of the hardiness of their cultivars.

In 1962 Siminovitch, Therrien, Wilner, and Gfeller (130) realized that leachates might come from nonliving wood or fiber tissues as well as from those injured by low temperatures. This could affect the conductivity attributable to winter injury. They rationalized that diffusible organic compounds such as amino acids and carbohydrates would be released only from cells killed by low temperatures in addition to the electolytic salts, and therefore devised a sensitive test for amino acids. Their theory proved valid with alfalfa, wheat, and apple, because the concentration of amino acids in the leachate increased in reverse proportion to the extent of survival following increased severity of freezing.

Protection against cold

298

Crops frequently are destroyed in Canada as elsewhere by damaging frosts, which may occur either in the spring during flowering or in the fall during harvest. In 1967 Siminovitch, Ball, Desjardins, and Gamble (128) of the Plant Research Institute, Ottawa, had the idea that a fire-fighting foam could be used to cover highly valued crops and protect them from frost damage. By adding a stabilizer to the foam they found that strawberries in the spring and tomatoes in the spring and fall could be protected overnight from temperatures as low as $-7°C$. The following morning the foam was either washed away with water or simply permitted to dissipate without damage to the plant. Later Siminovitch, Rheaume, Lyall, and Butler (129) learned that both tomatoes and strawberries could survive up to eight applications of foam. The treatment was judged by the Department of National Health and Welfare to be free from toxic ingredients. However, its expense has so far prevented its general use.

Further tests for cold hardiness

Roberts and Grant (120) at Lethbridge found that winter wheat changed its resistance to low temperatures during germination and development. In growth chambers, winter wheat had two periods of maximum resistance—the first when it was dry or just freshly moistened, the second when plants had four to six leaves. From these studies they were able to make fairly reliable predictions for field survival of individual varieties, which were useful in their winter wheat breeding program.

Winter damage is not always caused directly by freezing. Lebeau (81) noted that the fungus *Fusarium nivale* caused pink snow mold on winter wheat and several grass species following severe storms of wet snow. The condition could be avoided by growing varieties resistant to attack by snow mold.

Until 1973 only single killing temperatures had been used in testing winter wheats for winter injury. Fowler, Siminovitch, and Pomeroy (37) at Ottawa, noted that although such controlled testing saved time and afforded the opportunity of repeating a given test, these tests only related to field conditions when varieties with a wide range of frost hardiness were studied. They therefore devised a test that provided data as the temperature was lowered. By using a series of temperatures, they identified the temperature at which one-half the plants in a population were killed (LD^{50}). They found that the difference in temperature between no plants being killed and all plants being killed was usually less than 4 C degrees; that cold acclimation is controlled by many environmental variables of which temperature is perhaps the most important; and that ratings of varieties based upon lethal dose temperatures provided a more exact measure of hardiness than other methods. The following year Andrews, Pomeroy, and de la Roche (5) found that some varieties, although having hardened satisfactorily in the fall, declined in hardiness and vigor under ice. The Ottawa team continued to study the effects of winter flooding followed by ice encasement. They showed that cold hardiness decreased as plants were exposed to low temperature flooding of increased severity and duration. The loss of hardiness was attributed by Andrews and Pomeroy (4) to anaerobic processes. Andrews, Seaman, and Pomeroy (6) demonstrated that overwintering potential of winter wheat is reduced when it is cut in the fall before it has hardened and is further stressed by late winter flooding and icing.

Returning to peaches at Harrow, in 1975 Quamme, Layne, Jackson, and Spearman (112) used micro thermocouples (electrical thermometers) to determine the temperature at which peach buds were killed. They monitored the drop in temperature of buds as they were cooled and found that at a different temperature for each kind of bud the temperature rose suddenly, then dropped just as suddenly. This exothermic reaction coincided approximately with the temperature at which flower buds were killed. By using this method, W.G. Ronald at Morden, Manitoba, cooperating with H.A. Quamme and R.E.C. Layne at Harrow determined (113) the northern limit of growth for several cultivated and native *Prunus* species. From their data they were able to draw several conclusions that were of use in their peach breeding program.

The subject of low temperature research is not complete without mentioning the research of R.W. Salt at Lethbridge. For more than 20 years he studied the ways in which plants and insects survived extremely low temperatures. He found that insects, in particular (123), withstood temperatures below $-20°C$ because their body liquids supercooled. The physical mechanisms are complicated, and Salt explained a great deal as a result of his delicate detailed experimentation. One interesting discovery he made was that some insects produce glycerol, one of the antifreezes used to protect cooling systems in today's automobile!

References

1. Allan, J.R.; McDonald, S.; Hall, N.W. 1969. Carbon dioxide pressurized aquatic herbicide sprayer. Can. J. Plant Sci. 49:639–641.

2. Allen, W.R. 1972. Ontario superior fruit stock program. Can. Nurseryman (November), pp. 12–13.

3. Allen, W.R.; Dias, H.F. 1977. Properties of the single protein and two nucleic acids of tomato ring-spot virus. Can. J. Bot. 55:1028–1037.

4. Andrews, C.J.; Pomeroy, M.K. 1981. The effect of flooding pretreatment on cold hardiness and survival of winter cereals in ice encasement. Can. J. Plant Sci. 61:507–513.

5. Andrews, C.J.; Pomeroy, M.K.; de la Roche, I.A. 1974. Changes in cold hardiness of overwintering wheat. Can. J. Plant Sci. 54:9–15.

6. Andrews, C.J.; Seaman, W.L.; Pomeroy, M.K. 1984. Changes in cold hardiness, ice tolerance and total carbohydrates of winter wheat under various cutting regimes. Can. J. Plant Sci. 64:547–558.

7. Andrews, J.E. 1958. Controlled low temperature tests of sprouted seeds as a measure of cold hardiness of winter wheat varieties. Am. Soc. Hortic. Sci. 38:1–7.

8. Anonymous. 1944. Stockmen organize to combat plague menacing herds. Lethbridge Herald, 19 February, p. 11.

9. Atkinson, T.G.; Neal, J.L., Jr.; Larson, Ruby I. 1974. Root rot reaction in wheat: Resistance not mediated by rhizosphere or laimosphere antagonists. Phytopathology 64:97–101.

10. Bagnall, R.H.; Wetter, C.; Larson, R.H. 1959. Differential and serological relationships of potato virus M, potato virus S, and carnation latent virus. Phytopathology 49:435–442.

11. Bagnall, R.H.; Young, D.A. 1972. Resistance to virus S in the potato. Am. Potato J. 49:196–201.

12. Baker, Whiteford L. 1972. Eastern forest insects. U.S. Dep. Agric. For. Serv. Misc. Publ. 1175. 642 p.

13. Balch, R.E.; Bird, F.T. 1944. A disease of the European spruce sawfly, *Gilpinia hercyniae* (Htg.), and its place in natural control. Sci. Agric. 24:65–80.

14. Balch, R.E.; Reeks, W.A.; Smith, S.G. 1941. Separation of the European spruce sawfly in America from *Gilpinia polytoma* (Htg.) (Diprionidae, Hymenoptera) and evidence of its introduction. Can. Entomol. 73:198–203.

15. Bassett, I.J. 1967. Taxonomy of *Plantago* L. in North America: Sections *Holopsyllium* Pilger, *Palaeopsyllium* Pilger, and *Lamprosnatha* Decne. Can. J. Bot. 45:565–577.

16. Bradley, R.H.E. 1956. Effects of depth of stylet penetration on aphid transmission of potato virus Y. Can. J. Microbiol. 2:539–547.

17. Bradley, R.H.E.; Wade, C.V.; Wood, F.A. 1962. Aphid transmission of potato virus Y inhibited by oils. Virology 18:327–328.

18. Broadfoot, W.C. 1934. Studies on foot and root rot of wheat. IV. Effect of crop rotation and cultural practice on the relative prevalence of *Helminthosporium sativum* and *Fusarium* spp. as indicated by isolations from wheat plants. Can. J. Res. 10:115–124.

19. Cameron, A.E. 1918. Some blood-sucking flies of Saskatchewan. The Agric. Gaz. Can. 5:556–561.

300

20. Carson, Rachel. 1962. Silent spring. The Riverside Press, Cambridge, Mass. 368 p.

21. Chepil, W.S. 1946. Germination of weed seeds. I. Longevity, periodicity of germination, and vitality of seeds in cultivated soil. Sci. Agric. 26:307–346.

22. Chinn, S.H.F.; Tinline, R.D. 1964. Inherent germinability and survival of spores on *Cochliobolus sativus*. Phytopathology 54:349–352.

23 Conners, I.L., editor. 1972. Plant pathology in Canada. Can. Phytopathol. Soc., Univ. Man., Winnipeg. 251 p.

24. Cordukes, W.E.; Wilner, J.; Rothwell, V.T. 1966. The evaluation of cold and drought stress of turfgrasses by electrolytic and ninhydrin methods. Can. J. Plant Sci. 46:337–342.

25. Criddle, Norman. 1912. Wild oats and false wild oats: Their nature and distinctive characters. Dep. Agric. Bull. S-7. 11 p.

26. Daubeny, H.A.; Freeman, J.A.; Stace-Smith, R. 1975. Effect of tomato ringspot virus on drupelet set of red raspberry cultivars. Can. J. Plant Sci. 55:755–759.

27. Daubeny, H.A.; Stace-Smith, R.; Freeman, J.A. 1978. The occurrence and some effects of raspberry bushy dwarf virus in red raspberry. J. Am. Soc. Hortic. Sci. 103(4):519–522.

28. Davidson, T.R.; Sanford, G.B. 1954. Effect of age of potato foliage on expression of leaf-roll symptoms. Can. J. Bot. 32:311–317.

29. De Gryse, G.G. 1941. A brief note on the so-called "mysterious" disease of the spruce sawfly. For. Chron. 17:43–44.

30. De Pauw, R.M.; McBean, D.S.; Buzinski, S.R.; Townley-Smith, T.F.; Clarke, J.M.; McCaig, T.N. 1982. Leader hard red spring wheat. Can. J. Plant Sci. 62:231–232.

31. Dias, Humberto F. 1970. Transmission of cucumber necrosis virus by *Olpidium cucurbitacearum* Barr & Dias. Virology 40:828–839.

32. Dyer, E.D.A.; Wright, K.H. 1967. Striped ambrosia beetle *Trypodendron lineatum* (Oliv.). *In* Davidson, A.G.; Prentice, R.M. Important forest insects and diseases of mutual concern to Canada, the United States and Mexico. Dep. For. Rural Dev., Ottawa. 248 p.

33. Ewen, Al. B.; Mukerji, M.K. 1980. Evaluation of *Nosema locustae* (Microsporida) as a control agent of grasshopper populations in Saskatchewan. J. Invertebr. Pathol. 35:295–303.

34. Farstad, C.W. 1938. Thelyotokous parthenogenesis in *Chephus cinctus* Nort. (Hymenoptera: Cephidae) Can. Entomol. 70:206–207.

35. Fletcher, James. 1897. Weeds. Can. Dep. Agric. Bull. 28. 43 p.

36. Fletcher, Jas. 1897. Report of the entomologist and botanist. Pages 229–231 *in* Experimental Farms Reports for 1896. Queen's Printer. Ottawa.

37. Fowler, D.B.; Siminovitch, D.; Pomeroy, M.K. 1973. Evaluation of an artificial test for frost hardiness in wheat. Can. J. Plant Sci. 53:53–59.

38. Fredeen, F.J.H. 1956. Black flies (Diptera: Simuliidae) of the agricultural areas of Manitoba, Saskatchewan, and Alberta. Proc. 10th Int. Congr. Entomol. Montreal. 1956:3, 819–823.

39. Fredeen, F.J.H. 1969. Outbreaks of the black fly *Simulium arcticum* Malloch in Alberta. Quaest. Entomol. 5:341–372.

40. Fredeen, F.J.H. 1975. Effects of a single injection of methoxychlor black-fly larvacide on insect larvae in a 161-km (100-mile) section of the North Saskatchewan river. Can. Entomol. 107:807–817.

301

41. Fredeen, F.J.H.; Arnason, A.P.; Berck, B.; Rempel, J.G. 1953. Further experiments with DDT in the control of *Simulium arcticum* Mall. in the North and South Saskatchewan rivers. Can. J. Agric. Sci. 33:379–393.

42. Fredeen, F.J.H.; Rempel, J.G.; Arnason, A.P. 1951. Egg-laying habits, overwintering stages, and life-cycle of *Simulium arcticum* Mall. (Diptera: Simuliidae). Can. Entomol. 83:73–74.

43. Freeman, Jack A.; Finlayson, D.G. 1968. Evidence of incompatibility between a herbicide and insecticides in direct-seeded cabbage. Can. J. Plant Sci. 48:415–416.

44. Freeman, Jack A.; Finlayson, D.G. 1977. Response of rutabaga to combinations of herbicides and insecticides. J. Am. Soc. Hortic. Sci. 102:29–31.

45. Freeman, J.A.; Mellor, Frances C. 1962. Influences of latent viruses on vigor, yield and quality of British Sovereign strawberries. Can. J. Plant Sci. 42:602–610.

46. Freeman, J.A.; Stace-Smith, R. 1965. Systemic invasion of the root system of raspberry plants by the component viruses of red raspberry mosaic. Can. J. Plant Sci. 45:295–297.

47. Freeman, J.A.; Stace-Smith, R. 1968. Effects of tomato ringspot virus on plant and fruit development of raspberries. Can. J. Plant Sci. 48:25–29.

48. Freeman, J.A.; Stace-Smith, R.; Daubeny, H.A. 1975. Effects of tomato ringspot virus on the growth and yield of red raspberry. Can. J. Plant Sci. 55:749–754.

49. Freyman, S.; Palmer, C.J.; Hobbs, E.H.; Dormaar, J.F.; Schaalje, G.B.; Moyer, J.R. 1982. Yield trends on long-term dryland wheat rotations at Lethbridge. Can. J. Plant Sci. 62:609–619.

50. Fitzpatrick, R.E.; Stace-Smith, R.; Mellor, F.C. 1954. Heat inactivation of some strawberry and raspberry viruses. Proc. Can. Phytopathol. Soc. 22:13.

51. Gage, S.H.; Mukerji, M.K. 1978. Crop losses associated with grasshoppers in relation to economics of crop production. J. Econ. Entomol. 71:487–498.

52. George, J.A.; Davidson, T.R. 1963. Pollen transmission of necrotic ring spot and sour cherry yellows viruses from tree to tree. Can. J. Plant Sci. 43:276–288.

53. Gibson, Arthur. 1951. Entomology in Canada, with special reference to developments within the Dominion Department of Agriculture. Manuscript located at Agriculture Canada Library, Ottawa.

54. Gilpatrick, J.D.; Weintraub, M. 1952. An unusual type of protection with the carnation mosaic virus. Science 115:701–702.

55. Güssow, H.T. 1909. A serious potato disease occurring in Newfoundland. Dep. Agric. Bull. 63. 10 p.

56. Groh, Herbert. 1923. A survey of weed control and investigation in Canada. Sci. Agric. 3:415–420.

57. Hadwin, S. 1923. Insects affecting livestock. Can. Dep. Agric. Bull. 29, n.s. 32 p.

58. Hamilton, R.I.; Nichols, C. 1978. Serological methods of detection of pea seed-borne mosaic virus in leaves and seeds of *Pisum sativius*. Phytopathology 68:539–543.

59. Handford, R.H. 1964. The identification of nymphs of the genus *Melanoplus* of Manitoba and adjacent areas. Sci. Agric. 26:147–180.

60. Hansen, A.J. 1979. Inhibition of apple chlorotic leaf spot virus in *Chenopodium quinoa* by ribavirin. Plant Dis. Rep. 63:17–20.

61. Hansen, A.J. 1984. Effect of ribavirin on green ring mottle causal agent and necrotic ringspot virus in *Prunus* species. Plant Dis. 68:216–218.

62. Haufe, W.O. 1980. Control of black flies in the Athabasca River. Alberta Environ., Edmonton. 38 p.

63. Hardman, J.M.; Smoliak, J. 1980. Potential economic impact of rangeland grasshoppers (Acrididae) in southeastern Alberta. Can. Entomol. 112:277–284.

64. Harris, C.R.; Svec, H.J.; Tolman, J.H.; Tomlin, A.D.; McEwan, F.L. 1981. A rational integration of methods to control onion maggot in southwestern Ontario. Proc. Br. Crop Prot. Conf. 3:789–799.

65. Harris, P. 1984. *Carduus nutans* L., nodding thistle and *C. acanthoides* L., plumeless thistle (Compositae). Pages 115–124 *in* Kelleher, J.S.; Hulme, M.A., editors. Biological control programmes against insects and weeds in Canada 1969–1980. Commonwealth Agricultural Bureau, Slough, England.

66. Harris, P.; Wilkinson, A.T.S.; Myers, J.H. 1984. *Senecio jacobaea* L., tansy ragwort (Compositae). Pages 195–201 *in* Kelleher, J.S.; Hulme, M.A., editors. Biological control programmes against insects and weeds in Canada 1969–1980. Commonwealth Agricultural Bureau, Slough, England.

67. Harris, P.; Wilkinson, A.T.S.; Neary, M.E.; Thompson, L.S. 1971. *Senecio jacobaea* L., tansy ragwort (Compositae). *In* Biological control programmes against insects and weeds in Canada 1959–1968. Commonw. Inst. Biol. Control Tech. Commun. 4:97–104.

68. Harris, P.; Zwölfer, H. 1968. Screening of phytophagous insects for biological control of weeds. Can. Entomol. 100:295–303.

69. Hay, J.R. 1963. Experiments on the mechanism of induced dormancy in wild oats, *Avena fatua* L. Can. J. Bot. 40:191–202.

70. Hearle, Eric. 1938. Insects and allied parasites injurious to livestock and poultry in Canada. Can. Dep. Agric. Publ. 604. 108 p.

71. Jacoli, G.G. 1974. Translocation of mycoplasma-like bodies through sieve pores in plant tissues infected with aster yellows disease. J. Ultrastruct. Res. 46:34–42.

72. Katznelson, H. 1965. Nature and importance of the rhizosphere. Pages 187–207 *in* Ecology of soil-borne plant pathogens, 1st International Symposium on Factors Determining the Behavior of Plant Pathogens in Soil, Berkley, Calif. 1963.

73. Kemp, H.J. 1934. Studies on solid-stemmed wheat varieties in relation to wheat-stem sawfly control. Sci. Agric. 15:30–38.

74. Khan, M.A. 1968. Extermination of cattle grubs (*Hypoderma* spp.) on a regional basis. Vet. Rec. 83:97–101.

75. Khan, M.A.; Connell, R.; Darcel, C. le Q. 1960. Immunization and parenteral chemotherapy for the control of cattle grubs *Hypoderma lineatum* (De Vill.) and *H. bovis* (L.) in cattle. Can. J. Comp. Med. 26:177–180.

76. Klein, Kurt K. 1979. The economics of warble fly control. Can. Farm Econ. 14(4):20–27.

77. Klinkowski, M. 1970. Catastrophic plant diseases. Annu. Rev. Phytopathol. 8:37–60.

78. Koch, L.W. 1955. The peach replant problem in Ontario. I. Symptomatology and distribution. Can. J. Bot. 33:450–460.

79. Kuntz, S.E.; Drummond, R.D.; Weintraub, J. 1984. A pilot test to study the use of the sterile insect technique for eradication of cattle grubs. Prev. Vet. Med. 2:523–527.

80. Lapins, K. 1960. Artificial freezing of 1-year-old shoots of apple varieties. Can. J. Plant Sci. 4:381–393.

81. Lebeau, J.B. 1968. Pink snow mold in southern Alberta. Can. Plant Dis. Surv. 48(4):130–131.

303

82. Ledingham, R.J. 1970. Effects of straw and nitrogen on common root rot of wheat. Can. J. Plant Sci. 50:175–179.

83. Ledingham, R.J.; Atkinson, T.G.; Horricks, J.S.; Mills, J.T.; Piening, L.J.; Tineline, R.D. 1973. Wheat losses due to common root rot in the prairie provinces of Canada, 1969–71. Plant Dis. Surv. 53:113–122.

84. Ledingham, R.J.; Sallans, B.J.; Wenhardt, A. Influence of cultural practices on incidence of common rootrot in wheat in Saskatchewan. Can. J. Plant Sci. 40:310–316.

85. Leefe, J.S. 1968. Effect of soil pH on the phytotoxicity of simazine to strawberries. Can. J. Plant Sci. 48:424–425.

86. LeRoux, E.J. 1957. Importance of second-generation larvae of the codling moth, *Carpocapsa pomonella* (L.) (Lepidoptera: Tortricidae), on apple in Quebec. Ann. Entomol. Soc. Que. 3:75–80.

87. LeRoux, E.J. 1976. Biological control. Pages 287–295 *in* Fisher, D.V.; Upshall, W.H., editors. History of fruit growing and handling in United States of America and Canada 1860–1972. Regatta City Press Ltd., Kelowna, B.C.

88. Lott, T.B. 1943. Transmissible "twisted leaf" of sweet cherry. Sci. Agric. 23:439–441.

89. Lott, T.B.; Keane, F.W.L. 1960. The host range of the virus of twisted leaf of cherry. Plant Dis. Rep. 44:240–242.

90. McGugan, B.M.; Coppel, H.C. 1962. Biological control of forest insects— 1910–1958. *In* A review of the biological control attempts against insects and weeds in Canada. Commonw. Inst. Biol. Control Tech. Commun. 2:25–127.

91. Minshall, Wm. Harold. 1967. Effects of nitrogenous materials on translocation and stump exudation in root systems of tomato. Can. J. Bot. 46:363–376.

92. Moore, R.J.; Frankton, C. 1967. Cytotaxonomy of foliose thistles (*Crisium* spp. aff. *C. foliosum*) of western North America. Can. J. Bot. 45:1733–1749.

93. Mountain, W.B.; Boyce, H.R. 1985. The peach replant problem in Ontario. V. The relation of parasitic nematodes to regional differences in severity of peach replant failure. Can. J. Bot. 36:125–134.

94. Mountain, W.B.; Boyce, H.R. 1985. The peach replant problem in Ontario. VI. The relation of *Pratylenchus penetrans* to the growth of young peach trees. Can. J. Bot. 36:135–151.

95. Mukerji, M.K.; Ewen, Al. B.; Craig, C.H.; Ford, R.J. 1981. Evaluation of insecticide-treated bran baits for grasshopper control in Saskatchewan (Orthoptera: Acrididae) Can. Entomol. 113:705–710.

96. Mukerji, M.K.; Gage, S.H. 1978. A model for estimating hatch and mortality of grasshopper egg populations based on soil moisture and heat. Ann. Entomol. Soc. Am. 71:183–190.

97. Mukerji, M.K.; Gage, S.H.; Randell, R.L. 1977. Influence of embryonic development and heat on population trend of three grasshopper species (Orthoptera: Acrididae) in Saskatchewan. Can. Entomol. 109:229–236.

98. Munroe, E.G. 1971. Status and potential of biological control in Canada. *In* Biological control programmes against insects and weeds in Canada 1959–1968. Commonw. Inst. Biol. Control Tech. Commun. 4:213–255.

99. Neilson, M.M.; Morris, R.F. 1964. The regulation of European spruce sawfly numbers in the maritime provinces of Canada from 1937 to 1963. Can. Entomol. 96:773–784.

100. Newton, W.; Edwards, H.I. 1936. Virus studies. I. The production of antisera in chickens by inoculation with potato X. Can. J. Res. 14:412–414.

101. Patrick, Z.A. 1955. The peach replant problem in Ontario. II. Toxic substances from microbial decomposition products of peach root residues. Can. J. Bot. 33:461–486.

102. Peschken, V.P. 1984. *Cirsium arvense* (L.) Scop., Canada thistle (Compositae). Pages 139–146 *in* Kelleher, J.S.; Hulme, M.A., editors. Biological control programmes against insects and weeds in Canada 1969–1980. Commonwealth Agricultural Bureau, Slough, England.

103. Pickett, A.D.; Putman, Wm.L.; LeRoux, E.J. 1956 (1958). Progress in harmonizing biological and chemical control of orchard pests in eastern Canada. Tenth Int. Congr. Entomol. Proc. 3:169–174.

104. Pickett, A.D. 1976. Integrated control. Pages 292–295 *in* Fisher, D.V.; Upshall, H.W., editors. History of fruit growing and handling in United States of America and Canada 1860–1972. Regatta City Press, Kelowna, B.C.

105. Pickford, R. 1960. Survival, fecundity, and population growth of *Melanoplus bilituratus* (Wllk.) (Orthoptera: Acrididae) in relation to date of hatching. Can. Entomol. 92:1–10.

106. Platt, A.W. 1941. The influence of some environmental factors on the expression of the solid-stem character in certain wheat varieties. Sci. Agric. 21:139–151.

107. Platt, A.W.; Farstad, C.W. 1946. The reaction of wheat varieties to wheat stem sawfly attack. Sci. Agric. 26:231–247.

108. Platt, A.W.; Larson, Ruby. 1943. An attempt to transfer solid-stem from *Triticum durum* to *T. vulgare* by hybridization. Sci. Agric. 24:214–220.

109. Prebble, M.L.; Graham, K. 1957. Studies of attack by Ambrosia beetles in softwood logs on Vancouver Island, British Columbia. For. Sci. 3:90–112.

110. Proudfoot, K.G.; Olsen, O.A. 1976. Pink Pearl and Mirton Pearl: Two wart resistant potato cultivars for Newfoundland. Can. J. Plant Sci. 56:763–764.

111. Proverbs, M.D.; Newton, J.R.; Logan, D.M. 1969. Codling moth control by release of radiation-sterilized moths in a commercial apple orchard. J. Econ. Entomol. 62:1331–1334.

112. Quamme, H.A.; Layne, R.E.C.; Jackson, H.O.; Spearman, G.A. 1975. An improved exotherm method for measuring cold hardiness of peach flower buds. Hortic. Sci. 10:521–523.

113. Quamme, H.A.; Layne, R.E.C.; Ronald, W.G. 1982. Relationship of supercooling to cold hardiness and the northern distribution of several cultivated and native *Prunus* species and hybrids. Can. J. Plant Sci. 62:137–148.

114. Ragetli, H.W.J.; Weintraub, M. 1964. Immuno-osmophoresis, a rapid and sensitive method for evaluating viruses. Science 144:1023–1024.

115. Raine, J. 1967. Leafhopper transmission of witches' broom and clover phyllody viruses from British Columbia to clover, alfalfa, and potato. Can. J. Bot. 45:441–445.

116. Randell, R.L.; Mukerji, M.K. 1974. A technique for estimating hatching of natural egg populations of *Melanoplus sanguinipes* (Orthoptera: Acrididae). Can. Entomol. 106:801–812.

117. Rankin, W.H.; Hockey, J.F. 1922. Mosaic and leaf curl (yellows) of the cultivated red raspberry. Phytopathology 12:253–264.

118. Reeks, W.A. 1953. The establishment of introduced parasites of the European spruce sawfly *Diprion hercyniae* (Htg.) (Hymenoptera: Diprionidae) in the Maritime Provinces. Can. J. Agric. Sci. 33:405–429.

305

119. Rich, G.B. 1965. Systemic treatments for control of cattle grubs *Hypoderma* spp. in an isolated range herd. Can. J. Anim. Sci. 45:165–172.

120. Roberts, D.W.A.; Grant, M.N. 1968. Changes in cold hardiness accompanying development in winter wheat. Can. J. Plant Sci. 48:369–376.

121. Robertson, R.H. 1980. Antibody production in cattle infected with *Hypoderma* spp. Can. J. Zool. 58:245–251.

122. Saidak, W.J.; Marriage, P.B. 1976. Response of Canada thistle varieties to amitrole and glyphosate. Can. J. Plant Sci. 56:211–214.

123. Salt, R.E. 1969. The survival of insects at low temperatures. *In* Dormancy and survival. Symp. Soc. Exp. Biol. XXIII:331–350.

124. Sallans, B.J. 1948. Interrelations of common root rot and other factors with wheat yields in Saskatchewan. Sci. Agric. 28:6–20.

125. Sexsmith, J.J. 1969. Dormancy of wild oat seed produced under various temperature and moisture conditions. Weed Sci. 17:405–407.

126. Shemanchuk, Joseph A. 1981. Repellant action of permethrin, cypermethrin and resmethrin against black flies (*Simulium* spp.) attacking cattle. Pestic. Sci. 12:412–416.

127. Shepard, James F.; Claflin, Larry E. 1975. Critical analyses of the principles of seed potato certification. Annu. Rev. Phytopathol. 13:271–293.

128. Siminovitch, D.; Ball, W.L.; Desjardins, R.; Gamble, D.S. 1967. Use of protein-based foams to protect plants against frost. Can. J. Plant Sci. 47:11–17.

129. Siminovitch, David; Rheaume, Bernard. 1969. Stable nontoxic protein-based foam protects field plantings of crops from damage by frost. Cell Biol. Res. Inst. Contrib. 653. Ottawa.

130. Siminovitch, D.; Therrien, H.; Wilner, J.; Gfeller, F. 1962. The release of amino acids and other ninhydrin-reacting substances from plant cells after injury by freezing; a sensitive criterion for the estimation of frost injury in plant tissues. Can. J. Bot. 40:1267–1269.

131. Simmonds, P.M.; Ledingham, R.J. 1937. A study of the fungous flora of wheat roots. Sci. Agric. XVIII:49–59.

132. Simmonds, P.M.; Sallans, B.J. 1933. Further studies on amputations of wheat roots in relation to diseases of the root system. Sci. Agric. XIII:439–448.

133. Simmonds, P.M.; Sallans, B.J.; Ledingham, R.J. 1950. The occurrence of *Helminthosporium sativum* in relation to primary infections in common rootrot of wheat. Sci. Agric. 30:407–417.

134. Singh, R.P. 1970. Seed transmission of potato spindle tuber virus in tomato and potato. Am. Potato J. 47:225–227.

135. Singh, R.P.; Bagnall, R.H. 1968. Infectious nucleic acid from host tissues infected with the potato spindle tuber virus. Phytopathology 58:696–699.

136. Singh, R.P.; Clark, M.C. 1971. Infectious low-molecular weight ribonucleic acid from tomato. Biochem. Biophys. Res. Comm. 44:677–1083.

137. Sinha, R.C. 1984. Transmission mechanism of mycoplasma-like organisms by leafhopper vectors. Pages 93–109 *in* Harris, K.F., editor. Current topics in vector research. Vol. 2. Praiger Scientific, N.Y.

138. Sinha, R.C.; Chiykowski, L.N. 1984. Purification and serological detection of mycoplasma-like organisms from plants affected by peach eastern X-disease. Can. J. Plant Pathol. 6:200–205.

139. Slykhuis, J.T. 1953. Wheat streak mosaic in Alberta and factors related to its spread. Can. J. Agric. Sci. 33:196–197.

306

140. Slykhuis, J.T.; Yorston, J.; McMullen, R.D.; Li, Thomas S.C. 1980. Current status of little cherry disease in British Columbia. Can. Plant Dis. Surv. 60(4):37–42.

141. Smith, Allan E. 1982. Herbicides and the soil environment in Canada. Can. J. Soil Sci. 62:433-460.

142. Smith, D.S. 1972. Crowding in grasshoppers. II. Continuing effects of crowding on subsequent generations of *Melanoplus sanguinipes* (Orthoptera: Acrididae). Environ. Entomol. 1:314–317.

143. Timmons, F.L. 1970. A history of weed control in the United States and Canada. Weed Sci. 18:294–307.

144. Townshend, J.L.; Cline, R.A.; Dirks, V.A.; Marks, C.F. 1984. Assessment of turfgrasses for the management of *Pratylenchus penetrans* and *Pratylenchus projectus* in orchards. Can. J. Plant Sci. 64:355–360.

145. Tremaine, J.H.; Willison, R.S.; Allen, W.R. 1963. Three virus antigens associated with sour cherry necrotic ring spot. Phytopathol. Mediterr. 2:191–192.

146. Turnbull, A.L.; Chant, D.A. 1961. The practice and theory of biological control of insects in Canada. Can. J. Zool. 39:697–752.

147. Twinn, C.R. 1952. A review of studies of blood-sucking flies in northern Canada. Can. Entomol. 84:22–28.

148. Verma, P.R.; Chinn, S.H.F.; Crowle, W.L.; Spurr, D.T.; Tinline, R.D. 1981. Effect of imazalil seed treatment on common root rot and grain yields of cereal cultivars. Can. J. Plant Pathol. 3:239–243.

149. Ward, G.M.; Durkee, A.B. 1965. The peach replant problem in Ontario. III. Amygdalin content of peach tree tissues. Can. J. Bot. 34:419–422.

150. Weaver, G.M.; Jackson, H.O.; Stroud, F.D. 1968. Assessment of winterhardiness in peach cultivars by electric impedance, scion diameter and artificial freezing studies. Can. J. Plant Sci. 48:37–47.

151. Weintraub, J. 1961. Inducing mating and oviposition of the warble flies *Hypoderma bovis* (L.) and *H. lineatum* (De Vill.) (Diptera: Oestridae) in captivity. Can. Entomol. 43:149–156.

152. Weintraub, J.; Rich, G.B.; Thompson, C.O.M. 1959. Timing the treatment of cattle with trolene for systemic control of the cattle grubs *Hypoderma lineatum* (De Vill.) and H. bovis (L.) in Alberta and British Columbia. Can. J. Anim. Sci. 39:50–57.

153. Weintraub, M.; Ragetli, H.W.J.; Leung, Esther. 1976. Elongated virus particles in plasmodesmata. J. Ultrastruct. Res. 56:351–364.

154. Welsh, Maurice F.; Nylund, George. 1965. Elimination and separation of viruses in apple clones by exposure to dry heat. Can. J. Plant Sci. 45:443–454.

155. Wensley, R.N. 1956. The peach replant problem in Ontario. IV. Fungi associated with replant failure and their importance in fumigated and nonfumigated soils. Can. J. Bot. 34:967–981.

156. Westdal, P.H.; Barrett, C.F.; Richardson, H.P. 1961. The six-spotted leafhopper, *Macrosteles fascifrons* (Stål) and aster yellows in Manitoba. Can. J. Plant Sci. 41:320–331.

157. Wilks, J.M.; Welsh, M.F. 1955. Sweet cherry foliage indicator hosts for the virus that causes little cherry. Can. J. Agric. Sci. 35:595–600.

158. Wilner, J. 1955. Results of laboratory tests for winter hardiness of woody plants by electrolytic methods. Proc. Am. Soc. Hortic. Sci. 66:93–99.

159. Wilner, J. 1960. Relative and absolute electrolytic conductance tests for frost hardiness of apple varieties. Can. J. Plant Sci. 40:630–637.

160. Wilner, J.; Kalbfleisch, W.; Mason, W.J. 1960. Note on two electrolytic methods for determining frost hardiness of fruit trees. 1960. Can. J. Plant Sci. 40:563–565.
161. Wright, K.H.; Lejeune, R.R. 1967. Douglas-fir beetle *Dendroctonus pseudotsugae* Hopk. *In* Davidson, A.G.; Prentice, R.M. Important forest insects and diseases of mutual concern to Canada, the United States and Mexico. Dep. For. Rural Dev., Ottawa. 248 p.
162. Wright, N.S.; MacCarthy, H.R.; Forbes, A.R. 1970. Epidemiology of potato leaf roll virus in the Fraser River delta of British Columbia. Am. Potato J. 47:1–8.
163. Young, D.A.; Davies, H.T.; Johnston, G.R. 1979. Jemseg: A new early high-yielding potato variety with high resistance to virus Y and immune to virus X. Am. Potato J. 56:325–328.

308

Chapter 19
Engineering Research

Engineering research, in partnership with biological research, plays a role of ever increasing importance to the solving of agricultural problems. This chapter provides a few illustrations of the interplay and shows how engineering research directly affects farmers as they strive to produce food of world-class quality at competitive prices. Where appropriate, some examples of collaborative work by engineers appear in other chapters.

INSTRUMENTATION AND MECHANIZATION FOR RESEARCH

When Experimental Farms started in 1886, field work was accomplished with readily available horse-drawn farm machinery and gardeners' hand tools. Designs of field and laboratory experiments were unsophisticated; replication and randomization of treatments at one location had not been introduced. Sir Charles Saunders and his father before him were quite aware that experiments should be repeated at each of the experimental farms involved and over several years, in order to sample various weather and soil conditions. Both Saunders used plots of many kinds of cereals as part of their breeding programs. Some plots consisted of only 30 or 40 plants, each plant derived from an individual head, whereas other plots, ranging in size from one-tenth of an acre to one acre (0.04 to 0.4 ha), consisted of plants grown to increase seed from a desirable selection. Seeds were counted or weighed by hand.

Sir Charles Saunders was probably the first scientist in Experimental Farms to design and fabricate equipment specifically for his own research. After chewing his wheat samples to determine their gluten content, he ground flour from those selected, and, to test the wheat's quality, he baked small loaves of bread in an oven of his own design. (Laboratory equipment was available commercially for the chemists, plant pathologists, and entomologists.)

R.A. Fisher in Great Britain was the first scientist to apply statistical methods to agricultural experiments; he started in 1920. C.H. Goulden of Winnipeg studied under Fisher at Rothamsted for the summer of 1930 and introduced statistics to Experimental Farms and Science Service in 1931 (24). He was followed by G.B. Oakland in 1950. Because of the variation found within biological material, agriculture showed the way for others by devising mathematical systems to measure the extent of the variation and to identify differences among treatments. This required that each treatment or variety be repeated from three to six times at each location. The wheat, barley, and oat breeding programs at experimental farms increased as new stations were established. Hand methods of preparing, seeding, harvesting, and threshing small samples could no longer cope with the increased volume. Mechanization was imperative.

Field plot equipment

H.J. (Shorty) Kemp, an agronomist at the Swift Current Experimental Station, led the way in mechanization. Immediately following his graduation in agriculture he worked at the Experimental Farm, Brandon, for a year, then spent a year at Indian Head before going to Swift Current in 1921 to work in the Field Husbandry Section.

A rod row (5 m) was firmly established in 1925 by cerealists in Canada and the United States as the standard plot length. Swift Current had adopted it in 1924. The rod was a convenient measure because only about 250 seeds were required to plant a row and results were converted to bushels per acre by multiplying plot yields in kilograms by 130. Cerealists had difficulty, however, because germination was not uniform, harvesting with a sickle, sometimes even a jackknife, was tedious, and threshing the small lots was done by hand.

In 1926 Kemp devised a machine capable of placing seed within a standard plot. His first seeder had a funnel made of sheet metal and designed to distribute seed along one row of the plot as the seeder was moved forward. The machine marked succeeding drill rows, opened furrows at an even depth, deposited a specific quantity of seed along the row, covered the seed, and firmed the soil. Although it improved plant stands and saved time and labor, it was not perfect. Seed tended to clump in the hopper and thus in the row, causing bunching and weak plants, and leaving spaces in which weeds could grow. As an improvement, Kemp devised and introduced the V-belt seeder in 1930 (16), which became the standard plot seeder and is still used. It had a continuous molded rubber belt V-shaped in cross section that was stretched over two pulleys to permit seed to be distributed in a continuous ribbon along the bottom of the belt. The rear pulley was geared to the machine's ground wheels and, as the seeder was moved over a plot the belt rotated, uniformly depositing seed into a furrow. Hand-pushed single-row seeders greatly increased the accuracy of plot work, reduced the labor needed, and were useful in placing fertilizer for small plot trials.

The first attempt to mechanize seeders was at the Scott Experimental Station in 1952, when Kusch, Keys, and Nadan (19) mounted four V-belt units on the back of a riding garden tractor. A.D. Smith at Lethbridge constructed a similar machine; however, it suffered from engine vibration, in addition to which Alberta's chinook winds blew seed from the V-belt. D.T. Anderson therefore designed and built the first trailer-type four-row V-belt seeder with a cab enclosure to protect two operators and the four seeders. It could function in winds far in excess of those that the single-row hand-pushed rod-row seeder would tolerate. Similar seeders were built with 6-inch-row (15-cm-row) spacings for grass and legume plots.

Andrews and coworkers (1) observed that those rows packed by the tractor or the trailer wheels (usually the outside rows of a four-row plot) produced up to twice the yield of other rows. A.D. Smith therefore devised a powered four-row seeder, using four pneumatic drive-wheels in the front and four smaller pneumatic wheels at the back, each centered on a seeded row (30), which resulted in remarkably uniform stands. Its construction was improved upon at the Swift Current workshops and it became available to all other stations.

W.J. Cherewick, a plant pathologist in the Rust Laboratory at Winnipeg, wanted to be able to seed up to 5 inches (12.5 cm) deep in order to favor the development of smut. Kemp's V-belt seeder was not designed for such depths and so Cherewick made several modifications to it. However, these modifications made the machine top-heavy, resulting in vibration of the V-belt and agitation of the seed, which, in turn, changed the spacing of plants in the row. To solve the problem, Cherewick devised a new method of distributing seed in a furrow opened by a double disc (8). Up to 55 cc of seed was poured into a small metal container machined to fit snugly over the tip of a metal cone. Lifting the container slightly permitted the seed to flow uniformly down the outside of the cone onto an attached circular plate. The cone and plate were geared to rotate once for each row length to permit seed to be fed uniformly down a tube into an opened furrow. Seed on the cone and plate was protected from wind and vibration by a plastic guard. The cone principle was as revolutionary in 1953 as the V-belt had been in 1930. By 1960 the advantages of the new seeder were recognized and it soon became the standard seeder for experimental work throughout the world.

Both the Kemp V-belt and Cherewick cone seeders are used to place fertilizer and other granular products onto field plots. However, with the V-belt seeder uniform distribution of granular products is dependent upon the skill of the operator, and with the Cherewick seeder grit often lodges in the machined interface between the cone and the plate. These two problems motivated F.B. Dyck of the Swift Current Research Station to invent the cone-belt seeder (10) in 1976. It combines the advantages and eliminates the disadvantages of each of the previous seeders by having seed or fertilizer delivered by the cone to a flat belt running on the vertical face of the cone. This forms a vee to cradle the product until it falls through an aperture to the placement mechanism. The Dyck cone-belt has been widely adopted by plant scientists.

Seeding was only one of three operations that Kemp mechanized. He also developed rod-row harvesters, which cut and bundled grain plants ready to sack and thresh. The first one consisted of a short length of a hay-mower cutter bar powered by a small gasoline engine, but in 1925 small gasoline engines were still too cumbersome. The second one was made from a pair of hedge shears with extra long handles, the weight of which, when a handle was dropped, cut the grain. The third and successful machine was a rotary shear mower, which was powered through gears linked to the two ground wheels of the harvester that one operator could easily push. Two overlapping rotary knives rotated in opposite directions and cut standing grain, laying it into a sheet-metal basket. The rotary harvester was replaced in 1948 by the sickle-bar mower, which was attached to a garden tractor and could cut four rows (the standard plot width) at a time. Kemp continued to improve his harvesters as commercial garden tractors improved.

Kemp's third innovation was a plot threshing machine. He realized the importance of cleaning grain out of the machinery to prevent mixing one variety with another and of having the threshed seed free from all straw and chaff before weighing. He designed and built a thresher that could be powered by either a gasoline or an electric motor. It was eventually replaced by an American design that could be cleaned more easily. Kemp also designed and made a smaller thresher for single heads. Before the development of this thresher, seed had to be rubbed out by hand—an onerous task, particularly when thousands of heads were involved.

The Vulcan Iron Works of Winnipeg manufactured and sold hundreds of Kemp's harvesters and threshers in both Canada and the United States. In 1970 some of the harvesters and threshers were superseded by a self-propelled plot combine manufactured in West Germany. The Kemp V-belt and the Cherewick cone seeders have been improved upon by engineers at Swift Current and at Ottawa and mounted as self-propelled four-row seeders. They are being manufactured commercially in both Swift Current and Winnipeg.

Shorty Kemp moved to the Experimental Station, Saanichton, in 1946 and continued developing harvesters and threshers. He also widened his interests and designed seed cleaners, graders, and dehullers. He could not resist the lure of Saanichton's horticultural crops, so he designed and built small-fruit cultivators, sprayers, bulb-treating equipment, pea shellers, plant growth chambers, and several pieces of laboratory equipment. Kemp retired in 1959 secure in the knowledge that he had helped hundreds of scientists accelerate their research.

Today the rod row has gone the way of other Imperial measurements, not because of the Metric Commission but rather because of the requirements of international scientific journals. Long before metrification in Canada, the scientific community used and reported data measured in metric units and therefore the rod row has been replaced by the 5 m row and, later, the 3 m row. It is interesting that the distance between rows is still measured in inches—6, 7, or 9—because this is the standard row spacing on commercial seeders used by most farmers, and it is desirable that field experimentation follow field practice as closely as possible.

Seed counters

In 1936 when W.J. Mason started to work at the Central Experimental Farm his first job was to count seeds of wheat, oats, and barley and package them ready to seed rod rows, and to count 1000 seeds for the standard 1000 seed-weight measurement. He went overseas to serve in the armed forces in 1939 and shortly after his return in 1945, C.H. Goulden moved from Winnipeg to Ottawa as Dominion Cerealist. Mason had devised mechanical gadgets to speed seed counting and Goulden thought they could be improved upon.

The first Goulden and Mason counter (13) that was both reliable and reasonably priced used a commercially available electrically vibrated bowl with an inclined ramp around its inner periphery to separate seeds. The seeds vibrated their way up the incline, then fell through a plastic tube and struck a light metal plate. The plate was attached to a shaft that vibrated when a seed struck, and generated an electric current that caused a count to be made. As far as possible Goulden and Mason used commercially available parts, modifying them to their needs. The vibrating bowl was an industrial component designed to deliver small objects for packaging. To ensure that the seeds being counted were aligned end to end, they modified the bowl to provide a narrower spiral ramp. They eventually used custom bowls of cast aluminum to increase friction when counting slippery seeds such as flax. Once this problem was solved and seeds of different sizes and shapes could be separated reliably, Goulden and Mason had little trouble with the remaining part of the equipment. They were able to accurately count seeds at the rate of 200 to 300 per minute, varying with wheat, oats, barley, or flax. By the early 1960s the Goulden and Mason counter was marketed worldwide by an Ottawa firm.

The technology of seed counters has been developed dramatically since 1947 by branch engineers at Ottawa (23, 27). The original Goulden–Mason principles have not changed, but improvements have paralleled developments in the electronics industry. The original electrical system has been replaced by an electronic one, circuits have been miniaturized, tubes have been replaced by silicon chips, and physical contact has been replaced with photoelectronic detectors, using light and laser beams. Today, seeds ranging in size from 0.5 to 10 mm in diameter can be counted at rates of up to 60 000 per minute, the limiting factor now being the rate of single-file delivery of about 1100 seeds per minute. The vibrating bowl is still standard for delivering single seeds, although some counters use a vacuum that sucks single seeds into recessed holes, because vibratory separation of small seeds such as tobacco is not possible. With today's technology, Mason would be able to count his seeds in about 1/200 the time taken in 1958.

Growth cabinets

Plant scientists were continually frustrated by natural variations in temperature and light when growing their experimental material. Greenhouses partly solved the problem of temperature; however, plants near outside walls were often cooler than those in the center of a house, even though heating devices were usually placed adjacent to walls. Variation in light quantity and quality could partly be compensated for by adding incandescent, fluorescent, and other types of artificial light to extend day length and supplement daylight during a heavy overcast. By 1933 automatically controlled air-conditioned greenhouses were available, but when Went (37) devised the first plant growth room at the California Institute of Technology in 1943, a new day dawned.

Following World War II and Went's developments, most Canadian experimental stations and plant pathology laboratories constructed growth rooms, frequently in the basements of their office–laboratories. Their use gave improved control over plant growth and provided cereal breeders with the opportunity of producing two and sometimes three generations each year to speed their programs. Friend, Helson, and Fisher (11) of the Plant Research Institute found that both fluorescent and incandescent lights were needed to optimize the growth of wheat, and that at least 1700 foot candles (18 kilolux) of visible light were required at plant level (the sun delivers about 11 000 foot candles (118 kilolux) on a clear day). The need for incandescent lamps was for the red wavelengths that caused stem elongation and hence flowering.

With this basic information, engineers such as W. Kalbfleisch, K.R. Scott, and P.W. Voisey designed and built improved growth rooms (29) and made their greatest contribution with electronic light and temperature control devices (35). A key issue was that plant scientists demanded ever more light in the small chambers, which created heat control problems. Furthermore, standard industrial controls were designed to maintain constant conditions, not the diurnal variations required for plant growth. A breakthrough occurred when hot gas bypass controls were incorporated into the refrigeration system, that is, an early version of the heat pump. With this device temperature control improved from a variation of 3 to 0.1 C degrees. This advance permitted scientists to eliminate the effects of short cycle changes in temperature within artificial diurnal cycles.

Scientists at research stations located in areas such as Winnipeg, Saskatoon, and Lethbridge where there were large and important plant breeding programs were anxious to either develop or purchase growth rooms or growth cabinets. R.H. Cunningham, A.E. Hannah, and A.B. Campbell at Winnipeg worked closely with the Fleming Pedlar Refrigeration Company and by 1957 had a self-standing growth cabinet that controlled light, temperature, and humidity within close tolerances over a wide spectrum of growing conditions. The cabinets were also used by entomologists to rear insects and, under very special conditions, by scientists to keep both small and large animals in a known, reproducible, environment.

Since 1957 hundreds of growth cabinets have been manufactured by Canadian companies for domestic and foreign use. Another Canadian industry, setting world standards, has been established as a direct result of a need by Canadian scientists and the response of Research Branch in developing reliable controls for the required equipment. In 1984 Canada marketed $7.7 million worth of growth cabinets, 75 percent of which was exported.

Miniature flour mill

The final test of a new variety of wheat is the evaluation of its milling and baking qualities. It requires several kilograms of grain to produce flour for even the smallest commercial flour mill available. Therefore, a plant breeder must maintain and increase each agronomically outstanding selection through several generations to obtain sufficient seed. After each milling and baking test part of the breeder's seed has been lost to the miller, following which another year is needed to increase stocks.

Sir Charles Saunders produced flour and baked bread in order to evaluate the varieties from his wheat breeding program. He used an Allis-Chalmers, and later, a Wolf experimental flour mill (28), which had two pairs of rollers and three sieves, and required four breaks with the rollers closed slightly more for each succeeding break. It would mill a minimum of 500 g of wheat. He mixed 50 g of flour by hand into a dough on a heated glass table and baked 12 experimental loaves each day in his specially designed oven. His system worked well, but it was slow.

Between 1943 and 1957 a number of laboratory mills were developed in the United States, using burr and roll grinders. These were unsatisfactory because they ground rather than milled wheat, which makes separation of bran difficult. Adaptations of small commercial mills were expensive and difficult to clean. The solution for Canadian breeders came in 1961 when J.G. Kemp (the son of H.J. Kemp) and others from various research institutes (17) developed a micro flour mill. It was designed to mill four 100-g samples of wheat at a time; however, in practice, Kemp found that samples as small as 25 g could be milled successfully. Hard and soft wheats were milled at rates of between 15 and 18 samples per hour. The micro flour mill was well accepted by breeders and wheat chemists, because it was easily cleaned, produced excellent flour, and accommodated small samples of grain. It was manufactured commercially for several years and is still used in Canadian wheat breeding programs.

Dough tests

One test for deciding if a new variety of wheat produces good flour for bread making is to record the torque required to mix a flour–water dough. Cereal chemists agree that the greater the resistance of the dough to mechanical mixing, the stronger the flour and the better the bread. A "strong" flour retains the shape of a risen loaf when baked. Two machines used to measure this torque are the mixograph, requiring 35 g of flour, and the farinograph, requiring 50 g of flour, both of which measure the torque on a mixing paddle or a pin when turning through the dough and mechanically record it on a graph.

In 1966, Voisey, Miller, and Kloek (36) designed a tool for cereal breeders by converting the mechanical recording system to an electronic one, thus permitting data to be gathered in various ways. Tests with different strengths of flour confirmed that dough strength recorded on a strain gage dynamometer were accurate and permitted further miniaturization to handle as little as 4 g of flour in the mixograph or 10 g of flour in the farinograph. Sufficient flour could be obtained for testing much earlier in a breeding program, and this advantage has significantly reduced the time needed to develop new varieties. Once data were digitized, it was a simple matter to record them in a form suitable for direct analysis by a computer.

Data capture

Breeders and plant physiologists frequently need to measure the length or the area of roots or leaves, and animal research scientists often need to measure the area of hides and muscles or the length of hair. All such measurements are slow, tiresome, and subject to human error. Recently, Buckley (6) devised a system to automatically measure and record lengths or areas of such products. It consists of lenses through which an image of the object is projected onto a series of sensor cells. The number of cells activated is then converted to length or area and recorded digitally. Buckley found that by using his technique area measurements had an error of less than one in 300 000, a truly remarkable degree of precision!

The electronic counter-recorder, another device of Buckley and colleagues at Saint-Jean (7), can be mounted on a commercial sorter and automatically count apples or other spherical objects after they are sorted as to size. This counter-recorder, which is capable of counting objects of up to seven different sizes, saves time and improves accuracy.

Engineers have provided an amazingly wide range of new specialized instruments to increase the accuracy and efficiency with which scientists perform their experiments, the depth in which they study their subjects, and the speed with which they collect and analyze their data. Instrumentation for cereal breeding has been exploited because of the important part cereals play in Canadian agriculture. Similar and equally striking innovations have been made in livestock, soil, forage, and horticultural research.

FARM APPLICATIONS

G.N. Denike at Swift Current first recognized the importance of farm mechanization to the production of cereals and helped with its implementation. During the first 10 years of that experimental station many new cultivating and harvesting power machines were tested, all on steel or solid rubber rims. Denike, an agriculture graduate from Manitoba, worked at Swift Current as a summer student and joined the staff in 1929. Gray (14) records that Denike, after watching an airplane land on rubber tires realized that rubber tires could also withstand the pounding of a tractor. In 1933 he introduced pneumatic rubber tires for tractors, which saved fuel, reduced repairs, and produced fewer irascible operators. Denike became superintendent of the Swift Current Experimental Station in 1948, following L.B. Thomson.

For many years it was the policy of Experimental Farms, Science Service, and Research Branch to leave the development of farm machinery to manufacturing companies. The policy was probably appropriate for the machinery required to handle major crops and livestock enterprises where hundreds and even thousands of the same kind of machine such as large cultivators, seeders, and harvesters were used, but it was not an appropriate policy for the machinery required to handle minor crops such as vegetables, fruits, and ornamentals. Farmers who produced a minor crop and wished to improve their efficiency through the use of machinery either had to import and adapt the machinery themselves or design, build, and perfect them and, at the same time, compete with the foreign producers who often had machinery readily available. Some agricultural scientists accepted the challenge 20 or 30 years ago and worked with companies to the mutual benefit of both farmer and manufacturer. Since 1973 and the advent of contracting research to the private sector, scientists in the private sector, together with Research Branch, have developed a number of machines that have made Canadian producers more competitive with their international counterparts.

Machinery for miniature vegetables

Canada was importing both baby carrots and small red beets in quantity, even though they commanded a higher price than locally grown conventional-sized produce. Varieties that mature when small were available in Canada, but the machinery for their seeding and harvesting had not been tested under Canadian conditions.

Eventually G.B. Hergert of the Engineering and Statistical Research Institute (15) did test the seeding and harvesting machinery, in cooperation with R. Bernier and A. Liptay of the research stations at Saint-Jean, Quebec, and Harrow, Ontario, as well as others from the Quebec and Ontario ministries of agriculture. Following modifications, a commercial bed harvester from Holland was used on both organic and mineral soils. Toppers needed no modification. Bed formers and seeders were another matter. A bed former was developed to produce an elevated, flat, even bed that worked well under several different soil conditions. Three seeders were built and tested. The first one broadcast seed into grooves that were covered and flattened by a roller. The second one formed eight shallow seed openings, then covered and packed the soil. In both designs, seed was dispensed with cone research-type seeders. Later, a third seeder similar to the second design but with 18 double-disc adjustable openers proved to be the most satisfactory.

Horticulturists observed that small red beets and baby carrots growing on the edges of beds were larger than those growing in the center. They therefore developed markets for vegetables of both sizes. They also found yields per hectare of small red beets and baby carrots weighed less than those of conventional size. Calculations showed that the premium paid for baby carrots and small red beets needed to be greater than 90 percent above the price for conventional-sized produce. Canadian farmers can now successfully grow the small vegetables in competition with imports.

Cauliflower machinery

There are two types of cauliflower: self-blanching and standard. The first tends to shield itself from sunlight by producing leaves over the curd but has a lower yield and poorer quality. To prevent yellowing of standard cauliflower, leaves must be tied above the curd. Research Branch decided a California tying machine would adapt to Canadian conditions, and contracted with Industrial Equipment Manufacturing, Leamington, Ontario, to do so.

Growing cauliflower is labor intensive because it requires selective harvesting more than three times during the season. A machine was needed to identify and cut mature heads but pass over immature plants without damage. Fortunately, a prototype selective harvester was available and Research Branch contracted with Univerco Hydraulic Inc., Napierville, Quebec, to construct a Canadian version.

In 1982 the tying machine was tested by G.B. Hergert and M. Pelletier of the Engineering and Statistical Research Institute in cooperation with Nancy Smits of the Smithfield Experimental Station, Ontario, and R. Thériault of the Saint-Jean Research Station, Quebec. The first trials were disappointing because the machine did not have enough rigidity to be precise. By the end of the season, however, after repeated modifications, the machine tied 90 percent of the plants accurately. The knotter tied the leaves at a rate of 45 plants per minute as the machine moved forward. In 1983, a skillful operator could obtain 94 percent accuracy.

The harvester was more complicated because a mechanism was needed to softly squeeze each plant in order to judge if a curd was ready to harvest. Once this was accomplished there was little mechanical problem in cutting the plant at its base and elevating it into a bulk bin. Thériault improved upon the machine and by 1982 it was harvesting at the rate of 0.13 ha/hour. A lighter model needs to be designed before it is practical for commercial use. Both tying and harvesting machines preserve the quality of cauliflower for processing purposes.

Strawberry machinery

Strawberries and other small fruit such as raspberries and gooseberries require labor intensive input. Canada imported over $9 million worth of frozen strawberries from Poland, Mexico, and the USA in 1980. There is no biological reason why Canada cannot grow processible strawberries. The stumbling block is the high cost of harvesting. Canners Machinery Ltd., Simcoe, Ontario, in cooperation with Michigan State University developed a strawberry harvester suited to solid field culture. The same company also produced a dehulling device, with financial help from the Canadian Department of Industry, Trade, and Commerce. As recently as 1981, however, neither machine had been used in Canada! Growers were reluctant to try a solid field culture until they had seen the whole system demonstrated. So the Research Branch initiated cooperative work with strawberry growers in Simcoe and Blenheim, the Ontario Ministry of Food and Agriculture, the Canners Machinery Inc., Farm Lane Foods Inc., and Cedar Springs Cherry Growers Cooperative to develop a complete mechanically harvested strawberry operation. Together they formed the Processing Strawberry Research Committee. Funding was arranged through Agriculture Canada's contracting programs as well as through the Ontario provincial government's programs.

Cultural practices were refined to the point where several growers planted test areas. Processing equipment was installed and tested in two plants, using berries from 8 ha of land. Finally, some Ontario food manufacturers used several different varieties of mechanically harvested strawberries; their reactions were favorable under Ontario conditions. Testing will now go forward in the Maritime Provinces. We can look toward having more Canadian-grown strawberries for frozen and processed desserts as a result of engineering perseverance.

Low-volume air-blast spraying

J. Marshall, officer in charge of the Entomology Laboratory, Summerland, and his staff can take credit for introducing concentrate-sprayers into Canadian orchards. Classically, orchard sprays of low-concentrate insecticides and fungicides were applied in large volumes until spray dripped from the trees. The procedure was expensive because of the weight of water needed in the sprayer, time-consuming because two spray operators as well as a tractor driver were needed, and damaging because concentrations of spray built up in the soil along the drip-line. The first low-volume sprayers were tried by Marshall, McMechan, and Williams (21) in 1946. After 3 years these sprayers were recommended to orchardists because they were cheaper to operate, needed only one tractor driver, used much less water, produced no drip, and were equal to or more effective than high-volume sprayers. During the next 20 years McMechan improved the air-blast sprayer by increasing the air velocity, lowering the pressure, and adding surfactants to the spray mixture to provide for uniform deposits on the trees. He was also able to operate the new machine with power from the tractor and could therefore discard the sprayer motor.

REMOTE SENSING

Ever since wheat has been grown in large quantities on the Great Plains, grain merchants and later the Canadian Wheat Board have estimated its quantity and, insofar as possible, its quality months before harvesttime. Such estimates are prepared to assess how well the supply will meet the demand and to provide sufficient storage and transportation. The Dominion Bureau of Statistics asked key farmers to give periodic yield estimates starting almost at seeding time. Grain companies and the Canadian Wheat Board had their own skilled estimators, and the Winnipeg Free Press had the renowned Miss E. Cora Hind, a world authority on grain and livestock estimating. From the air such knowledgeable people could distinguish a high-yielding crop from a low-yielding one and a diseased crop from a healthy crop. They could differentiate among crops such as corn, cereals, grapes, and apples.

Remote sensing is not new. Today's technology had its beginnings when photographs were taken from aircraft at the turn of the century. Military personnel recognized the potential value of remote sensing and were quick to develop it. By the late 1930s air photography was standard practice and of tremendous use to land and coastal surveyors. World War II interrupted the civilian use of remote sensing but furthered its development by the military.

An indication that both crop yield and crop quality could be accurately measured from the air came in 1967 when E.J. Brach (5) of the Engineering

Research Service developed a portable spectrophotometer that would measure light intensities and wavelengths reflected from various crops. Such an instrument divides wave energy into various segments. Rain drops act in the same manner, dividing light from the sun into the colors of a rainbow. In a laboratory in 1968, using spectrophotometry, Brach was able to differentiate wheat, oats, and barley when the plants were only 10 days old. This was such a momentous development that the branch allocated $2 million over a 10-year period to support the research. By 1970, Brach had developed a continuous-integrating spectrophotometer that measured the total energy falling on an apple, for instance, during its growth. In 1972, the National Aeronautical and Space Administration (NASA) of the United States launched its first Earth Resources Technology Satellite (ERTS-1), which gave a tremendous boost to the development of remote sensing for agricultural purposes. NASA used some of Brach's findings when deciding upon which wavelengths should be scanned.

By 1976, Brach and Mack (3) had learned how to differentiate corn, wheat, oats, and barley by comparing the ratios of reflected blue and green wavelengths with the reflected red wavelength. These ratios could be mathematically quantified and, provided the measuring instrument was close to the crop, the system worked well. There was doubt concerning its practical application when used from an aircraft or a satellite. During the next few years the universities of Manitoba, Winnipeg, and Montreal used the technique for graduate students' theses, with Brach as an adviser. Glick (12) at Manitoba learned how to distinguish varieties of wheat, oats, and barley when they were in bloom (anthesis) on the basis of their reflectance patterns. Again at Manitoba, Berard (2) was able to determine if the protein content of a standing wheat crop was high, medium, or low, based on near infrared reflectance. In cooperation with J.J. Jasmin at Saint-Jean, and J.M. Molnar at Ottawa, Brach (4) found remote spectroscopy to be useful in determining the maturity of lettuce. Desjardins of the Land Resource Research Institute and coworkers at McGill University (9) were able to measure the exchange of CO_2 (carbon dioxide) above a cornfield, a forest, and a lake from a low-flying (30 m), slow-moving aircraft, using an entirely new CO_2 analyzer developed by Brach. With the aid of the analyzer, they were able to monitor crop performance over large areas; for example they could determine the percentage of damage to a corn crop after a severe storm.

The development of remote sensing both from aircraft and from satellites has accelerated rapidly, thanks to the research of scientists in Research Branch and elsewhere. We can now accurately predict quantity and quality of grain crops, assess damage done to crops by storms, and determine the maturity of crops before harvesting them, all valuable information for growers and marketing agencies. Further advancements are anticipated.

CANADA PLAN SERVICE

Plans for the construction of farm buildings kept pace with Canadian farming through the drafting service available at the Central Experimental Farm, started in 1924. This service was the forerunner of the Canada Plan Service. Thousands of standard plans for constructing and repairing farm buildings were made available to farmers. The Division of Field Husbandry, Soils and Agricultural Engineering began studies in the mid-1940s on farm building design. By 1956, the division, provincial departments, and faculties of agriculture were all involved. The architect's office continued to supply the drafting service, but by 1959 it could no longer meet the demand; therefore some provinces established their own farm plan service, which frequently overlapped those of other provinces and of the Experimental Farm.

The National Committee on Agricultural Engineering (NCAE), sponsored jointly by the National Research Council of Canada and the Experimental Farm Service, recommended coordination among the various agencies at its inaugural meeting in 1944 (18). In 1945, NCAE organized a Farm Buildings Sub-Committee, which included representation from provincial departments of agriculture, faculties of agriculture, the new National Housing Administration, and the Experimental Farm Service. It met at the Swift Current Experimental Station in 1945 and decided to issue plans for 13 farm buildings that would include houses, workshops, barns, and farm equipment. The architect's office agreed to redraft the selected plans. By 1949, 11 of the 13 plans were published under the joint auspices of the Prairie Rural Housing Committee and Central Mortgage and Housing Corporation.

The eastern section of NCAE, under Professor C.G.E. Downing, Ontario Agricultural College (now the University of Guelph), collected sketches for over 400 farm structures. These were presented to the second conference in 1952, which selected the most useful ones. Drafting services at the Ontario Agricultural College, the Experimental Station at Swift Current, and the Central Experimental Farm prepared working drawings. Downing, on behalf of NCAE, proposed to the National Advisory Committee on Agricultural Services, which was made up of deputy ministers, that a national farm building plan service should be started. The deputies agreed and the following year Downing presented them with a sample copy of a beef cattle housing catalog that had been planned, prepared, printed, and financed by the Ontario Agricultural College. By April 1954, the deputies decided the federal government should edit and publish plan catalogs in English and French and the provinces could distribute them. After 10 years' of preliminary work the Canadian Farm Building Plan Service (CFBPS) was finally a reality. In 1969 the word "Canadian" was changed to "Canada" and in 1973 the current shorter name of "Canada Plan Service" (CPS) was accepted.

CPS is truly national in scope with representatives from each province. The three drafting centers were consolidated into one and located at Guelph, remaining there from 1958 until 1968, when the center was moved to the Central Experimental Farm, Ottawa, and directed by J.E. Turnbull. About 80 new or revised master sheets are sent to all provinces each year. Provinces reprint from the master sheets as needed. The Canada Plan Service is an internationally recognized archetype of a smoothly administered, cooperative, countrywide service to farmers.

321

FARM STRUCTURES

A national plan service justifies its existence when its buildings function properly. Research Branch engineers have developed and tested innovative construction methods, ventilation systems, and manure removal–storage systems. In most instances, these new methods have been used on buildings to house, feed, or handle livestock wastes.

Due to climatic extremes in Canada, the design of livestock housing that has good ventilation and optimum heat retention is a major challenge. Poultry and swine grow best when the temperature of their building is controlled within narrow limits. Frequently, such buildings require more than just animal heat. Furthermore, greatly increased ventilation is needed to keep temperatures close to outdoor conditions during summer months. Cattle, however, are more cold-tolerant and can be housed in colder buildings with natural ventilation during winter months. They are rarely housed in the summer. Therefore, each type of livestock must have its own building design.

Turnbull and Coates (31) found that slotted inlets near ceilings improved the distribution of air in poultry houses over open-ended ducts. Munroe and co-workers (25) showed that for cold-tolerant cattle, a porous insulated ceiling inletting from a well-ventilated attic performed better in cold weather than a conventional airtight construction. Even with outside temperatures of $-40°C$, Munroe found the porous ceiling system kept inside temperatures between $-1°C$ and $-10°C$ and still provided adequate ventilation. Larkin and Turnbull (20) adapted air-to-air type heat exchangers to reclaim heat normally lost in winter through exhaust fans. Standard ventilation recommendations for various species and ages of livestock were calculated and published (26).

All farm livestock enterprises must store manure prior to spreading it on fields. Turnbull, Phillips, and Hore (33), found that where suitable clay was available, earth-banked storages were the least expensive for liquid and semi-solid manures. For solid manure with bedding, aboveground paved concrete slab storages with low curb walls, were the next best. Where high water-tables existed, concrete-walled silos for liquid manures were the only solution. Although a roof or cover is desirable in regions of heavy precipitation, its cost is high. Automatic moving sprinklers have been developed that satisfactorily spread liquid manure, even when it contains a reasonably high level of solids. Tanker injection of liquid manure into the soil provides the best possible conservation of manure nutrients and control of odors, but it requires stone-free soil and much tractor power to pull the tanker-injector.

Designs issued by the Canada Plan Service were based upon construction standards developed by the Canadian Farm Building Code and other authorities, such as the Canadian Standards Association, but they were untested for farm construction. Turnbull and Todd (34) have improved the strength of nailed wood joints and more recently Turnbull, Lefkovitch, and Low (32) have shown that steel and plywood gussets developed several planes of shear through each nail, thus further improving roof trusses. The net result is that half the number of nails are now required to frame trusses for a given roof. Other tests on trusses, and the publication of standard truss designs, have changed the shape of modern farm buildings. Virtually all farm buildings are now clear-span structures without interior posts to interfere with function.

In 1980 the Engineering and Statistical Research Institute made a valuable test facility that simulated snow and wind loads on roof trusses, rigid frames, and ceiling diaphrams in order to verify the safety of components designed for farm buildings. One of the first tests was to find the stiffening inherent in metal-clad buildings against sidesway caused by wind. Massé, Turnbull, and Williams (22) showed that additional screws could be used where roof and siding sheets lap, thus eliminating the more expensive bracing formerly used. This concept is still at the introductory stage, its impact yet to be measured. Safe, easy-to-build, and economical designs are important to farmers when constructing buildings. The Canada Plan Service, supported by the experimental capacity of the Engineering and Statistical Research Institute, fills their need.

Biologists possessing mechanical aptitudes such as Kemp and Goulden, have added immeasurably to the ease, speed, and accuracy of executing research in agriculture. Engineers such as Kalbfleisch, Thériault, Downing, Voisey, and Dyck have found solutions to hundreds of problems and stand tall beside other great innovators at home and abroad.

References

1. Andrews, J.E.; Smith, A.D.; Grant, M.N.; Wells, S.A. 1959. Variation among rows in power-seeded cereal variety trials. Can. J. Plant Sci. 39:25–29.
2. Berard, Raymond Gilbert. 1982. The feasibility of detecting the protein content of a standing wheat crop by near infrared reflectance spectroscopy. M.Sc. Thesis, Univ. Man., Winnipeg.
3. Brach, E.J.; Mack, A.R. 1976. Differentiation of selected annual field crops throughout the growing season by their spectral reflectance properties. Can. J. Remote Sensing 3:55–65.
4. Brach, E.J.; Molnar, J.M.; Jasmine, J.J. 1977. Detection of lettuce maturity and variety by remote sensing techniques. J. Agric. Eng. Res. 22:45–54.
5. Brach, E.J.; Wiggins, B.W.E. 1967. A portable spectrophotometer for environmental studies of plants. Lab. Pract. 16:302–309.
6. Buckley, D.J. 1975. Solid state, image sensing arrays: Tools for dimension and area measurement of agricultural products. 1976. Can. Agric. Eng. 18:26–30.
7. Buckley, D.J.; Rousselle, G.L.; Thériault, R.; St. Amour, G.; Nicholls, C.F. 1983. A microcomputer-based system for automatically counting and weighing pre-sorted apples. Am. Soc. Agric. Eng. 26:1849–1853.
8. Cherewick, W.J. 1954. An improved experimental plot seeder. Can. J. Agric. Sci. 34:642–643.
9. Desjardins, R.L.; Brach, E.J.; Alvo, P.; Schuepp, P.H. 1982. Aircraft monitoring of surface carbon dioxide exchange. Science 216:733–735.
10. Dyck, F.B. 1976. A cone-belt distributor for seed and fertilizer. 4th Int. Conf. Mech. Field Explor. 97–100. Ames, Ia.
11. Friend, D.J.C.; Helson, V.A.; Fisher, J.E. 1960. The influence of the ratio of incandescent to fluorescent light on the flowering response of Marquis wheat grown under controlled conditions. Can. J. Plant Sci. 41:418–427.
12. Glick, Harvey L. 1979. Species and cultivar discrimination of cereals by field spectroscopy. M.Sc. Thesis, Univ. Man., Winnipeg, Man.
13. Goulden, C.H.; Mason, W.J. 1958. An electronic seed counter. Can. J. Plant Sci. 38:84–87.
14. Gray, James H. 1978. Man against the desert. Western Producer Prairie Books, Saskatoon. 250 p.
15. Hergert, G.B. 1983. Production techniques for baby carrots and small red beets. Can. Agric. Eng. 25:33–37.
16. Kemp, H.J. 1935. Mechanical aids to crop experiments. Sci. Agric. 15:488–506.
17. Kemp, J.G.; Whiteside, A.G.O.; MacDonald, D.C.; Miller, H. 1961. Ottawa micro flour mill. Cereal Chem. 38:51–59.
18. Kenney, C. Hamilton. 1978. Canada plan service 25 anniversary. Hist. Ser. 12. Agric. Can. and Ont. Minist. Agric. and Food. Ottawa. 26 p.
19. Kusch, A.G.; Keys, C.H.; Nadan, H.J. 1950. A four-row power plot seeder. Sci. Agric. 30:201–208.
20. Larkin, B.S.; Turnbull, J.E. 1979. The economics of heat recovery systems for animal shelters. Can. Agric. Eng. 21:53–59.
21. Marshall, J.; McMechan, A.D.; Williams, K. 1963. Low-volume air-blast spraying in British Columbia orchards 1957–1962. Can. Dep. Agric. Publ. 1191. 18 p.
22. Massé, D.I.; Turnbull, J.E.; Williams, C.J. 1983. Screwed connections for corrugated steel diaphragm ceilings in farm buildings. Can. Agric. Eng. 25:95–99.

23. McLaughlin, N.B.; Giesbrecht, J.; Blight, D.F. 1976. Design and performance of an electronic seed counter. Can. J. Plant Sci. 56:351–355.
24. Morse, Pamela A. 1984. A history of the Statistical Research Service, Ottawa. Contrib. I-643, Eng. Stat. Res. Serv., Res. Br., Agric. Can., Ottawa. 32 p.
25. Munroe, J.A.; Wolynetz, M.S.; Turnbull, J.E.; Darisse, J.P.F. 1981. Cold weather conditions in a free-stall barn fitted with a porous ceiling. Can. Agric. Eng. 23:17–21.
26. NRC Standing Committee on Farm Buildings. 1977. Canadian farm building code. Nat. Res. Counc. Can. 15564. Ottawa.
27. Reid, W.S. 1974. Seed counters aid research. Can. Agric. 19(1):24–27.
28. Saunders, Charles E. 1921. Researches in regard to wheat, flour and bread. Bull. 97. Dom. Dep. Agric., Ottawa. 56 p.
29. Scott, K.R.; Kalbfleisch, W. 1963. Plant growth building with evaporative cooling. Can. Agric. Eng. 5:32–33.
30. Smith, A.D.; Bergen, H. 1962. Note on a power plot seeder. Can. J. Plant Sci. 42:548–550.
31. Turnbull, John E.; Coates, James A. 1971. Temperatures and air-flow patterns in a controlled-environment, cage poultry building. Trans. Am. Soc. Agric. Eng. 14:110–120.
32. Turnbull, J.E.; Lefkovitch, L.P.; Lowe, D. 1981. Multi-laminated nailed truss connections. Can. Agric. Eng. 23:113–119.
33. Turnbull, J.E.; Phillips, P.A.; Hore, F.R. 1977. Design concepts and construction costs of eight different manure storages. Can. Agric. Eng. 19:65–70.
34. Turnbull, J.E.; Todd, D.M. 1970. The effect of nailing patterns on the tensile strength of plywood gussets. Can. Agric. Eng. 12:42–44.
35. Voisey, Peter W. 1963. Note on the modification of an electronic temperature controller to provide diurnal temperature variations in plant growth chambers. Can. J. Plant Sci. 43:111–112.
36. Voisey, P.W.; Miller, H.; Kloek, M. 1966. An electronic recording dough mixer. 2. An experimental evaluation. Cereal Chem. 43:420–432.
37. Went, F.W. 1943. Plant growth under controlled conditions. 1. The air-conditioned greenhouses at the California Institute of Technology. Am. J. Bot. 30:157–163.

325

Chapter 20
Preserving, Processing, and Packaging Food

Except for tomatoes and cucumbers grown in greenhouses, Canada must import its fresh vegetables from November to April and must store or preserve its excess production for use after the growing season. Livestock are slaughtered, dairy cows produce milk, and chickens lay eggs all year round, although production of these products may decline when fresh feed is not available. Some form of preservation is needed for excess milk, eggs, and other commodities. Food scientists and engineers in the Food Research Institute, Ottawa, and in fruit and vegetable processing laboratories and meat laboratories of research stations at Kentville, Nova Scotia; Morden, Manitoba; Lethbridge and Lacombe, Alberta; and Summerland, British Columbia, have solved a number of preservation problems and improved many processes, thus bringing increased returns to primary producers and greater variety of higher quality foods to Canadian consumers.

STORAGE OF FRESH FOODS

Apples, pears, and root vegetables such as turnips and potatoes have been stored for centuries in root cellars. These dark, damp, partially buried storages are built into the side of a hill or sunk into a dry, well-drained part of the farmyard. Those parts not buried are covered with several metres of soil and sod. They are often constructed such that a team of horses with wagon, or a truck, can be driven through the cellar for off-loading and pickup. Suitable varieties of fruits and vegetables remain in good condition in a root cellar until February or March, neither freezing nor overheating.

In 1931 a low-temperature laboratory for studying controlled storage of fruits and vegetables was built at the Division of Horticulture. It consisted of six 30-m^3 chambers. One chamber was kept at a temperature of 0°C, the others were kept at temperatures above 0°C. The facility was expanded to 21 chambers in 1936, including an additional two for freezing purposes. W.R. Phillips soon learned that apples and pears could be kept longer under temperatures near the freezing point of water than they could in root cellars where temperatures were 5 to 10 degrees higher. The rate of cooling, the relative humidity, and the levels of oxygen and carbon dioxide formed in the storage area as a result of respiration all affected the durability and keeping qualities of fruits. Phillips worked on the mechanical aspects of freezing and packaging. Following these experiments, which were successful, the department assisted financially with the construction of commercial cold storages.

S.W. Porritt and D.V. Fisher of Summerland, and C.A. Eaves of Kentville, also investigated problems of fruit storage. Shortly after World War II, scientists at Summerland demonstrated that apple varieties such as Delicious and Winesap stored well at temperatures between -1 and $0°C$, and that the varieties McIntosh and Newtown developed core flush, scald, and soft flesh at the same temperatures. Porritt started to investigate the use of controlled-atmosphere storage shortly after it was developed in England in 1953 (20). Fruit continues to respire (use oxygen and produce carbon dioxide) while in storage, the higher the temperature the faster the respiration. In controlled-atmosphere storage, the level of carbon dioxide is increased to about 5 percent from 0.03 percent and the level of oxygen is reduced such that the sum of carbon dioxide and oxygen equals about 21 percent of the atmosphere. This alteration lowers the rate of respiration and so increases the storage life of the fruit. Under these conditions Porritt found that temperatures could be raised to $2°C$, preventing core flush, scald, and soft flesh in Newtown and McIntosh varieties.

Initially, levels of carbon dioxide in controlled-atmosphere storages were maintained by simple venting to the ambient air. As the technique was refined, carbon dioxide needed to be controlled independently of oxygen. Scrubbers containing caustic soda (sodium hydroxide dissolved in water) were used to remove carbon dioxide from the storage air before it became too concentrated. However, these scrubbers were expensive to build and hazardous to operate. In the mid-1950s a water scrubber was developed. Although costly to build, this type of scrubber was not hazardous.

At about this time, fruit companies in the Annapolis Valley were constructing new buildings in which to house controlled-atmosphere storages. In one storage being observed by C.A. Eaves of the Experimental Station, Kentville, the operators noticed that the level of carbon dioxide remained low, although the storage was full of apples. After preliminary studies, Eaves concluded that the lime in the newly poured concrete floor was removing the carbon dioxide as it was being produced by the apples. From this original observation in 1959, Eaves and his technician, H.J. Lightfoot, designed and built the world's first dry lime scrubber to remove carbon dioxide. As they used the technique they found that unopened bags of fresh lime could be stacked in an auxiliary airtight room and connected to the storage by ducts. Air movement through the ducts was created by the fans of the evaporator coils needed in all cold storages to lower temperatures. Eaves had invented the inexpensive, safe, and effective Kentville dry lime scrubber that was tested in several countries. To this day, the dry lime scrubber continues to be used in every major apple-producing region of the world. It still is cost competitive with even the latest technological advances in carbon dioxide removal.

Later in 1982 Lidster, Lightfoot, and McRae (18) at Kentville demonstrated that if the oxygen level in cold storages was reduced to about 1 percent, apples that had kept well for 10 months could be stored for a year, or even longer. In addition, these apples would have outstanding quality and could be displayed in retail stores twice as long as before. Several large commercial storages in Nova

Scotia have adopted the new recommendations to their complete satisfaction. Canadian apples now compete with high-quality imports from the southern hemisphere, even during our off-season.

In the Okanagan, when trees of the newly introduced Spartan apple variety started to produce fruit, packing houses found that under certain conditions the apples developed an internal breakdown frequently associated with a lack of calcium. Porritt, Lidster, and Meheriuk (21) and many other scientists determined that this breakdown could be reduced by dipping the fruit into a calcium chloride solution prior to storage at a low relative humidity.

During 1985 the most recent development in fruit and vegetable storage was made by a team of scientists from Research Branch and two universities in Nova Scotia. P.D. Lidster at Kentville, working with chemists C.M. Elson and D.H. Davies of Saint Mary's University and E.R. Hayes of Acadia University, modified the chitin molecule so that it can be sprayed onto fresh produce. A biodegradable film forms, slowing respiration. By converting naturally occurring, insoluble chitin, the main component of lobster and crab shells, to a soluble form, the scientists (9) now have a product that uses fishing industry waste to prolong the storage life of high-quality fruit and vegetables. Treated produce retains its freshness for many months in normal commercial cold storage or even in a home refrigerator, because a modified atmosphere develops within the film as the produce respires.

IRRADIATION

Canada supplies 90 percent of the Cobalt 60 used in the world. Cobalt 60 is a by-product of the CANDU reactor. It has been known since 1947 that gamma rays produced from Cobalt 60 are useful in preserving stored foods.

At the Winnipeg Research Station, F.L. Watters and K.F. MacQueen (26) demonstrated that gamma rays killed some kinds of stored-product insects and sexually sterilized others, but they had no effect upon the stored grain. In 1961 V.W. Nuttall and colleagues (19) at the Genetics and Plant Breeding Research Institute in cooperation with Atomic Energy of Canada, Limited, subjected onions to five different levels of radiation prior to storage. They found that of those stored without radiation 75 percent sprouted, whereas all irradiated onions were sprout free following 300 days in storage. H.B. Heeney and associates (12) at Smithfield, Ontario, irradiated fresh strawberries and were able to keep them 3 weeks at 4°C without any molding or loss of quality. Low level radiation of fresh foods such as meat, fish, poultry, and cereals delays their spoilage by reducing the incidence of fungi and bacteria.

In 1960, Canada became one of the first countries to permit irradiation of food products, when the Department of Health and Welfare authorized its use on potatoes to inhibit sprouting. A commercial operation was established in Saint-Hilaire, Quebec, which, for many years, produced sprout-free potatoes for table use. In 1963, the United States approved the use of gamma rays for destroying insects in wheat and wheat products. In 1969, Canada's Department of Health and Welfare followed the American example, but did not extend its approval beyond potatoes and wheat. Despite clearance by the World Health Organization, the Food and Agriculture Organization, and the International Atomic Energy Agency in 1981 stating that food irradiated to a dose of one million rad (10 000 Gy) was toxicologically safe, a number of countries including Canada and the United States still have not finalized their regulatory process for food irradiation. Canada has now undertaken to amend the regulations and approve the irradiation of other foods.

Under no circumstance will food that has been irradiated emit radiation itself. No harm is caused to people handling and eating foods that have been irradiated and it seems likely that renewed effort will result in this efficient, effective, and harmless food preservation technique being made available to Canadian processors and consumers. In anticipation of the event, Research Branch is working with Diversified Research Laboratories, Toronto, on the efficacy of irradiating whole, parts, and mechanically separated poultry meat for salmonella control; the Food Research Centre being built at Saint-Hyacinthe, Quebec, will have a pilot plant facility for irradiation of food. Commercial installations are being considered elsewhere in Canada for the treatment of food products.

DEHYDRATION

The first Canadian experiments on home dehydration of fruits were started at the Horticulture Division during the latter part of World War I. In May 1923, the House of Commons expressed concern about the large imports into Canada of dehydrated foods and appointed a Dehydration Committee to study Canadian commercial methods (3). The following year, through the efforts of C.S. McGillivray, Canada's chief canning inspector, and E.S. Archibald, Director, Experimental Farms, dehydration laboratories were established at Grimsby, Ontario, and Penticton, British Columbia. Five years later, a processing laboratory was added to the horticulture building at the Experimental Station, Summerland, British Columbia, with F.E. Atkinson appointed as its head in June 1929. At about the same time, C.C. Eidt of the Kentville Experimental Station undertook to improve the dehydration process and, in 1938, he published a reference book on the subject, which for many years was regarded as the standard. Two apple dehydration plants designed to process fruit based upon the methods researched at Kentville were opened in Ontario.

Apples to be dehydrated were sliced, treated with sulfur dioxide to prevent mold growth, and then dried. At first, the dehydrated product turned a nut brown, but by adjusting the concentration of sulfur dioxide a creamy whiteness was achieved. Sulfuring became a requirement under the Meat and Canned Foods Act of Canada.

To facilitate overseas shipments of vegetables during World War II, further work was done on their dehydration. Mary MacArthur at Ottawa discovered that vegetables needed blanching to inactivate enzymes before dehydration.

Kentville used its own plant to produce experimental lots of dehydrated vegetables that were then stored and tested at Ottawa. MacArthur had a large dehydration tunnel built at Ottawa in 1942 in which she conducted more than 2000 experiments during the last 4 years of World War II. As a result of her work the appearance and nutritional value of commercial dehydrated cabbages, carrots, potatoes, and turnips improved markedly.

FROZEN FOODS

The patriarch of the frozen food industry in Canada is W.H. (Bill) Heeney, who was a kraft paperbox manufacturer in Philadelphia. In the late 1920s, Dr. Clarence Birdseye needed boxes in which to package his frozen food, and contacted Heeney. Heeney was so intrigued with this new concept of preserving food that, with the urging of Birdseye, he returned to his home city, Ottawa, to start a quick-frozen-food company. He had the cooperation of the Horticulture Division and worked with his sister Lilian, who was a graduate in food science from Macdonald College. In 1933 they froze 10 000 pounds (4500 kg) of strawberries and raspberries, held them until Christmas, then had milkmen sell them on their delivery routes. Refrigeration in the horse-drawn milk carts was not necessary during an Ottawa December. The response was so great that in 1934 he packaged 300 000 pounds (136 000 kg) of strawberries grown under contract at Simcoe, Ontario. By 1935 he had outlets in Montreal, Quebec City, and Toronto. He sold his interests in 1956 after developing his company into a large national organization, *Zer-O-Pack*, that froze all types of fruit and vegetables.

Scientists working in the low-temperature laboratory of the Horticulture Division strived to improve methods of handling, packaging, and storing frozen foods and later verified their results on a pilot plant scale. The division was in a unique position to assess which varieties were best for freezing in both commercial and home freezers because they also had extensive varietal trials. The work on freezing was interrupted by World War II when the laboratory was needed urgently for research on dehydrating vegetables. Research on freezing was reactivated in 1944, and in 1945 the Consumer Section, Marketing Service of the department, published the first booklet on home freezing, based upon the research of the Horticulture Division. By 1949 Mary MacArthur was confident enough of her results to summarize them and publish them in a booklet for commercial use.

INSTANT MASHED POTATOES

In 1960 Asselbergs, Saidak, and Hamilton (2) of the Plant Research Institute studied commercially available potato granules both microscopically and organoleptically. They found granules of all brands to be about half the size of cells from freshly cooked potatoes and none equaled the fresh product in taste. They therefore devised a method of making potato flakes, or crystals, that more nearly resembled those of freshly cooked potatoes by boiling, drying, and

331

mashing french fried slices, adding a small amount of skim milk solids, then putting the mash through a double drum drier developed by J.G. Kemp of the Engineering Research Service. The distance between the drums was adjusted carefully to equal the size of freshly cooked potato cells. When reconstituted and taste-tested, the crystals were greatly superior to any other available dehydrated potato; the process was patented. It is used commercially by several Canadian manufacturers of instant potato flakes.

FRUIT JUICES AND WINES

The initial work at Summerland on developing brandy liqueurs from cull apples led to investigations of the production of juices from apple and other fruits. At the time, sweet apple juice contained heavy pulp sediments, which detracted from its appearance. Atkinson was able to clarify the juice by using tannin and gelatin. He devised a flash pasteurizer to inactivate the polypheno oxidase enzymes, which discolored the juice during slow pasteurization. Fortunately, acid-resistant steel cans were newly available as containers. The clarified apple juice met with success on the market and in 1953 it grossed nearly two million dollars for Okanagan fruit growers.

Meanwhile C.C. Strachan, a recent graduate from the Massachusetts Institute of Technology, studied the effects of adding ascorbic acid (vitamin C) to clarified apple juice with the intent of making it nutritionally equivalent to citrus juice (22). He found that ascorbic acid improved the apple flavor by preventing oxidation. R.P. Walrod, manager of B.C. Fruit Processors, encouraged Atkinson and Strachan to develop the idea and extend it to unclarified juice, which resulted in "opalescent apple juice." It received immediate acceptance on the market and remains a popular product.

Cranberry juice in combination with juices of other fruit is widely used as a fruit cocktail. The major advantage of cranberry juice, other than its tangy flavor, is its bright color, and therefore any process that will increase the yield and color is of commercial importance. D.F. Wood at Kentville (28) found that by adding a very small amount (1:1000) of a pectinase enzyme to cranberries at the time of pressing, both the depth of color and the amount of juice increased by about 60 percent over the normal commercial extraction method.

Wines have been made in various food processing laboratories, primarily in support of enological grape-breeding programs, but food scientists also have devised methods of making wines from other fruits. One such method is that devised by Tibor Fuleki of Kentville, who successfully made commercially-acceptable wine from blueberries (11). He found that when appropriate strains of yeast were used at lower than normal temperatures, a wine was produced that, to experienced palates, was not different from wine made of good-quality grapes. Blueberry wine now is manufactured in Moncton, New Brunswick, and in the Lac-Saint-Jean area of Quebec, where it is used as a base for several types of aperatif wines.

Because loganberries were in surplus supply on the Pacific Coast in 1929, growers formed a wine company and produced more than 3000 hL (hectolitres) of loganberry wine. However, the wine was unmarketable because of slime-producing organisms. The advice of W. Newton, a plant pathologist at Saanichton, was sought, and by filtering and sterilizing the wine he was able to salvage most of it. Newton continued to advise the company on many of its operations.

Apples, too, can be manufactured into an acceptable wine or cider, according to Atkinson, Bowen, and MacGregor (5). Lacking the usual European varieties high in tannin, they used nonsaleable dessert varieties. By carefully selecting an appropriate yeast from 30 isolates, they found one which produced superior sparkling cider when fermented at 15°C for 4 days. The system worked well and produced a commercially valuable product.

Wines manufactured from grapes have been made in Canada for several generations. Until recently, however, only the hardy American grape, *Vitis labrusca*, had been used. This species produces a wine with a characteristic foxy flavor. Summerland Research Station and the Ontario Department of Agriculture have made excellent progress in finding and developing varieties of wine grapes (*V. vinifera*) and hybrids hardy under Canadian conditions. Bowen, MacGregor, and Fisher (8) at Summerland made wine from 25 *V. vinifera* and hybrid varieties grown in the station vineyards and recommended some that were both hardy and produced good wine. Variety trials of grapes started at Kentville in 1913, although only table varieties were identified until as recently as 1971. Some of these were planted in a nearby vineyard, but only when they were replaced with *V. vinifera* in 1977 were wineries able to produce a high-quality grape wine.

CANNED FOODS

It was in 1929, when a new horticulture building was erected on the Central Experimental Farm, that adequate space became available for extensive tests on preserving fruits and vegetables. This research was reported first in 1930 by Ethel W. Hamilton who compared many varieties of raspberries, using four methods of canning, and concluded that the water bath method was best for the home. In addition, she canned varieties of seven different vegetables, six of other fruits, and originated recipes for pickles and relishes.

Canned ready-to-use fruit pie fillings were introduced to the Canadian market in 1946. However, homemakers and those operating commercial kitchens were wary of the product, which varied from canner to canner and from year to year because no standard recipe had been developed. At Summerland, C.C. Strachan, A.W. Moyls, and F.E. Atkinson cooperated with Dorothy Britton, home economist, in developing commercial processes for many kinds of fruit pie fillings (23). Apple was their principal fruit, but they also used apricots, plums, peaches, cherries, and many kinds of berries, either singly or in mixes.

The fresh-tasting, firm pieces of apple in today's pies are no accident. Several kinds of gases fill the intercellular spaces in many fruits and vegetables. These gases, of which oxygen may be present in relatively large amounts, caused problems by exploding when heated, thus breaking down the fruit's structure, corroding the tin plate of the packing can, and producing off-flavors. Canners first attacked the problem by soaking peeled apple slices in a salt brine, which replaced the gas. They then washed away most of the salt with water. Another method was to draw a vacuum on the product to remove the gases, then release the vacuum with salt brine, hot water, syrup, or pickling solution. Both methods were time-consuming and expensive. Kitson (16) changed all this in 1961 when he invented a continuous-vacuum process at the Summerland Research Station. The system is elegant in its simplicity, for it places the vacuum chamber 10 m above the inflow pump and outlet gate valve. Within the chamber is a continuous-moving belt to keep the apple slices in motion during evacuation of gases. The weight of apple slices and liquid flowing to and from the chamber via two 10-m pipes is sufficient to maintain a partial vacuum there and to permit a continuous process. The resulting product is so superior that it has been the industry standard since about 1965. The equipment is manufactured commercially in Vancouver and sold worldwide.

SMALL CANNERY EQUIPMENT

At the request of fruit growers in the early 1930s, several fruit and vegetable processing laboratories at experimental stations and the Horticulture Division, fabricated and tested equipment such as evaporators, heat exchangers, belt cookers, concentrators, and can fillers to serve small private canneries. Many made top-quality, sometimes unusual products, which commanded a premium price.

In the Okanagan, sulfur dioxide to dehydrate apples and manufacture maraschino cherries was required. Atkinson at Summerland designed a cannery absorption tower that used raw sulfur at only 10 percent of the cost of the compressed gas formerly used.

After designing equipment such as preparation tables, draper belts, and cooking retorts complete with safety valves, Atkinson and Strachan wrote one of the more popular Department of Agriculture bulletins, *Small canneries* (4), with details for canneries of three different sizes. By following the advice given in the bulletin entrepreneurs were able to successfully develop a number of small canneries, which have operated for 30 to 40 years.

BLANCHER

Many fruits and vegetables contain enzymes that can cause spoilage unless they are inactivated before the product is frozen and stored. Untreated enzymes alter flavor, color, and texture, frequently rendering the fruit or vegetable unpalatable. Heat above 85°C is used to inactivate enzymes.

In the home kitchen, blanching is accomplished by plunging the vegetable into boiling water. Commercial blanching is done by using steam at atmospheric pressure or by using hot water. Both methods have the following disadvantages: (1) high energy consumption, (2) waste water polluted from organic matter due to leaching, and (3) the degradation of vitamin C as the surface of the product becomes overheated while its center is being brought to the required temperature.

One evening in 1970 three scientists meeting in Kansas City had an unplanned brain-storming session on the blanching process. They were Robert Stark from Kentville, John Kitson from Summerland, and Gordon Timbers from the Engineering and Statistical Research Institute, Ottawa. Upon Timbers' return to Ottawa some ideas gelled. Several concepts were engineered into a system where the product was subjected to just enough heat to bring the whole piece up to temperature, held, and then cooled rapidly—the heat-hold-cool concept, which had not been used commercially. By 1975, Timbers and J.C. Caron had built a prototype and sent it to Kentville for preliminary tests. The new machine forced steam up through a moving belt and surrounded the pieces of vegetable with steam for a short time, then held them in an insulated zone until the temperature equilibrated throughout. They then were cooled rapidly and frozen. Uncondensed steam was recirculated to conserve energy and reduce the volume of waste water.

Between 1975 and 1978, Stark and Dan Cumming tested the experimental prototype and were delighted with the results. For instance, peas required only 35 seconds to be blanched in steam rather than 90 seconds by the old method. They then required 55 seconds in the holding chamber. The retention of vitamin C was much improved.

Up to this point the project had cost $7000 in materials and a lot of time and thought. The next step, the commercialization of the equipment, would be much more costly, although financially rewarding. The Kentville Research Station arranged a contract with the Atlantic Bridge Company (ABCO) of Lunenburg, Nova Scotia, to make a commercial prototype, and in 1979 the M.W. Graves Company in Kentville tested it in its commercial operation. They processed 1.2 million kg of different kinds of vegetables and found the equipment could handle between three and eight times more volume than conventional systems, using the same amount of steam—even better results than those achieved with the experimental prototype. There was only one-tenth the volume of waste water from the new process and thus less pollution of the company's sewage system. As a final and very important bonus, vegetables were more nutritious, broccoli, for instance, retaining over 50 percent more vitamin C!

Canadian Patents and Development Ltd., a Crown corporation, patented the principle both in Canada and in the United States (see U.S. patent number 4 387 630, 14 June 1983). ABCO build four sizes of blanchers capable of handling from 1000 kg/hour to 13.5 thousand kg/hour. The Food Machinery Corporation (FMC) of California market them. By mid-1984 four machines had been sold, three in Canada and one in the United States. Prototype units owned by ABCO and FMC are on lease or demonstration to firms in Canada, the United

States, Europe, and New Zealand. Firms that have an ABCO blancher find they can use it for cooking as well as for a number of other innovative processes.

The Engineering and Statistical Research Institute, the Kentville Research Station, and ABCO Ltd., received the Gordon Royal Maybee Award from the Canadian Institute of Food Science and Technology, the Industrial Achievement Award from the United States Institute of Food Technologists, and the Food Processing Award from the Putnam Press of the United States for the development of the new blancher. The cost of research and development will be repaid many times by creating new jobs in manufacturing and sales and more nutritious frozen vegetables for consumers. Research is paying high dividends to food processors and the Canadian taxpayer.

ROLLTHERM COOKER–COOLER

J.A. Kitson at Summerland found that food processed in cans in a cooking retort held its heat for a long time, even though steel baskets containing the cans were plunged into cold water. He therefore devised a system of belts to rotate the cans as they moved through a cold water bath. This gentle agitation removed heat much more quickly; therefore Kitson reasoned that heat could be added quickly, using the same process. It was then that he, Dugal R. MacGregor of Summerland, and Darrell F. Wood of Kentville devised and patented (17) a cooker–cooler, using the rolling principle. The equipment was described in a French food industries magazine in 1968 and won for the department top award for the invention of a food processing technique and machine. Additional advantages of the Rolltherm Cooker–Cooler are that fruit and vegetables have better flavor, nutrition, and texture, and that the machine is cheaper to build than other automatic cookers.

INSPECTION AND GRADING

Canadians enjoy one of the best food grading systems in the world. The Department of National Health and Welfare keeps a close watch on the purity of foods, whereas the Department of Agriculture inspects for grade. A product, such as apples, must meet standards for size and color, and be free from blemish to be classed as Canada No. 1. For meat, inspection determines whether a side of beef is Canada A1 … D4.

Carcass research with beef and pork, initiated by the Lacombe Research Station in 1948, has been one of the outstanding success stories of the branch. The index system for pork introduced in 1968 and the grading system for beef adopted in 1972 are acclaimed as the most advanced carcass classification procedures in the world, and the principle they embody, specifically the method of ranking carcasses based on lean content, has been copied widely.

The Lacombe investigation began when H.T. Fredeen and A.H. Martin started to search for reliable measurements of live animals that would predict carcass merit and could be put to use when selecting breeding stock. They recognized that rate of gain and feed conversion were only two of the indicators of potential economic merit. They wanted to measure tissue composition and used techniques such as visual appraisal, fat probes, radiography, and an electrical device known as the "lean meter" to predict lean-to-fat ratios.

Field tests conducted in 1960 introduced the fat probe to hog producers and within a short time it was used across Canada. Later, Lacombe demonstrated an accuracy equivalent to the fat probe with an electronic echo sounder, which replaced the probe in the Canadian Record of Performance (ROP) program, in the early 1970s.

Although live animal probing opened the doors to meaningful genetic improvement of breeding stock, it was clear that it would have little impact on the reduction of fat unless commercial pig producers could see a financial return. Fredeen therefore established a simple indexing procedure that linked carcass value in the marketplace directly with carcass weight and backfat. In 1967 he tested more than 5000 carcasses across Canada, which fully validated the indexing proposal, and the scheme was officially adopted the following year. By 1974 trimmable fat had been reduced by an average of 0.5 kg per carcass, and by 1983 it had been reduced by 1.0 kg per carcass. In practical terms, this means that today's annual hog slaughter yields 6 million kg more trimmed retail meat than would have been marketed from hogs of 1968 breeding.

Intrigued by this progress, the beef industry in 1968 asked Fredeen to investigate the feasibility of revising their carcass classification system. Research began in 1970 with financial support from the Alberta Department of Agriculture and the Western Stock Grower's Association. The results were presented to the industry the following year but much heated debate ensued. Processors and retailers, even though recognizing the need for change, envisaged it taking place over several years. Finally, it was a cattle producer, Johnathon Fox, Jr., of Lloydminster, who asked, "When docking a dog's tail do you cut it an inch at a time?" The parallel was well taken and the entire industry voted to make the change. The new grading standards based on lean quantity and meat quality were implemented in September 1972. Within a week top market prices were being paid for quality carcasses in the minimum fat category. The chain of research was completed when the Engineering Research Service developed a simple device for inspectors to measure fat thickness.

The long-standing problem of excessively fat cattle vanished almost overnight. Time, feed, and money were no longer being wasted on producing, trimming, and disposing of excess quantities of low-value fat. Finally, consumers had an abundant supply of the low-fat beef they favored.

While Fredeen made progress at Lacombe, E.J. Brach at the Engineering and Statistical Research Institute found that he could objectively measure color, size, and texture to determine grade of food products. He knew that smooth surfaces reflect diffused light poorly, whereas coarse surfaces reflect it well. By converting the energy of reflected light into an electrical current he could determine if an animal had been finished on grass or on grain. By measuring color he could ascertain the amount of fat in the meat. The grading of eggs, meat,

337

apples, cherries, potatoes, and other products normally is done by comparing a reading of the product to be graded with a similar reading of a product of known grade, usually called a standard. In 1984, Brach, together with well-qualified inspectors, tested several prototypes of his instrument in abbatoirs. Meter readings were taken and by referring to a table of standards the inspectors could determine the grade of each carcass. Once the prototype is fully tested, Brach expects to make the instruments more sophisticated by having them display the grade directly. The first step toward an automatic inspection system has been taken.

Texture contributes to the eating pleasure of all foods. The chewing quality of a steak and the crispness of an apple are important characteristics. Two approaches have been used to study texture: sensory analyses (taste tests), which are subjective in nature, and an instrument that objectively measures the force required to deform and rupture a food. Sensory analyses require a taste panel of several people who can measure only a few samples at one sitting because one's taste buds quickly become fatigued. Taste panels are sometimes expensive and cumbersome; nevertheless, they are irreplaceable as the final test of any new product. Until 1970 there was one type of instrument to measure texture for industrial quality control and another type of instrument for laboratory use in research. This necessitated a recalibration of its equipment before a cannery or meat-packing plant could adopt a new research method.

In 1970 P.W. Voisey of the Engineering Research Service completed the development of the Ottawa Texture Measuring System (OTMS), which has such versatility it can be used in both research and industrial laboratories (25). One part of the instrument is a container in which the test material is placed and the other part exerts forces on it. The OTMS has a motorized, variable speed, screw-operated press. The simple, versatile product cell is unique. It has solid sides and parallel wire grid bottoms of selected spacings through which samples are pressed. Graphs of time and force can be compared with different samples of the same product (peas, fruits, meats, pastas) and objective decisions can be made concerning their relative textures. In tests with various classes of products, Voisey and others found the OTMS clearly differentiated among varieties of cooked soybean, ripeness of canned pears, grades of canned peas, and the tenderness of boiled scallops and cod fillets.

License to manufacture the OTMS was granted to Canners Machinery Ltd., Simcoe, Ontario, and machines now are being used by both research and industrial laboratories in many countries. Australia and Israel have adopted the OTMS to develop standards against which their peas are graded. A sophisticated new technology has been incorporated into another, more reliable standard of food inspection.

PRESERVING EGGS

As early as 1898 F.T. Shutt, the Dominion Chemist, searched for an improved method of preserving eggs. At that time chickens usually laid fewer than 100 eggs each a year, most of them during the spring and summer months. For the balance of the year production was low and prices were high. To compensate, people preserved eggs by "putting them down" in water glass, which was a 10-percent solution of sodium silicate. Shutt found that a concentrated solution of lime (calcium oxide) preserved fresh eggs as well as or better than water glass. It was also less expensive and more easily used. During World War II, C.K. Johns of the Bacteriology Division, Science Service (see Chapter 8) worked on bacterial counts in powdered eggs. In 1969 two Research Branch scientists markedly improved the quality of frozen eggs.

Because of the demand for frozen eggs, a large industry has been built both in Canada (10 million kg/year) and in the United States (230 million kg/year). Egg shells are broken and the contents homogenized and pasteurized before freezing in 15-kg pails. Eggs frozen this way have many disadvantages. Because of the large pail volume, freezing takes from 36 to 72 hours, and thawing from 24 to 36 hours. Even though pasteurized at 60°C, viable organisms can remain and continue to grow throughout the lengthy freeze and thaw periods. In the defrosting process, whites defrost first, then the yolks, even though they are homogenized; hence a whole pail must be thawed and remixed before use. The product is used in bakeries and dairies, by canners and confectioners, and in hotel, restaurant, and institutional kitchens.

M.M. Aref of the Food Research Institute dealt with these problems in 1969. He knew that semen was frozen rapidly by immersion in liquid nitrogen at $-196°C$. Would homogenized and pasteurized liquid eggs respond to the same treatment? They did, and came out looking like popcorn. Canadian Patents and Development Ltd., applied for patents (1) the same year.

In 1970 G.E. Timbers of the Engineering Research Service devised a method (24) of quickly freezing egg droplets. This method resulted in a particulate or granular frozen, light-yellow egg that was easily poured and measured, and could be more efficiently packaged than the popcorn-like pieces. It thawed rapidly without separating, and was ready to use within 15 minutes. Timbers found he could package the product in clear plastic bags of any premeasured size. Chefs could remove exactly the quantity they needed from a bulk container without any previous thawing. Because freezing was done rapidly at very low or cryogenic temperatures and the end product was granulated, Aref and Timbers called the new frozen egg product Cryogran. Taste panel experts could not distinguish between omelettes and other products made from Cryogran and those made from fresh eggs.

In 1972 Timbers took the cryogenic freezer to the world's leading food and agricultural competition in Paris (Industries alimentaires et agricoles) where the process won the Technique Prize. Many other liquid food products such as concentrated fruit juices, custards, creams, and purees can be frozen equally well by the new process. Probably the largest user will be industrial chemical companies who need to store highly perishable liquids in a frozen state. The new system did not meet with industrial acceptance until 1984, when the first commercial (500 kg/hour) system was built and used at Rodney, Ontario (6).

339

Many other processors of food, yeasts, bacteria, and other liquids now are showing an increased appreciation of the equipment. One scientist said that its potential is limited only by our imagination.

DAIRY RESEARCH

Clause 7(*b.*) of the Experimental Farm Station Act (*see* Appendix III) charged officers of each farm station with examining the economic questions involved in the production of butter and cheese. So it was that the Central Experimental Farm started to make and sell butter by 1900. The manufacture of soft cheese was started in 1911, followed by cheddar cheeses in 1913. Also in 1913, dairies were completed at both Lacombe and Agassiz, where dairymaids tested new methods of churning butter and making cheese and then advised local dairies of their findings. Dairy personnel at the Central Experimental Farm prepared and sent out countless bulletins and pamphlets to farmers across Canada advising them of improved methods of manufacturing dairy products and warning them of many pitfalls.

One of the early major contributions to the improvement in quality of butter was made by Bouchard (7) in 1907, who discovered that impure water used to wash lactic acid from butter imparted off-flavors. E.G. Hood and A.H. White in the 1930s found that washing cheese frequently resulted in off-flavors caused by bacterial infection. Investigations continued and in 1956 White, Beattie, and Riel (27) found, after studying 170 churnings from 29 Canadian dairies, that there was no difference in flavor between washed and unwashed butter when it was made from sweet cream rather than from sour cream. Furthermore, washing slightly reduced the yield. Therefore, today butter is not washed and the extra yield represents savings to the butter industry of several million dollars per year.

Johns (13) determined that the growth of bacteria in raw milk was virtually eliminated by the methods of production and efficient cooling developed in the 1950s. Thus, tests on freshly taken samples of cooled milk frequently did not detect poor dairy practices and latent infection when they should have. He therefore devised a method of incubating samples before tests were made, giving any bacteria present time to multiply and become detectable.

In 1972 D.B. Emmons revived the work Johns had done during World War II on grading skim milk powder. The logistics of many dairy plants sampling and shipping to a central laboratory for analyzing had meant that results were delayed by about 4 weeks. Emmons developed a system whereby official grades could be issued based on analyses done in the commercial laboratories where the powder was produced. By devising strict analytical methods and a system of comparisons and controls, dairy plants now analyze their own powder quickly, and grades are assigned immediately by telegraph. This saves the industry about two million dollars annually in inventory and storage costs, and provides for the sale of a fresher product. The system has served as a model for other products in the food inspection services of Agriculture Canada.

Emmons and coworkers also solved a starter culture problem in the manufacture of cottage cheese. Dairies sometimes were unable to obtain curds from apparently healthy milk to which starter cultures had been added; a granular sludge formed instead. After extensive investigation the scientists discovered (10) that some starter cultures had antibodies which agglutinated (formed clusters), onto which casein from the milk precipitated. Having found the cause, the solution since 1963 has been to grow and select suitable cultures that avoid agglutination and produce the desired curd.

CANOLA MEAL

The success of the Canadian rapeseed program has resulted from the skill of plant breeders in developing varieties suitable for Canadian conditions, and the persistence of food chemists in manipulating the seed and its products. Oil from canola is used as human food and the meal by-product is used as a high-protein supplement for livestock feed. Profitable canola production depends upon both oil and meal being marketable.

In its raw state, meal from old rapeseed varieties contains goitrogenic substances (glucosinolates), which must be removed or inactivated before the meal can be fed to livestock. When fed raw meal, the growth of rats is reduced to nearly zero and the thyroid gland increases in size by a factor of four or five. Detoxified meal from the newly developed canola, when fed to animals, results in growth essentially the same as feeding casein (the standard protein). The high fiber content in canola meal is a deterrent to its use as human food. Its seed coat fragments are an objectionable color and are indigestable to all animals but cattle. A third problem is that zinc in the meal is nutritionally unavailable. When fed as the only source of protein supplement, rats suffered a decreased appetite, a rapid loss of weight, a reduction in live-born pups, and a lighter weight of pup. Careful investigation showed that zinc was bound by phytates in the meal. The condition was easily corrected for rats by adding zinc to their drinking water. On the positive side, the balance of amino acids, which are the components of proteins, are superior in canola meal to any other known oilseed meal.

In 1978 J.D. Jones and I.R. Sibbald of the Food Research and Animal Research institutes fractionated canola seed by various methods in an attempt to obtain meal suitable for human and animal use (15). They found that removing hulls, which are high in fiber and low in energy, leaves a material containing more energy than does soybean meal. At about the same time, Jones (14) devised a method of extracting protein from canola that made an acceptable food for humans as well as for farm livestock. He and others patented the process in both Canada and the United States. Instead of the old method of flaking seed, cooking at temperatures of up to 130°C, pressing to remove part of the oil, and dissolving the remaining oil in hexane, Jones dehulled the seed, reduced its moisture content, and then extracted its oil directly with hexane solvent. The meal then was detoxified according to methods outlined in the patents. The process has many advantages because the improved oil contains no gums. The

341

meal, without any objectionable hulls, is suitable for human consumption to replace or extend other sources of protein. Also, it is a valuable supplement for all classes of livestock. To date, no Canadian oil extracting company is using the process. There is, however, considerable interest on the part of European processors.

The fruit, vegetable, dairy, and meat laboratories of the six research stations involved have been particularly successful in gaining the confidence of their local processing industries. By so doing, they have been able to provide optimum assistance. This type of advisory work has been only alluded to in the preceding paragraphs, because it has involved the interpretation and application of already known technologies.

Much more could have been said about many other food research studies of practical significance such as the quality of butter and cheese; the development of milk by-products and starter cultures; the fractionation of oats; the restructuring of meats; the properties of proteins; the biochemistry of muscle; extrusion cooking; the chemistry of sugars; the concentration of captured fruit juice aromas; and the improvement of hospital meals. Most have had an immediate commercial application, and have benefited Canadian farmers by assisting processors to either widen their product line or improve their efficiency, and should have enabled processors to pay more for the raw product.

During some periods in the history of Research Branch, the policy has been for its research to stop at the farmer's gate. This is not so today. There is a growing involvement on the part of branch scientists with the total food system, a system that goes beyond the farmer and the processor to the distributor and the retailer. The thinking is that since Canadian farmers can and do grow high-quality food, the Canadian consumer should have the opportunity of buying high-quality food, food that has been properly preserved, properly packaged, and properly processed.

References

1. Aref, M.M.; Timbers, G.E. 1975. Frozen food substance from egg; its production and apparatus therefore. Can. Pat. No. 964921.
2. Asselbergs, E.A.; Saidak, P.; Hamilton, H. 1961. Studies on processed potato products. Can. Food Ind. 32:30–33.
3. Atkinson, F.E. 1954. Food technology achievements at Summerland, B.C. Agric. Inst. Rev. 9(4):37–40.
4. Atkinson, F.E.; Strachan, C.C. 1950. Small canneries. Dom. Can. Dep. Agric., Publ. 828. 83 p.
5. Atkinson, F.E.; Bowen, J.F.; MacGregor, D.R. 1959. A rapid method for production of a sparkling apple wine. Food Technol. 13:673–675.
6. Barratt, Bob. 1984. "Cryogran" freezes liquid foods in pellet form. Food Can. 44(7):25–27.
7. Bouchard, J.G. 1907. Sweet-cream butter. Part II. Directions for the manufacture of butter from sweet or unripened cream. Can. Dep. Agric. Bull. 13. Dairy and cold storage series. 15 p.
8. Bowen, J.F.; MacGregor, D.R.; Fisher, D.V. 1972. Wine grape varieties for British Columbia. Can. Inst. Food Sci. Technol. 5:44–49.
9. Elson, C.M.; Lidster, P.D.; Hayes, E.R. 1985. Development of the differentially permeable fruit coating and *Nutri Save* for the modified atmosphere storage of apples. *In* Proc. 4th Controlled Atmos. Res. Conf., Raleigh, N.C. (In press.)
10. Emmons, D.B.; Elliot, J.A.; Beckett, D.C. 1963. Agglutination of starter bacteria, sludge formation, and slow acid development in cottage cheese manufacture. J. Dairy Sci. 46:600.
11. Fuleki, Tibor. 1965. Fermentation studies on blueberry wine. Food Technol. 19:105–108.
12. Heeney, H.B.; Rutherford, W.M.; MacQueen, K.F. 1964. Some effects of gamma radiation on the storage life of fresh strawberries. Can. J. Plant Sci. 44:188–193.
13. Johns, C.K. 1959. Application and limitations of quality tests for milk and milk products. J. Dairy Sci. 42:1625–1650.
14. Jones, J.D. 1979. Rapeseed protein concentrate preparation and evaluation. J. Am. Oil Chem. Soc. 56:716–721.
15. Jones, J.D.; Sibbald, I.R. 1979. The true metabolizable energy values for poultry of fractions of rapeseed (*Brassica napus* cv. Tower). Poult. Sci. 58:385–391.
16. Kitson, J.A. 1961. Continuous vacuum unit simplifies apple process. Food Eng. 33(9):94–95.
17. Kitson, John A.; MacGregor, Dugal R.; Wood, Darrell F. 1971. Continuous atmospheric roll cooker. Can. Pat. No. 883762.
18. Lidster, P.D.; Lightfoot, H.J.; McRae, K.B. 1983. Fruit quality in respiration of 'McIntosh' apples in response to ethylene, very low oxygen, and carbon dioxide storage atmospheres. Scientia Hortic. 20:71–83.
19. Nuttall, V.W.; Lyall, L.H.; McQueen, K.F. 1961. Some effects of gamma radiation on stored onions. Can. J. Plant Sci. 41:805–813.
20. Porritt, S.W. 1963. Progress of C.A. storage in British Columbia. Wash. State Hortic. Assoc. Proc. 59:152–154.
21. Porritt, S.W.; Lidster, P.D.; Meheriuk, M. 1975. Postharvest factors associated with the occurrence of breakdown in Spartan apple. Can. J. Plant Sci. 55:743–747.
22. Strachan, C.C. 1942. Factors influencing ascorbic acid retention in apple juice. Dom. Can. Dep. Agric., Publ. 732. 31 p.

343

23. Strachan, C.C.; Moyles, A.W.; Atkinson, F.E.; Britton, Dorothy. 1960. Commercial canning of fruit pie fillings. Can. Dep. Agric. Publ. 1062. 24 p.
24. Timbers, G.E. 1974. Frozen particulate liquid products. Can. Inst. Food Sci. Technol. J. 7:68–71.
25. Voisey, Peter W. 1971. The Ottawa texture measuring system. Can. Inst. Food Technol. J. 4:91–103.
26. Watters, F.L.; MacQueen, K.F. 1967. Effectiveness of gamma irradiation for control of five species of stored-product insects. J. Stored Prod. Res. 3:223–234.
27. White, A.H.; Beattie, D.M.; Riel, R.R. 1956. Washed and nonwashed butter. I. Flavor quality and curd content. J. Dairy Technol. 39:261–267.
28. Wood, D.F. 1967. Anthocyanin and juice yields from frozen cranberries. Agric. Can. Res. Stn Technol. Rep., Kentville, N.S. 11 p.

344

PART III
EXTRAMURAL ACTIVITIES

Research Branch
Agriculture Canada
1886–1986

Chapter 21
Cooperation

To complete this reflection on the life of the Research Branch, Canada Department of Agriculture, recognition must be given to the significant activities of its staff while away from their regular places of work. In this chapter some of the ways in which staff have made contributions extramurally are examined.

COMMUNITY INVOLVEMENT

Four of the original five Experimental Farms were established in small communities. They soon became, and some still are, the largest employers within their localities. The experimental farms also attracted to their service, individuals from outside their communities. These staff members possessed skills and talents beyond those for which they were hired. Consequently, they became community leaders in numerous social and business endeavors. As additional experimental stations were established, most were located near small rural towns; thus the community influence of experimental farms expanded.

When tragedies such as flood and fire struck, Research Branch staff proved to be capable leaders. In 1948, when the Lower Fraser River overflowed its banks and flooded the valley, W.H. Hicks, Superintendent, Experimental Farm, Agassiz, was appointed by the government of British Columbia as coordinator of flood relief for the Agassiz–Harrison Valley (part of the Lower Fraser Valley). Hicks made the safety of people his main concern. When assured that no human life was endangered, he proceeded to move livestock to higher land and arrange for their feed. This was no easy task, as rail and road connections had been severed. Hicks assigned all available Experimental Farm staff and equipment to duties in the community. The 1948 scare has given the Experimental Farm (now Research Station), incentive to keep its emergency measure techniques updated. Periodically, action plans are reviewed to ensure readiness in the event of natural, chemical, or general disaster.

Another flood, similar to that experienced in the Lower Fraser Valley, occurred in 1973 when the St. John River swelled, overflowing its banks. G.M. Weaver, Director, Research Station, Fredericton, cooperating closely with the New Brunswick Deputy Minister of Agriculture and with Department of National Defence personnel from Gagetown, removed all cattle from flooded areas to the research station. The cattle were fed and cared for by the staff of the station until the waters receded and the flooded areas were again habitable.

This spirit of cooperation is returned by the farming community. In 1959, for instance, when the main dairy barn was destroyed by fire at the Research Station, Lethbridge, Alberta, neighboring farmers came to the rescue with accommodation and milking machines for the dairy herd.

Agricultural fairs also enhanced a cooperative link between the community and the staff of experimental farms. Often the superintendents themselves, in addition to many staff of experimental farms and stations, served as members of boards of directors of local fair associations. For example, in 1888 James Fletcher, Chief, Entomology and Botany Division, became the first experimental farm officer to be appointed to the Central Canada Exhibition Association, Ottawa. He has been succeeded by Experimental Farm and Research Branch staff members including W. Saunders, J.H. Grisdale, E.S. Archibald, E.S. Hopkins, C.H. Goulden, J.C. Woodward, T.H. Anstey, and J.J. Cartier. In addition, several were elected to the presidency or served on the board of directors of the Ottawa Winter Fair.

To help formulate plans for the World's Columbian Exposition to be held in 1893 in Chicago, Saunders, in January 1892 was appointed Executive Commissioner for Canada. The Canadian exhibit was judged to be the finest agricultural display in the building. Special seedings of grain and vegetables were made in the spring of 1893 at experimental farms in order to provide excellent examples of produce throughout the duration of the fair. This is the fair at which the Great Canadian Mammoth Cheese from Perth, Ontario, was featured. The 1900 World Exposition in Paris (see Chapter 3) received 1200 jars of preserved fruit and a continuous supply of fresh fruit and vegetables from experimental farms to demonstrate the production capabilities of Canadian farmland and its farmers.

Breed, production, and other types of agricultural societies closely associated with farmers receive attention from experimental farm staff. Their help may be in the form of technical consultation, service on executive councils, or the provision of meeting rooms and support services. Recipient organizations include the Canadian Seed Growers' Association, the Canadian Cattlemen's Association, the Nova Scotia Fruit Growers' Association, 4-H Clubs, and the Agricultural Improvement Associations (see Chapter 13).

Experimental Farm, Science Service, and Research Branch men and women have been staunch supporters of many local, national, and international service clubs. They have a sincere sense of civic responsibility, and the *esprit de corps* that they enjoy in the workplace extends to community activities resulting in numerous youth and adult leaders coming from "the farm." Staff sometimes accept duties as reeves and municipal councillors, members of school, university, and hospital boards as well as organizers of civic, church, and charitable functions. Some employees have resigned or retired in order to enter provincial or federal politics.

FEDERAL–PROVINCIAL COOPERATION

With provincial departments of agriculture

The Canadian Conference of Agricultural Instruction, organized by the Dominion Department of Agriculture, held on 24–25 March 1914, brought together representatives of provincial departments of agriculture. The attendees included five provincial ministers of agriculture, eight provincial deputy ministers of agriculture, and 19 representatives of departments of education. It was convened by the Honourable Martin Burrell, federal Minister of Agriculture, as a forum to exchange ideas and to discuss plans under the provisions of the 1913 Agricultural Instruction Act. The conference proved to be a landmark in uniting federal and provincial departments of agriculture. By 1922, the program was completed. It had financed the development of technical skills of teachers in agriculture, helped the provinces organize agricultural extension services, encouraged the instruction of agricultural topics in secondary schools, provided assistance to boys and girls clubs (now 4-H Clubs), and supplied agricultural fairs with prize money.

In 1932 the National Research Council of Canada (NRC) organized the Associate Committee on Agricultural Research. It included all deans of faculties of agriculture, a representative from the Canadian Society of Technical Agriculturists (now the Agricultural Institute of Canada), appointees from the Dominion Department of Agriculture, and representatives from NRC. The objective of the committee was to advise those organizations conducting agricultural research on how problems might be solved scientifically.

The NRC associate committee was dissolved in 1935 upon the formation by the Department of Agriculture of a National Advisory Committee on Agricultural Services (NACAS) chaired by Deputy Minister G.S.H. Barton. The committee still functions. Over time it has changed its organization and its name. Records are incomplete, but it seems that in 1960, when Deputy Minister J.G. Taggart was chairman, the committee regrouped its 12 parts into three and changed its name to the National Coordinating Committee on Agricultural Services (NCCAS). At about the same time, each province organized a parallel series of technical committees to deal with problems on a provincial level. In 1965, under Deputy Minister S.C. Barry NCCAS's name was changed to its current designation—the Canadian Agricultural Services Coordinating Committee (CASCC, pronounced "cask").

CASCC is mainly research oriented. Three groups of committees form its base: provincial agricultural services coordinating committees chaired by appropriate provincial deputy ministers of agriculture (the Atlantic Provinces as a group have one committee); six expert committees on engineering, animal production, crop production, food, land resources, and socio-economics; the general services section of deputy ministers, which deals with policy; and the Canadian Agricultural Research Council (CARC). In addition, Statistics Canada, the National Research Council of Canada, and the Agricultural Institute of Canada are members of CASCC. On occasion, representatives of the departments of Health and Welfare; Industry, Trade, and Commerce; and Consumer and Corporate Affairs attend as observers. Research Branch supplies the secretariat for both CASCC and CARC, as well as the operating expenses for CARC.

During the 1930s and after World War II (NACAS's first meeting following the cessation of hostilities was in 1946), soil survey and conservation, rape and sunflower seed production, weed control, grading of beef cattle, artificial insemination, and female sterility of dairy cattle were the main areas of concern. As the years passed, new subjects such as farm building plans (see Chapter 19), trace elements for plants and animals, performance testing of beef cattle and swine, and residues of insecticides and fungicides were brought to the attention of CASCC. The federal deputy ministers of Agriculture Canada (see Appendix II) chaired the meetings. To begin with, these were held in the fall immediately before the annual Dominion–Provincial Agricultural Conference (now called the Agricultural Outlook Conference) that reviews annually the following year's prospects for each group of agricultural commodities. This arrangement, however, detracted from the business of CASCC. In addition, committee members found that research and administrative funds for the following year had already been allotted, leaving managers little opportunity of responding to recommendations from CASCC. In 1959 the meeting date was changed to the spring in order to overcome this difficulty.

CASCC is described by Wasik (6) as "the pinnacle of a pyramid of committees representing virtually all our agri-food institutions." He estimates that 2500 people from across Canada form the pyramid. CASCC has been an important instrument in advancing the Canadian agri-food system.

The research section of CASCC, consisting of deans of faculties of agriculture and a few others, was organized by Deputy Minister Barry in 1965, the year in which CASCC was formed. In 1974, Deputy Minister S.B. Williams, upon the urging of B.B. Migicovsky, Director General, Research Branch, and that of C.M. Switzer, Dean, Faculty of Agriculture, and D.G. Howell, Dean, Ontario Veterinary College, both of the University of Guelph, established the Canadian Agricultural Research Council (CARC) as a body independent of CASCC. One of CARC's first tasks was to prepare an inventory of agricultural research. This inventory, which is updated each year, lists and categorizes all active agricultural research from universities, provincial governments, the federal government, and private industry. It is used as an aid by scientists and managers to avoid overlap when planning new programs. In 1977 the inventory reported 3700 research projects and 1760 full-time agricultural research scientists in Canada. By 1984 there were 4200 projects and 2350 scientists. In addition, since 1974, CARC has made special studies on energy, land resources, rapeseed meal, crop losses from weeds, agricultural engineering, biotechnology for agriculture, animal production, rural development, and the horticultural industry. It has also examined the need for research in aquaculture.

Provincial and federal agrologists who are at the operational level maintain a close, often informal, association with one another. Research Branch directors and scientists have always recognized the importance of making their research results available to farmers as quickly as possible. The first superintendents, therefore, were involved as much in extension activities as they were in research activities. This situation prevailed until each province appointed agricultural extension officers. Experimental Farm, Entomological Branch, and Science Service officers then willingly responded to calls from provincial district agriculturists (in some provinces called county agents) to assist at field days in the summer and at short courses in the winter.

The practice of providing office space on experimental stations to provincial extension officers started in the Maritime Provinces. Indeed, at Fredericton, the research station and the New Brunswick Department of Agriculture share a building on federal government property. Similar arrangements are now in effect at most federal research stations. The one reverse situation is at Vineland, Ontario, where the Research Branch Laboratory is the guest of the Ontario Horticultural Research Station. The advantages are many: early awareness by research scientists of new problems facing farmers in the field; rapid, direct transfer of research results to provincial extension specialists; and one-stop shopping for information by farmers in person, by telephone, or by letter.

With provincial departments of forestry

Arrangements with provincial forestry services similar to those with agriculture extension services, involved, in particular, the reporting of forest insects. By 1940 more than 2000 observers from provincial forestry services and the forestry industries stationed in all parts of Canada had made regular reports to the forest insect survey. From 1940 until 1960, when forest entomologists moved to the Department of Forestry, the results of the survey were published by Science Service. Laboratory facilities and person-years have been provided by the provinces of New Brunswick, Quebec, and Ontario in addition to those supplied by Science Service and Research Branch.

The Canadian Pulp and Paper Association granted scholarships in forest entomology, and the Canadian Lumbermen's Association has promoted important contacts between governments and the lumber industry.

With university faculties of agriculture

Relations between faculties of agriculture and Research Branch have always been positive. Many of the professional staff of Research Branch and its predecessors have been graduates of Canadian universities, which in itself provides a basis for close affiliation. In addition, K.W. Neatby was a strong advocate of placing some Science Service laboratories on the campuses of those Canadian universities with faculties of agriculture. Today, research stations function on the campuses of the universities of Laval, Western Ontario, Manitoba, Saskatchewan, and British Columbia. In some instances, staff from the research stations provide instruction at graduate courses and the students do their laboratory work at the research station.

The Division of Horticulture, in particular, encouraged interaction with departments of horticulture at the universities of Guelph (Ontario Agricultural College at the time), Manitoba, Saskatchewan, Alberta, and British Columbia. In these five instances, the Dominion Horticulturist, M.B. Davis, assigned a horticulturist to each university. It was the responsibility of the horticulturist to work closely with the university department in conducting fruit and vegetable variety trials. Also, one horticulturist was assigned to the Provincial Horticultural Station, Vineland, Ontario. Federal public service staff reductions, which started about 1960, terminated this arrangement.

NATIONAL RESEARCH COUNCIL OF CANADA

The contributions made by the ministers of agriculture, the Honourable Richard W. Motherwell, and the Honourable Robert Weir, from 1929 to 1939, the formative years of the National Research Council of Canada, (NRC) are mentioned in Chapter 5. NRC, Experimental Farms, and Science Service cooperated on many programs during World War II when resources were especially scarce.

In 1948 the Prairie Regional Laboratory (PRL) of NRC was formally opened on the campus of the University of Saskatchewan. PRL, as a research unit, was already several years old (3), having operated from the university laboratories until its building was complete. Gridgeman (4) describes the support given to the formation of PRL by the Honourable James Gardiner, Minister of Agriculture. The first research conducted by scientists at the new laboratory was on rapeseed and was done in conjunction with the Forage Crops Laboratory (see Chapter 17). Together they played an important part in bringing rapeseed to its current economic level and in developing hardboard from wheat straw. Cooperation and friendly competition between scientists at PRL and the research station continues, to the advantage of both organizations.

COMMONWEALTH AGRICULTURAL BUREAUX

In 1908 (5) the Colonial Office in London, England, recognized the problem of insects affecting people, crops, and livestock on the west coast of Africa. The office contacted a number of entomologists in British African colonies. In an attempt to find a solution, governments of the United Kingdom (UK) and the African colonies provided funds to form the Entomological Research Committee (Tropical Africa) in 1910. The Committee arranged for African entomologists to forward insect specimens to the care of the secretary, Sir Guy Marshall, at the Natural History Museum, London, for identification. He, in turn, published the *Bulletin of Entomological Research* to disseminate current, topical information to field entomologists. This was the start of what became the Commonwealth Agricultural Bureaux (CAB)—an organization consisting of 10 bureaus and four institutes (1) that exists to provide an international identification and information service.

In 1911 at the Imperial Conference, C. Gordon Hewitt, the Canadian Dominion Entomologist, firmly favored a proposal to expand the services of the Entomological Research Committee. In 1912, governments of all the major Commonwealth countries agreed to proceed and the Imperial Bureau of Entomology formally started to function in 1913. The UK, Canada, Australia, New Zealand, and India provided the required funds and do so still.

At the Imperial War Conference, in 1918, the Bureau of Mycology, headquartered at Kew Gardens, UK, was established. As with the Imperial Bureau of Entomology, the Bureau of Mycology was financed jointly by all Commonwealth countries.

The Imperial Agricultural Research Conference was held in London in 1927. Delegates from many Commonwealth countries recognized the value of the services performed by the Bureau of Entomology and by the Bureau of Mycology, and they urged that similar clearinghouses for information on soil science, animal health, animal nutrition, animal genetics, pasture and forage crops, horticulture, and agricultural parasitology be established. The idea was accepted by all governments in the Commonwealth and on 1 April 1929, the Executive Council of CAB held its inaugural meeting at which a representative from each member government attended. Canada, from the outset, has been an ardent supporter of CAB, funding having been supplied by Science Service and, subsequently, by Research Branch. Canadian representatives to the Executive Council of CAB have been H.L. Trueman of Science Service; G.M. Carman, R.M. Prentice, and R. Trottier, all of Research Branch.

At the working level, university, provincial, and federal scientists play an active role in Canada's support of the CAB. In particular, Canadian scientists (2) have a keen interest in the biological control of insects and weeds. Indeed, during World War II the Commonwealth Institute for Biological Control, established by Sir Guy Marshall as a subsidiary of the Commonwealth Institute of Entomology, was moved together with its director, W.R. Thompson, from England to the Dominion Parasite Laboratory in Belleville, Ontario. It was transferred to Ottawa in 1948 and then relocated to permanent quarters in Trinidad during 1962.

ORGANIZATION FOR ECONOMIC COOPERATION AND DEVELOPMENT

The involvement of E.J. LeRoux in the Organization for Economic Cooperation and Development (OECD) began in Washington, D.C., in the fall of 1980. He was attending a Tripartite meeting of the three directors of agricultural research from the United States, Great Britain, and Canada, when Deputy Minister Lussier telephoned him to ask if he would let his name stand for the chairmanship of the OECD Committee for Agriculture, with headquarters in Paris, France. LeRoux agreed to the nomination.

Research Branch had been involved with the subcommittees of Directors of Agricultural Research and Agricultural Policies for many years. Both were part of the Committee for Agriculture, but LeRoux would be the first person from the branch to be associated with the main committee. LeRoux, who speaks English and French fluently and is a Canadian, was experienced in chairing meetings at which there were differing, firmly held opinions. In accord with this nomination were his minister, the Honourable Eugene Whelan, his deputy minister, Gaetan Lussier, and the Department of External Affairs. He also had the support of the non-European member countries—the United States, Australia, New Zealand, and Japan.

The Department of External Affairs briefed LeRoux on four areas of endeavor. They were as follows: to reorient the agricultural programs from academic studies to those relating to urgent needs of member countries, to complete the policy adjustment study for presentation to the OECD Council, to complete a review of the economy of centrally planned countries, and to review the agriculture of developing countries. Formidable tasks for a chairman who had had no previous active association with OECD!

At LeRoux's first meeting in May 1978 the opening item of business was the election of a new chairman. The delegate from Denmark put forward a motion, which was seconded by the Spanish delegate, that E.J. LeRoux from Canada be chairman. Approval was unanimous, revealing the high regard the 23 participating countries had for Canada. M. DeBouverie from Belgium was reelected vice-chairman. LeRoux developed a close rapport with the Canadian Ambassador to the OECD, His Excellency Ronald Stuart MacLean, and also with Mr. Albert Simantov, the Executive Secretary of the Committee for Agriculture, and his staff. LeRoux's objectives were to perform his new duties in such a way that Canada's impact on the world scene would be properly recognized, and to fulfill the mandate entrusted to him by the Canada Department of External Affairs. He earned the respect of the delegates from other countries by conducting meetings in a professional and tactfully efficient manner. Several delegates commented that the committee discussions under his chairmanship were full but not tedious, were reactive to the wishes of ministers, and were challenging with the intent of producing results. The United States' delegates were pleased that North America had filled the chair so effectively.

Over a period of 3 years LeRoux chaired nine meetings and attended two others, one of Deputy Ministers of Agriculture and one of Ministers of Agriculture. He arranged for the Honourable John Wise, Canadian Minister of Agriculture, to chair the meeting of ministers in March 1980. Before it took place, however, the federal Conservative Government was defeated, so Mr. Jaime Lamo de Espinosa, the Minister of Agriculture for Spain, acted as chairman.

One of LeRoux's major accomplishments was to guide the study on Positive Adjustment Policies (PAP) through the various committee meetings and have it accepted by the OECD Council. This had been a difficult task because the representative of the European Economic Community (EEC) organization as well as individuals from France and Belgium continually challenged the study. Skill was needed to negotiate agreement between the dissenting countries and the remainder of the committee. He reorganized the working groups making them smaller and more effective. He contributed both in Paris and in Ottawa to the establishment of priorities for the World Food Council, which assembled in Ottawa in September 1979. The Honourable Eugene Whelan became its president in 1983. LeRoux led a study to develop an agenda for the World Conference on Agrarian Reform and Rural Development. Above all, he encouraged the Committee of Research Directors to act positively in organizing cooperative research in four areas of food production. During his chairmanship, the Prairie Agricultural Machinery Institute of Humboldt, Saskatchewan, was named an official OECD farm machinery testing station. There, Canadian manufacturers could obtain OECD approval for farm tractors without having to send prototypes to Europe. This meant that Canadian tractors could be readily marketed within the EEC.

LeRoux recommended that the Canadian delegate to the OECD Paris meetings in future come from Agriculture Canada, and receive support from other departments of the federal government rather than the reverse. He improved communications within Agriculture Canada, between his department and other departments concerned with OECD agricultural matters, and between provincial governments and the government of Canada on OECD activities in order that Canada could speak with a united voice. He urged that a qualified agrologist be recruited to the staff of the Canadian post to OECD.

Research Branch, through LeRoux's talents, raised Canada's profile while contributing to the development of the OECD Committee for Agriculture.

INTERNATIONAL COOPERATION

With developed countries

Science, and research especially, is an international activity. Except for subjects dealing with national defense, the results of scientific research are normally disseminated to the international community through publication in scientific journals. Scientists themselves have networks within which they pass prepublished information. They frequently work at one another's laboratories for better exchange of information.

Some developed countries recognized the need to structure this effective informal arrangement. Therefore, with the guidance of External Affairs, agreements for the pooling of technical information and for the exchange of agricultural scientists with Canada have been negotiated since 1970 between the Research Branch and Belgium, Brazil, Cuba, Czechoslovakia, France, Germany, Hungary, Israel, Japan, Poland, Romania, the USSR, and Yugoslavia. A number of scientists in several countries have undertaken international work transfers. Often benefit has been derived by both sides.

B.B. Migicovsky, Assistant Deputy Minister (Research), kept in close contact with T.W. Edminster, Chief Administrator of Agricultural Research, U.S. Department of Agriculture. There was already frequent interaction between Research Branch and the Chief Scientist, U.K. Ministry of Agriculture, Fisheries, and Food. Because each had a reduced scientific staff, the leaders of the three national agricultural research agencies met in Washington, D.C., in 1978, to devise ways of maintaining or even increasing their research effectiveness and avoiding duplication of effort. They accomplished their objectives by collaborating in areas of mutual interest through a planned exchange of scientists. The following year, E.J. LeRoux was the host and invited J. Poly, Director General, Institut national de la recherche agronomique (INRA), France, to join the group in Ottawa. Since then representatives from the four countries have met annually. From this cooperation has come joint research programs on nitrogen fixation, photosynthesis, biotechnology, and human nutrition. Since 1978, more than 100 Canadian scientists have exchanged with their counterparts in the United Kingdom, the United States, and France.

With developing countries

The International Institute of Agriculture, which Canada supported, held its inaugural meeting in Rome in 1905. Two actions were taken that were of particular significance to Experimental Farms. The first was the formation of the International Meteorological Commission. This action caused R.F. Stupart, Director of the Canadian Meteorological Service, who was a member of the commission, to hire three graduates from the Ontario Agricultural College, Guelph. They were to investigate the application of meteorology to agricultural requirements. At the same meeting, on Canada's behalf H.T. Güssow, Dominion Botanist, signed an agreement that created the International Phytopathological Commission. Sixty-five other countries signed the agreement, which fostered the definition of specific arrangements for the control of plant diseases and permitted regular international trade of nurserystock and other plant material. The actions of Stupart and Güssow have benefited Canadian farmers by improving the accuracy of forecasting weather conditions, and by making it possible for Canadians to import and export plant material with assurance of its freedom from disease and with minimum loss of time for inspection at ports of entry.

The International Institute of Agriculture was absorbed by the League of Nations in 1924 and then by the United Nations Organization through the Food and Agriculture Organization (FAO) in 1945. In fact, FAO had its beginning at the Quebec Conference on 16 October 1945. Because food production is one of the major subjects in development programs, agriculture has been of prime importance. Until 1970 Canada Department of Agriculture supported development programs on an individual project basis. The department granted leave of absence to employees to serve with FAO, United Nations Development Program (UNDP), WORLDBANK, and the International Atomic Energy Agency (IAEA). Technical help has been provided through the Canadian International Development Agency (CIDA) to assist in planning, operating, and assessing agriculturally related development programs. To expedite this process, the Research Branch supplied a liaison officer, T.G. Willis, whose time was divided between activities based at CIDA and those based at the department.

In 1970, after the Research Branch had for several years provided agrologists to other organizations undertaking development programs for CIDA, the branch, on behalf of the department, contracted directly with CIDA to execute two programs—a rainfed agricultural program in India and a wheat production scheme in Tanzania. Both programs emphasize the practical application of research. Both have consistently met their objectives. Both are still operative. J.E. Andrews, Director, Research Station, Swift Current, planned the India program, and following his move to Lethbridge, administered the project that has now grown to span 23 research stations and to employ 250 scientists in India. As a direct result of the Indo–Canadian project, rainfed crop yields have increased by 150 percent. The wheat production project in Tanzania involves one research station and a 20 000-ha area on which wheat is grown. The project is administered by J.S. Clark, Director, Land Resource Research Institute. Andrews and Clark have been honored with Public Service Merit Awards for their contributions to Canada's international development objective.

Since 1970 similar programs have been started for CIDA in Indonesia, Sri Lanka, Pakistan, Colombia, Haiti, and Brazil. Some, such as the Brazilian Wheat Breeding Program, have directly benefited the counterpart Canadian program by integrating research from each country. A more rapid progression of research is permitted, because two crops may be grown and studied each year—one in the northern hemisphere and one in the southern hemisphere. Since Research Branch first contracted with CIDA, 122 scientists from 21 research stations have served in 38 countries (see Table 21.1).

357

Table 21.1 Contributions of Research Branch to International Development

Year	Number of scientists	Number of countries	Number of person-years
1970	3	3	0.5
1971	17	16	7.5
1972	12	7	9.4
1973	12	7	6.0
1974	12	9	4.0
1975	5	2	4.6
1976	6	5	2.0
1977	7	6	4.0
1978	3	3	2.0
1979	10	6	1.5
1980	25	17	10.0
1981	20	10	10.0
1982	30	18	7.0
1983	38	19	6.5
1984	18	9	5.5
Total			80.5

International development has been promoted through other channels. Research Branch has seconded staff to several international agricultural research centers. One director, W.L. Pelton, because of his valuable experience with the India Rainfed Agricultural Program, was seconded for a year in 1976 to the Consultative Group on International Agricultural Research (CGIAR) to establish the International Centre for Agricultural Research in the Dry Areas (ICARDA) in the Middle East. Through cooperation with the Canadian-sponsored International Development Research Centre (IDRC), Research Branch has been involved in organizing and in executing training programs for scientists and research administrators from the Third World.

References

1. Anonymous. 1980. Perspectives in world agriculture. Commonwealth Agricultural Bureau, Slough, England. 532 p.
2. Anonymous. 1981. Biological control programmes against insects and weeds in Canada 1969–1980. Commonwealth Agricultural Bureau, Slough, England. 410 p.
3. Eggelston, Wilfrid. 1978. National Research in Canada. The NRC 1916–1966. Clarke, Irwin & Co. Ltd., Toronto. 470 p.
4. Gridgeman, N.T. 1979. Biological sciences at the National Research Council of Canada: The early years to 1952. Wilfrid Laurier University Press. 153 p.
5. Scrivenor, Sir Thomas. 1980. CAB—the first 50 years. Commonwealth Agricultural Bureau, Slough, England. 73 p., 9 append.
6. Wasik, R.J. 1985. CASCC—A national agri-food think-tank. Food in Canada 45(3):44.

APPENDIXES

Research Branch
Agriculture Canada
1886–1986

APPENDIX I

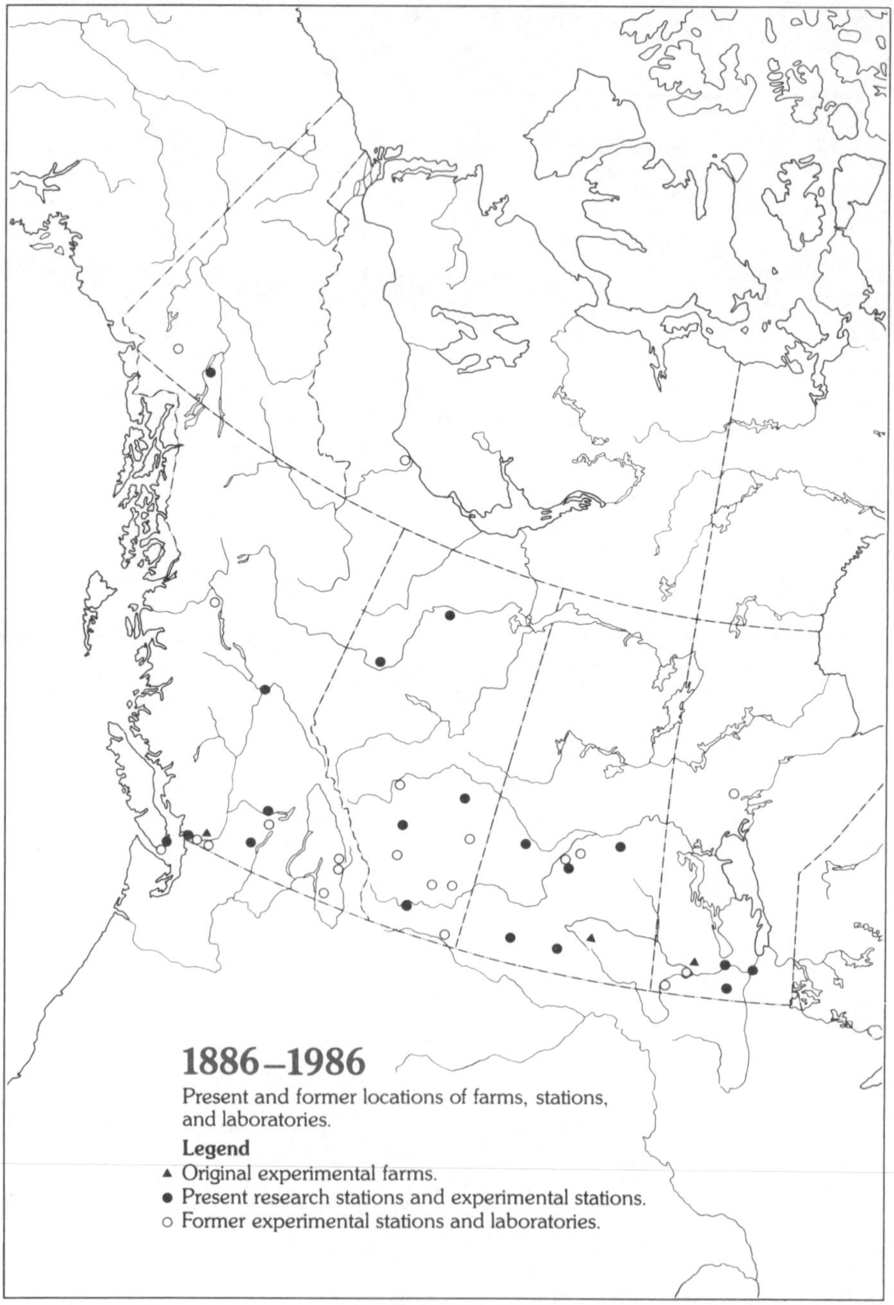

Present and former locations of farms, stations, and laboratories.

Legend

▲ Original experimental farms.
● Present research stations and experimental stations.
○ Former experimental stations and laboratories.

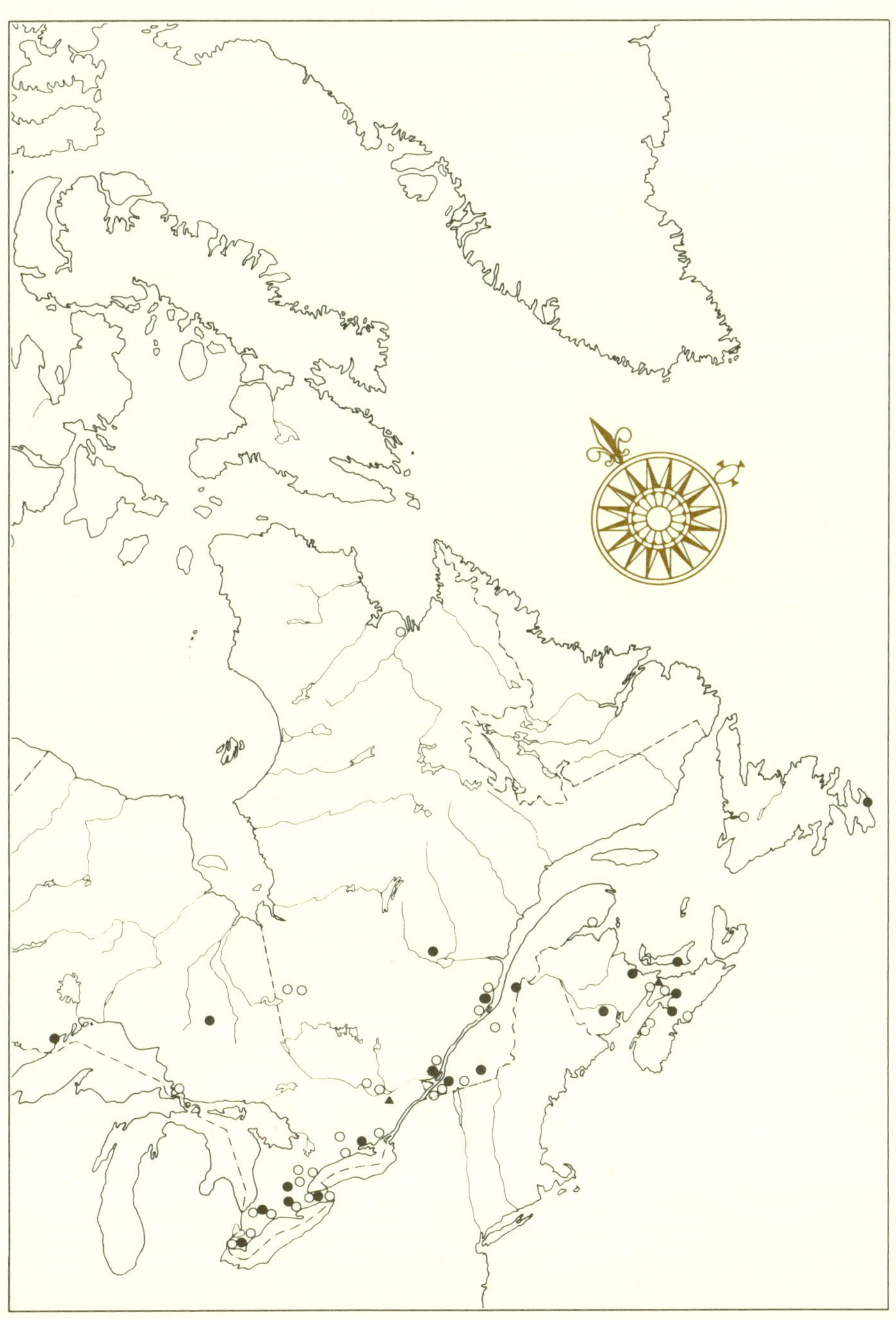

⚜

APPENDIX II
MINISTERS, DEPUTY MINISTERS, AND OFFICERS IN CHARGE

The following tables are arranged:
— by province, east to west, with headquarters first;
— chronologically by location within each province;
— chronologically within each location; and
— chronologically within each responsibility center (farm, station, or laboratory).

The five original farms are called experimental farms, all others are called experimental stations or laboratories regardless of what they may have been called from time to time. For 1959 and after, research stations and institutes are named.

Officers in charge are not designated as superintendents or directors except with the headquarters staff where directors, chiefs of divisions, and later, assistant deputy ministers and directors general are identified as such. Incumbents holding acting positions are included from the date they became acting if they were appointed to the position. If someone else was appointed, the name of the acting appointee has not been included. Only officers in charge who were in residence are listed.

The year of disposition is the final one for the last officer in charge, unless otherwise stated. If no location is given, then it was to the same location and organization. If the disposition was to a different location, service, branch, or department it is identified. When a new unit started, its source is not given unless it came from a service, branch, or department unrelated to Research Branch and therefore not identified elsewhere in Appendix II.

The information was obtained from various written sources. In the event of discrepancies, data from appropriate volumes of *American Men of Science* or some other source in which the contents were checked by the person concerned were used. Question marks indicate missing data, parentheses indicate approximate dates.

MINISTERS

Hon. John Henry Pope	1878–1885
Hon. John Carling	1885–1892
Hon. Auguste Real Angers	1892–1895
Hon. Walter Humphies Montague	1895–1896
Hon. Sidney Arthur Fisher	1896–1911
Hon. Martin Burrell	1911–1917
Hon. Thomas Alexander Crerar	1917–1919
Hon. Simon Fraser Tolmie	1919–1926
Hon. Richard William Motherwell	1926–1930
Hon. Robert Weir	1930–1935

Hon. Douglas C. Harkness	1957–1960
Hon. Alvin Hamilton	1960–1963
Hon. Harry Hays	1963–1965
Hon. J.J. Greene	1965–1968
Hon. H.A. Olson	1968–1972
Hon. Eugene Whelan	1972–1979
Hon. John Wise	1979–1980
Hon. Eugene Whelan	1980–1984
Hon. Ralph Ferguson	1984
Hon. John Wise	1984–

DEPUTY MINISTERS

Dr. J.C. Taché	1867–1888
Mr. J. Lowe	1888–1895
Mr. W.B. Scarth	1895–1902
Mr. G.F. O'Halloran	1902–1918
Dr. J.H. Grisdale[1]	1918–1932
Dr. G.S.H. Barton	1932–1949
Dr. J.G. Taggart[1]	1949–1959
Mr. S.C. Barry	1960–1966
Mr. S.B. Williams[1]	1967–1975
Mr. L.D. Hudon	1975–1977
Dr. G. Lussier	1977–1982
Mr. J.P. Connell	1982–

[1]From Experimental Farms Service or Research Branch.

HEADQUARTERS, DIVISIONS, and INSTITUTES

OTTAWA

EXPERIMENTAL FARMS SERVICE

Directors

W Saunders, CMG, LLD, FRSC, FAAAS	1886–1911
JH Grisdale, BAgr, DScA, FAIC, FAAAS, FAGA	1911–1919
ES Archibald, BA, BSA, LLD, DSc, FRSC, FAIC	1919–1951
ES Hopkins, BSA, MSc, PhD, FAIC	1951–1955
CH Goulden, BSA, MSA, PhD, LLD, FRSC, FAIC, FASA	1955–1959

Chiefs of Divisions

Agriculture
W Saunders, CMG, LLD, FRSC, FAAAS	1886–1890
JW Robertson, LLD	1890–1896
W Saunders, CMG, LLD, FRSC, FAAAS	1897–1898
JH Grisdale, BAgr, DScA, FAIC, FAAAS, FAGA	1899–1912
(to Animal Husbandry and to Field Husbandry)	

Cereal
  W Saunders, CMG, LLD, FRSC, FAAAS   1886–1902
  CE Saunders, Kt, BA, PhD, LLD, DSc, FRSC, FAIC   1903–1922
  LH Newman, BSA, DSc, FAIC   1923–1948
  CH Goulden, BSA, MSA, PhD, LLD, FRSC, FAIC, FASA   1948–1955
  DG Hamilton, BSc, MS, PhD, FAIC, FAAAS   1955–1959
  (to various research institutes)

Horticulture
  WW Hilborn   1886–1889
  J Craig   1890–1898
  WT Macoun, DSc, FAIC   1898–1933
  MB Davis, BSA, MSc, FAIC   1933–1955
  H Hill, BSA, MSA, PhD   1955–1959
  (to various research institutes)

Chemistry
  FT Shutt, BA, MA, DSc, FIC, FRSC   1887–1933
  CH Robinson, BA   1933–1937
  (to Science Service)

Entomology and Botany
  J Fletcher, LLD, FRSC, FLS, FESA   1887–1908
  (to Entomology and to Botany and Plant Pathology)

Poultry Husbandry
  AG Gilbert   1888–1913
  (to Poultry)

Entomology
  CG Hewitt, BA, PhD, DSc, FRSC, FESA   1909–1914
  (to Entomological Branch and to Bee Division)

Botany and Plant Pathology
  HT Güssow, FLS, FRHS, LLD   1909–1937
  (to Science Service)

Animal Husbandry
  ES Archibald, BA, BSA, LLD, DSc, FRSC, FAIC   1912–1919
  GB Rothwell, BSA   1919–1931
  GW Muir, BSA   1931–1951
  HKCA Rasmussen, BSA, MS, PhD, LLD, FAIC   1951–1958
  (to Animal and Poultry Science)

Field Husbandry (Field Husbandry, Soils and Agricultural
Engineering, 1940)
  JH Grisdale, BAgr, DScA, FAIC, FAAAS, FAGA   1912–1918
  ES Archibald, BA, BSA, LLD, DSc, FRSC, FAIC   1919–1920
  ES Hopkins, BSA, MSc, PhD, FAIC   1920–1946
  PO Ripley, BSA, MS, PhD, FAIC   1946–1959
  (to various research institutes)

Forage
  MO Malte, BA, MA, PhD   1912–1921
  GP McRostie, BSA, MSA, PhD   1922–1930
  LE Kirk, BA, BSA, MSA, PhD   1931–1937
  TM Stevenson, BSA, MSc, PhD, FAIC   1938–1959
  (to various research institutes)

Tobacco
  (from the Tobacco Branch)
  F Charlan   1912–1924
  CM Slagg, BS, MS   1924–1928
  NT Nelson, BSA, MSc, PhD   1928–1946
  NA MacRae, BA, MSc, PhD   1946–1959
  (to various research institutes)

Poultry
  FC Elford   1913–1937
  G Robertson   1937–1946
  HS Gutteridge, BSA, MSc   1946–1957
  (to Animal and Poultry Science, 1958)

Bee
  FWL Sladen   1914–1921
  CB Gooderham, BSA   1921–1949
  CA Jamieson, BSA, PhD   1949–1958
  (to Entomology Research Institute)

Extension and Publicity
  JF Watson   1914–1917
  WA Lang   1917–1921
  FC Nunnick, BSA   1921–1935
  (to departmental headquarters)

Illustration Stations
  J Fixter   1915–1927
  JC Moynan, BSA   1928–1953
  AE Barrett, BSA, MSc   1953–1959
  (to various research and experimental stations)

Fibre
  GG Bramhill, BSA   1917–1818
  RJ Hutchinson   1918–1952
  (to Field Husbandry)

Bacteriology
  AG Lochhead, BA, MSc, PhD, FRSC   1923–1937
  (to Science Service)

Animal and Poultry Science
  HKCA Rasmussen, BSA, MS, PhD, LLD, FAIC   1958–1959
  (to Animal Research Institute)

365

ENTOMOLOGICAL BRANCH

Directors

CG Hewitt, BA, PhD, DSc, FRSC, FESE	1914–1920
A Gibson, LLD, FRSC, FRES, FESA	1920–1937
(to Entomology Division, Science Service)	

Chiefs of Divisions

Field Crops and Garden Insects

A Gibson, LLD, FRSC, FRES, FESA	1914–1920
RC Treherne, BSA	1920–1923
HGM Crawford, BSA, MS	1925–1937
(to Entomology, Science Service)	

Forest Entomology

JM Swaine, BSA, MSc, PhD, FRSC	1914–1934
JJ de Gryse, PhCand	1934–1937
(to Entomology, Science Service)	

Plant Inspection

LS McLaine, BSc, MSc, FAAAS	1919–1938
(to Plant Protection, Production Division)	

Systematic Entomology

JH McDunnough, MA, PhD	1919–1937
(to Entomology, Science Service)	

Stored Products Insects

EH Gray, BSA, BS, MSc, PhD	1932–1937
(to Entomology, Science Service)	

SCIENCE SERVICE

Directors

JM Swaine, BSA, MSc, PhD, FRSC	1937–1946
KW Neatby, BSA, MSA, PhD, DSc, LLD, FRSC, FAIC	1946–1958
R Glen, OC, BSc, MSc, PhD, LLD, DSc, FAIC, FRSC, FESC, FESA	1958

Chiefs of Divisions

Bacteriology

AG Lochhead, BA, MSc, PhD, FRSC	1937–1938
(to Bacteriology and Dairy Research)	

Botany and Plant Pathology

HT Güssow, FLS, FRHS, LLD	1937–1944
JH Craigie, OC, AB, MSc, PhD, DSc, LLD, FRSC, FAIC, FRS	1945–1952
WF Hanna, OC, CBE, OLM, BA, BSc, MSc, PhD, LLD, FRSC, FAIC	1953–1958
(to various research institutes)	

Chemistry
 CH Robinson, BA   1937–1949
 JC Woodward, MC, BSA, MS, PhD, FCIC, FAIC, FBCS   1949–1955
 ARG Emslie, BSA, MSA, DSc, FCIC   1955–1959
 (to various research institutes and services)

Entomology
 A Gibson, LLD, FRSC, FRES, FESA   1937–1942
 LS McLaine, BSc, MSc, FAAAS   1942–1943
 HGM Crawford, BSA, MS   1943–1950
 R Glen, OC, BSc, MSc, PhD, LLD, DSc, FAIC, FRSC, FESC,
 FESA   1950–1957
 BN Smallman, BA, MA, PhD, FAAAS   1957–1959
 (to Entomology Research Institute)

Animal Pathology
 (from Health of Animals Branch)
 EA Watson, DVM   1937–1942
 CA Mitchell, VS, BVSc, DVM, FRSC, HARCVS   1942–1952
 (to Production Service)

Bacteriology and Dairy Research
 AG Lochhead, BA, MSc, PhD, FRSC   1938–1953
 (to Bacteriology)

Plant Protection
 (from Production Service)
 WN Keenan   1942–1956
 (to Production Service)

Forest Biology
 JJ de Gryse, PhCand   1951–1952
 ML Prebble, BScF, MSc, PhD   1952–1959
 (to Research Branch)

Bacteriology
 AG Lochhead, BA, MSc, PhD, FRSC   1953–1956
 H Katznelson, BSA, MSc, PhD, FAAAS, FRSC   1956–1959
 . (to various research institutes)

RESEARCH BRANCH

Assistant Deputy Ministers

 CH Goulden, BSA, MSA, PhD, LLD, FRSC, FAIC, FASA   1959–1962
 R Glen, OC, BSc, MSc, PhD, LLD, DSc, FAIC, FRSC, FESC,
 FESA   1962–1968
 JA Anderson, BSc, MSc, PhD, DSc, LLD, FCIC, FAIC, FRSC   1968
 JC Woodward, MC, BSA, MS, PhD, FCIC, FAIC, FBCS   1968–1974
 BB Migicovsky, BSA, MS, PhD, DSc, FCIC, FAIC   1975–1978
 EJ LeRoux, BA, MSc, PhD, DSc, FAIC   1978–

Directors General

R Glen, OC, BSc, MSc, PhD, LLD, DSc, FAIC, FRSC, FESC, FESA	1959–1962
JA Anderson, BSc, MSc, PhD, DSc, LLD, FCIC, FAIC, FRSC	1963–1968
BB Migicovsky, BSA, MS, PhD, DSc, FCIC, FAIC	1968–1974

Directorates

Operations
EJ LeRoux, BA, MSc, PhD, DSc, FAIC   1975–1978

Program Coordination
(renamed Planning and Coordination [1970], Planning and
Evaluation [1975], and Program Coordination [1980])

JC Woodward, MC, BSA, MS, PhD, FCIC, FAIC, FBCS	1959–1964
PO Ripley, BSA, MS, PhD, FAIC	1964–1965
JC Woodward, MC, BSA, MS, PhD, FCIC, FAIC, FBCS	1965–1968
HKCA Rasmussen, BSA, MS, PhD, LLD, FAIC	1968–1970
AE Hannah, BSA, MS, PhD	1970–1972
EJ LeRoux, BA, MSc, PhD, DSc, FAIC	1972–1975
DG Hamilton, BSc, MS, PhD, FAIC, FAAAS	1975–1977
JW Morrison, BSc, MSc, PhD	1978–1982
RL Halstead, BSA, PhD	1982–1985
WB Baier, Diplomlandwirt, DrAgr, MSc	1985–

Forest Biology Division
ML Prebble, BScF, MSc, PhD   1959–1960
(to Department of Forestry)

Institutes and Services

BB Migicovsky, BSA, MS, PhD, DSc, FCIC, FAIC	1964–1968
EJ LeRoux, BA, MSc, PhD, DSc, FAIC	1968–1972
WB Mountain, BSc, PhD	1973–1975
(to Central and Institutes Region)	
JW Morrison, BSc, MSc, PhD	1980–1985
RL Halstead, BSA, PhD	1985–

Administration

SB Williams, ON, BSA, MSc, FAIC	1959
DH Laughland, BSA, MA, PhD, FCIC	1959–1965
RA Ludwig, BSc, MSc, PhD, FAPS	1965–1977
JE Ryan, RIA	1977–1984
LR Radburn, BA	1984–

Research Services

Statistical

GB Oakland, BA, MA, PhD, FIS, FSS	1950–1960
P Robinson, BA, DipStat, PhD	1960–1972
LP Lefkovitch, BSc	1972–1977
(to Engineering and Statistical Research Institute)	

Analytical Chemistry
    RB Carson, BA, BSc, MS, FCIC   1959–1971
    (to Chemistry and Biology Research Institute)

Engineering
    W Kalbfleisch, BE, MS   1959–1967
    CGE Downing, BE, MSc, FASAE, FEIC   1967–1977
    (to Engineering and Statistical Research Institute)

Scientific Information (Research Program, 1973)
    GM Carman, BSA, MSc, PhD   1959–1965
    C Chaplin, BSc, MSc, PhD   1965–1976
    C Aubé, BSc, MSc, PhD   1976–1978
    S Plourde, BA, MA   1978–1979
    R Trottier, BSc, MSc, PhD   1980–1982
    J-C St-Pierre, BSc, MSc, PhD   1982–1984
    Y Bélanger, BSc   1985–

Ornamental
    AP Chan, BSc, MSc, PhD   1973–1976
    (to Ottawa Research Station, Ontario Region)

Regions

Eastern
    DG Hamilton, BSc, MS, PhD, FAIC, FAAAS   1964–1975
    (Ontario to Central and Institutes, 1975)
    JJ Cartier, BA, BSc, MSc, PhD, FESC   1975–1980
    (to Atlantic and to Quebec)

Western
    HKCA Rasmussen, BSA, MS, PhD, LLD, FAIC   1964–1968
    TH Anstey, BSA, MSA, PhD, FAIC   1969–1979
    AA Guitard, BSc, MSc, PhD, FAIC   1979–1981
    JE Andrews, BSA, MS, PhD, FAIC   1981–1983
    (to Prairie and to Pacific)

Central and Institutes
    WB Mountain, BSc, PhD   1975–1979
    (to Ontario and to Institutes and Services)

Atlantic
    EE Lister, BS, MS, PhD   1980–1985
                                       1985–

Quebec
    J-J Jasmin, BScAgr, MSc   1980–

Ontario
    JJ Cartier, BA, BSc, MSc, PhD, FESC   1980–

Prairie
    WL Pelton, BSA, MSA, PhD   1983–

369

Pacific
    SC Thompson, BSc, MSc, PhD   1983–

RESEARCH INSTITUTES

Animal
    ARG Emslie, BSA, MSA, DSc, FCIC   1959–1965
    RS Gowe, BSA, MS, PhD   1965–1980
    (to Animal Research Centre, Ontario Region)

Dairy Technology
    CK Johns, BSA, MSc, PhD, FAIC, FAPHA   1959–1963
    (to Food)

Entomology
    GP Holland, BA, MA, DSc, FRSC   1959–1969
    WB Mountain, BSc, PhD   1969–1973
    (to Biosystematics)

Genetics and Plant Breeding
    AWS Hunter, BSA, MSc, PhD   1959–1964
    (part to Food in 1962)
    (to Ottawa Research Station, Ontario Region)

Microbiology
    H Katznelson, BSA, MSc, PhD, FAAAS, FRSC   1959–1967
    (to Cell Biology)

Plant
    HA Senn, BA, MA, PhD, FRSC   1959–1960
    RA Ludwig, BSc, MSc, PhD, FAPS   1961–1965
    AP Chan, BSc, MSc, PhD   1965–1973
    (to Biosystematics, to Cell Biology, and to Ornamental
    Research Service)

Soils
    (renamed Land Resource Research Institute [1978], with part
    of program to Chemistry and Biology)
    PC Stobbe, BSA, MSc, PhD, FCSSS, FAIC   1959–1969
    JS Clark, BSA, MSA, PhD   1969–

Research Institute for Biological Control, see Belleville, Ontario

Pesticide Research Institute, see London, Ontario

Insect Pathology Research Institute, see Sault Ste. Marie, Ontario

Chemical Control
    JJ Fettes, BScF, PhD   1959–1960
    (to Department of Forestry)

Food
    RPA Sims, BSc, PhD, FCIC   1962–1973
    J Holme, BA, MA, PhD   1974–1983
    NW Tape, BSA, PhD   1983–

Cell Biology
    RM Hochster, BSc, PhD   1967–1971
    (renamed Chemistry and Biology)
    G Fleischmann, MA, PhD   1972–1974
    JG Saha, BSc, MSc, PhD   1974–1980
    IA de la Roche, BSc, MSc, PhD   1980–1985
    (to Plant Research Centre, see Ontario)

Biosystematics
    DF Hardwick, BA, MSc, PhD   1973–1978
    GA Mulligan, BSc   1978–

Engineering and Statistical
    P Voisey, AMIMechE   1978–

371

NEWFOUNDLAND

ST. JOHN'S WEST

Entomology Laboratory
    (from the government of Newfoundland)
    HA Butler, BSc   1949–1957
    RF Morris, BSA, MS   1957–1959
    (to Research Station)

Experimental Station
    IJ Green   1950–1956
    HWR Chancey, BSA, MSA, FAIC   1956–1959
    (to Research Station)

Plant Pathology Laboratory
    OA Olsen, BSA, MSc, PhD   1957–1959
    (to Research Station)

Research Station
    HWR Chancey, BSA, MSA, FAIC   1959–1984
    HR Davidson, BSA, PhD   1984–

Soil Survey
    HWR Chancey, BSA, MSA, FAIC   1952–1956
    JFG Millett, BSA, MSc, PhD   1956–1963
    PK Heringa, BSA, MSA   1963–1984
    F Hender, BSc   1984–

CORNER BROOK

Forest Biology Laboratory
    WJ Carroll, BSc, MSc   1952–1960
    (to Department of Forestry)

PRINCE EDWARD ISLAND

CHARLOTTETOWN

Experimental Station
JA Clark, BSA, MSA, DSc	1909–1947
RC Parent, BSA, MSc, FAIC	1947–1966
(to Research Station)	

Plant Pathology Laboratory
PA Murphy, BSc, PhD	1915–1920
JB McCurry, BSA	1920–1923
RR Hurst, BSA	1925–1939
(to Science Service Laboratory)	

Entomology Laboratory
FM Cannon, BSc, MSc	1936–1939
(to Science Service Laboratory)	

Science Service Laboratory
RR Hurst, BSA	1939–1960
(to Experimental Station)	

Soil Survey
GB Whiteside, BSA	1943–1962
JI MacDougall, BSc	1971–1984
C Veer, DipFor	1984–

Research Station
GC Russell, BS, MS, PhD	1966–1970
LB MacLeod, BSc, MSc, PhD, FCSSS	1970–

SUMMERSIDE

Experimental Fox Ranch
(Experimental Fur Ranch 19??)
   (from National Research Council Fox Ranch, Hull, Quebec)
GE Smith, BASc	1925–1941
CK Gunn, BSc, MSc, PhD	1941–1968
(to Research Station, Charlottetown, 1969)	

NOVA SCOTIA

NAPPAN

Experimental Farm
WM Blair	1887–1896
GW Forrest	1896–1897
R Robertson	1898–1913
WW Baird, BSA	1913–1952
SB Williams, ON, BSA, MSc, FAIC	1952–1959
TM MacIntyre, BSc, BSA, MSc, FAIC	1959–1979
FW Calder, BSc, MS	1979–

BRIDGETOWN

Entomology Laboratory
  GE Sanders, BSA   1910–1915
  (to Annapolis Royal)

KENTVILLE

Experimental Station
  JR Starr   1911–1912
  WS Blair, DSc   1912–1938
  A Kelsall, BSA, DCL   1938–1952
  CJ Bishop, BSc, AM, PhD, DSc, FRSC, FAIC, FASHS   1952–1958
  TH Anstey, BSA, MSA, PhD, FAIC   1958–1959
  (to Research Station)

Plant Pathology Laboratory
  JF Hockey, BSA, DSc   1924–1959
  (to Research Station)

Chemistry Laboratory
  FA Herman, BSc, FCIC   1936–1954
  RF Bishop, BSc, MSc, PhD   1954–1959
  (to Research Station)

Entomology Laboratory
  AD Pickett, BSA, MSc, DSc   1952–1959
  (to Research Station)

Research Station
  RA Ludwig, BSc, MSc, PhD, FAPS   1959–1961
  JR Wright, BSc, MS, PhD, FCIC   1961–1978
  GM Weaver, BSc, PhD   1979–

ANNAPOLIS ROYAL

Entomology Laboratory
  GE Sanders, BSA   1915–1921
  A Kelsall, BSA, DCL   1921–1938
  AD Pickett, BSA, MSc, DSc   1939–1952
  (to Kentville)

Chemistry Laboratory
  FA Herman, BSc, FCIC   1923–1936
  (to Kentville)

TRURO

Soil Survey
  GB Whiteside, BSA   1934–1943
  RE Wicklund, BSA, MSc, PhD   1943–1947
  DB Cann, BSA, MSc, PhD   1947–1964
  JI MacDougall, BSc, BScA   1964–1969

373

JL Nowland, BA, MSc   1969–1973
GJ Beke, BSA, MSc, PhD   1973–1981
KT Webb, BSc, MSc   1983–

HALIFAX

Forest Entomology Laboratory
FG Cuming, BSc, MSc   1949–1951
(to Debert)

DEBERT

Forest Biology Laboratory
FG Cuming, BSc, MSc   1951–1960
(to Department of Forestry)

NEW BRUNSWICK

 FREDERICTON

Entomology Laboratory
Field Crops Insects
JD Tothill, BSA   1911–1922
RP Gorham, BSc   1920–1946
JB Adams, BA, MSc   1946–1959

Fruit Insects
GP Walker   1917–1924
CWB Maxwell, BSA, MSc   1925–1959
(to Research Station)

Experimental Station
WW Hubbard   1912–1921
CF Bailey, BSA   1922–1947
SA Hilton, BSA, MS, FAIC   1947–1959
(to Research Station)

Plant Pathology Laboratory
GC Cunningham, BSA   1915–1923
JF Hockey, BSA, DSc   1923–1924
DJ MacLeod, BA, MA, PhD   1924–1959
(to Research Station)

Forest Entomology Laboratory
JD Tothill, BSA   1923–1924
JL Simpson   1924–1930
RE Balch, BSA, MS, PhD   1930–1951
(to Forest Biology Laboratory)

Soil Survey
PC Stobbe, BSA, MSc, PhD, FCSSS, FAIC   1938–1940
H Aalund, BScA   1940–1947
RE Wicklund, BSA, MSc, PhD   1947–1953

JFG Millette, BSA, MSc, PhD   1953–1961
KK Langmaid, BSc, MSc   1961–1975
C Wang, BSA, MSc, PhD   1975–1979
R Wells, BSc, MSc, PhD   1979–1981
HW Rees, BSc   1981–

Forest Pathology Laboratory
AJ Skolko, BScF, MA, PhD   1946–1948
VJ Nordin, BA, BScF, PhD   1949–1951
(to Forest Biology Laboratory)

Forest Biology Laboratory
RE Balch, BSA, MS, PhD   1951–1959
RM Belyea, BA, PhD   1960
(to Department of Forestry)

Research Station
SA Hilton, BSA, MS   1960–1966
F Whiting, BSc, MSc, PhD   1966–1971
G M Weaver, BSc, PhD   1971–1979
CS Bernard, BSA, MSc, PhD   1979–1984
YA Martel, BA, BSA, PhD   1984–

BUCTOUCHE

Experimental Station
JM Wauthy, BSc   1979–1982
R Rioux, BSc   1982–1985
                                                     1985–

QUEBEC

CAP ROUGE

Experimental Station
GA Langelier, DScA   1911–1933
CE Ste Marie, BSA   1933–1940
(closed)

SAINTE-ANNE-DE-LA-POCATIÈRE

Experimental Station
J Begin   1912–1922
JA Ste Marie, BSA   1921–1937
JR Pelletier, BSA, MA, MSc, DSc   1937–1960
(to Research Station)

Plant Pathology Laboratory
HN Racicot, BA   1923–1930
CJ Perrault, BSA, MSc, DSc   1930–1952
(to Science Service Laboratory)

Entomology Laboratory
    JA Duncan, BSc, MSc   1949–1952
    (to Science Service Laboratory)

Soil Survey
    RW Baril, LScA, MSc, FCSS, OMA   1947–1962
    (closed)

Science Service Laboratory
    CJ Perrault, BSA, MSc, DSc   1952–1960
    (to Research Station)

Research Station
    CJ Perrault, BSA, MSc, DSc   1960–1967
    (to Experimental Station, some staff to Sainte-Foy)

Experimental Station
    JM Girard, BS, BSA, MSA   1967–1968
    JE Comeau, BSc, MSc   1969–

COVEY HILL

Entomology Laboratory
    CE Petch, BSA   1912–1914
    (to Hemmingford)

FARNHAM

Experimental Station
    O Chevalier, INA   1912–1916
    JE Montreuil, BA, BSA   1919–1928
    R Bordeleau, BSA   1928–1946
    (closed)

HEMMINGFORD

Entomology Laboratory
    CE Petch, BSA   1914–1952
    (closed)

LENNOXVILLE

Experimental Station
    JA McClary   1914–1937
    JA Ste Marie, BSA   1937–1952
    E Mercier, BA, BSA, MSA, PhD   1952–1960
    GJ Brisson, BA, BSA   1960–1962
    PE Sylvestre, BA, BSA, MSc   1962–1965
    (to Research Station)

Research Station
    PE Sylvestre, BA, BSA, MSc   1965–1968
    CS Bernard, BSA, MSc, PhD   1968–1979

YA Martel, BA, BSA, PhD   1979–1984
J-C St-Pierre, BSc, MSc, PhD   1984–

LA FERME

Experimental Station
  P Fortier, BSA   1916–1932
  JCH Chabot, BS, BSA   1932–1936
  (closed)

FORT COULONGE

  Forest Insect Laboratory ?   1918–?

SAINT JOACHIM

Experimental Station (Horse Farm)
  GA Langelier, DScA   1920–1940
  (sold to the government of Quebec)

L'ASSOMPTION

Experimental Station
  JE Montreuil, BA, BSA   1928–1946
  R Bordeleau, BSA   1946–1962
  J Richard, BSA, MSc   1962–1971
  PP Lukosevicius, BSA, MSc, PhD   1971–1981
  JPF Darisse, BA, BSc, MSc   1981–

BERTHIERVILLE

Forest Entomology Laboratory
  JSL Daviault, BSA   1929–1942
  (closed)

SAINTE-ANNE-DE-BELLEVUE

Soil Survey
  PC Stobbe, BSA, MSc, PhD, FCSSS, FAIC   1934–1938
  DB Cann, BSA, MSc, PhD   1938–1947
  JGP Lajoie, BA, BSA, MSc   1947–1964
  (to Sainte-Foy)

Animal Pathology Laboratory
  WE Swales, BVSc, PhD   (1947)–1952
  (to Production Service)

NORMANDIN

Experimental Station
  JA Belzile, BSA   1935–1962
  JE Laplante, BSA, MSc   1962–1969

JPF Darisse, BA, BSc, MSc   1969–1981
JM Wauthy, BSc   1982–

MACAMIC

Experimental Station
  WA Montcalm   1936–1937
  A Courcy   1937–1943
  R Bernier, BA, BSc   ?
  (closed)

HULL

Animal Disease Research Institute
  (from Health of Animals Branch)
  EA Watson, DVM   1937–1942
  CA Mitchell, VS, BVSc, DVM, LLD, FRSC, HARCVS   1942–1952
  (to Production Service)

SAINT-JEAN-SUR-RICHELIEU

Entomology Laboratory
  JB Maltais, BSA, MSc   1940–1949
  AA Beaulieu, BSA, MSc   1949–1959
  (to Research Laboratory)

Plant Pathology Laboratory
  L Cinq-Mars, BA, BSA, MSc   1949–1959
  (to Research Laboratory)

Research Laboratory
  AA Beaulieu, BSA, MSc   1959–1962
  (to Research Station)

Research Station
  AA Beaulieu, BSA, MSc   1962–1972
  J-J Jasmin, BScAgr, MSc   1972–1980
  CB Aubé, BSc, MSc, PhD   1980–

SAINT-CHARLES-DE-CAPLAN

Experimental Station
  JGG Provencher, BA, BSc   1948–1952
  L Bellefleur, BSA   1952–1957
  JDR Bernier, BA, BSc   1957–1963
  JPF Darisse, BA, BSc, MSc   1963–1966
  M Hughes   1966–1970
  (sold to the government of Quebec)

SAINTE-FOY

Forest Entomology Laboratory
  JSL Daviault, BSA   1954–1955
  (to Forest Biology Laboratory)

Forest Pathology Laboratory
   HR Pomerleau, BSA, MSc, DSc, FRSC   1954–1955
   (to Forest Biology Laboratory)

Forest Biology Laboratory
   JSL Daviault, BSA   1955–1960
   (to Department of Forestry)

Research Station
   CJ Perrault, BSA, MSc, DSc   1967–1968
   SJ Bourget, BSc, MS, PhD   1968–

Soil Survey
   R Marcoux, BSc, MSc   1975–1979
   JM Cossette, BSc   1979–

SAINTE-CLOTHILDE

Experimental Station
   J-J Jasmin, BScAgr, MSc   1956–1962
   (none now in residence)

FORT CHIMO

Experimental Station
   RI Hamilton, BSc, MSA, PhD   1956
   H Gasser, BSc, MSc, PhD   1957–1959
   RI Hamilton, BSc, MSA, PhD   1960–1961
   BWA Parks   1962–1965
   (closed)

ONTARIO

VINELAND

Entomology Laboratory
   WA Ross, BSA   1911–1947
   GG Dustan, BSA, MS   1947–1959
   (to Research Station)

Research Station
   DA Chant, BA, MA, PhD   1960–1963
   WB Mountain, BSc, PhD   1964–1969
   GM Weaver, BSc, PhD   1969–1971
   AJ McGinnis, BSc, MS, PhD   1972–1980
   DR Menzies, BSc, MS, PhD   1981–

ST. CATHARINES

Plant Pathlogy Laboratory
   WA McCubbin, BA, MA, PhD   1912–1919

WH Rankin, AB, PhD   1919–1922
GH Berkeley, BA, MA, PhD   1923–1959
(to Research Stations, Vineland and Vancouver)

HARROW

Experimental Station
(from Tobacco Branch, Canada Department of Agriculture)
WA Barnet, BSA   1913–1915
DD Digges, MSc   1915–1926
HA Freeman, BSA, MSc   1926–1928
HF Murwin, BSA   1929–1959
(to Research Station)

Plant Pathology Laboratory
LW Koch, BA, MA, PhD   1938–1947
(to Science Service Laboratory)

Entomology Laboratory
WE van Steenburgh, OBE, ED, BS, MA, PhD   1938–1940
HR Boyce, BSA, MSA   1940–1947
(to Science Service Laboratory)

Science Service Laboratory
LW Koch, BA, MA, PhD   1947–1959
(to Research Station)

Research Station
LW Koch, BA, MA, PhD   1959–1969
GC Russell, BS, MS, PhD   1970–1975
JM Fulton, BSc, MSA, PhD   1975–1980
CF Marks, BSc, MSA, PhD   1981–

STRATHROY

Entomology Laboratory
HF Hudson, BSA   1913–(1932)
(disposition?)

WALKERVILLE

Experimental Station
GC Routt   1915–?
(disposition?)

KAPUSKASING

Experimental Station
JPS Ballantyne   1916–1945
ET Goring   1945–1949
FX Gosselin, BSA   1949–1964
JE Comeau, BSA, MSc   1965–1969

JM Wauthy, BSc   1969–1978
J Proulx, DVM   1978–

CHATHAM

Entomology Laboratory
   HGM Crawford, BSA, MS   1920–1925
   GM Stirrett, BSA, MS, PhD   1926–1948
   (biological control work to Belleville, 1929)
   GF Manson, BSc, MSc   1948–1967
   (to Harrow and London)

ST. THOMAS

Entomology Laboratory
   AB Baird, BSA, MS   1923–1929
   (to Belleville)

SIMCOE

Entomology Laboratory
   JA Hall, BSA   1928–1958
   A Hikitchi, BA, BSA   1958–1960
   (to Research Station, Vineland)

BELLEVILLE

Entomology Laboratory
   (Research Institute for Biological Control, 1959)
   AB Baird, BSA, MS   1929–1948
   A Wilkes, BSA, MSc, PhD   1948–1955
   BP Beirne, BSc, MA, MSc, PhD, FCSZ   1955–1967
   PS Corbet, BSc, PhD, DSc, FIB   1967–1972
   (closed, staff to Ottawa, Winnipeg, and Regina)

Commonwealth Institute for Biological Control
   WR Thompson, BSA, MSc, DSc, PhD, FRS, FRSC   1940–1948
   (to Ottawa)

DELHI

Experimental Station
   GL Haslam, BSA   1933–1935
   FA Stinson, BSA, MSc, PhD   1935–1949
   LS Vickery, BSA, MS   1949–1975
   (to Research Station)

Research Station
   CF Marks, BSc, MSc, PhD   1976–1981
   PW Johnson, BSA, MSc, PhD   1981–

GUELPH

Soil Survey
GA Hills, BSA, MSc 1935–1943
NR Richards, BSA, MSc, DSc 1943–1952
RE Wicklund, BSA, MSc, PhD 1953–1970
CJ Acton, BSA, MSc, PhD 1970–

Entomology Laboratory
DG Peterson, BA, MSc 1955–1964
(closed, staff to Saskatoon, Lethbridge, and Ottawa)

THUNDER BAY (formerly Fort William)

Experimental Station
JK Knights, BSA 1937–1958
WB Towill, BSA 1958–1978
J Wilson 1978–

MARMORA

Entomology Laboratory
GH Hammond, BSA, MSc 1939–1961
(closed)

OTTAWA

Poultry Pathology Laboratory
(from Health of Animals Branch)
AB Wickware, VS ..... (1941)–(1950)
(to Production Service in 1952?)

Forest Entomology Laboratory
EB Watson, BA 1945–1953
(became part of Forest Biology Headquarters)

Entomology Laboratory
WG Matthewman, BSA, MSc ..... (1954)–1959
(to Entomology Research Institute)

Research Station
AWS Hunter, BSA, MSc, PhD 1964–1971
FK Kristjansson, BSA, MS, PhD 1971–1976
T Rajhathy, IngAgr, MSc, DAgrSci, FRSC 1976–1985
(to Plant Research Centre)

Animal Research Centre
RS Gowe, BSA, MS, PhD 1980–

Plant Research Centre
AI de la Roche, BSc, MSc, PhD 1985–

SAULT STE. MARIE

Forest Biology Laboratory
    CE Atwood, BScA, MScA, PhD   1944–1945
    ML Prebble, BScF, MSc, PhD   1945–1952
    RM Belyea, BA, PhD   1952–1960
    WA Reeks, BSc, MSc   1960
    (to Department of Forestry)

Insect Pathology Laboratory
    (Research Institute, 1959)
    JW MacB Cameron, BSA, MSc, PhD   1950–1960
    (to Department of Forestry)

WOODSLEE

Experimental Station
    JW Aylesworth, BSA, MSc, PhD   1946–1980
    (to Research Station, Harrow)

TORONTO

Forest Pathology Laboratory
    JE Bier, NScF, MA, PhD   1947–1951
    LT White, BScF, BA, MA, PhD   1951–1953
    (to Maple)

KINGSTON

Insect Disease Laboratory
    GE Bucher, BA, MA, PhD, FESC   1948–1955
    (to Belleville)

LONDON

Science Service Laboratory
    H Martin, BSc, MSc, DSc, FRIC, FCIC   1951–1960

Pesticide Research Institute
    EY Spencer, BSc, MSc, PhD, FCIC   1960–1978
    HV Morley, BSc, PhD   1978–1980
    (to Research Centre)

Research Centre
    HV Morley, BSc, PhD   1980–

MAPLE

Forest Biology Laboratory
    LT White, BScF, BA, MA, PhD   1953–1960
    (to Department of Forestry)

SMITHFIELD

Experimental Station
    HB Heeney, BSc, MSc   1960–1979
    SR Miller, BSc, MSc, PhD   1980–

MANITOBA

BRANDON

Experimental Farm
SA Bedford	1888–1905
N Wolverton, BA	1906–1907
J Murray, BSA	1907–1911
WC McKillican, BSA	1911–1925
MJ Tinline, BSA	1925–1946
RM Hopper, BSA, MSc	1946–1960
(to Research Station)	

Entomology Laboratory
RD Bird, BSc, MSc, PhD	1933–1952
(to Winnipeg)	

Research Station
JE Andrews, BSA, MS, PhD, FAIC	1960–1965
WN MacNaughton, BSc, MSc, PhD	1965–1980
BH Sonntag, BSA, MSc, PhD	1980–

TREESBANK

Entomology Laboratory
N Criddle	1913–1933
(to Brandon)	

MORDEN

Experimental Station
EM Straight, BSA	1918–1921
WR Leslie, BSA	1921–1956
CC Strachan, BSA, MS, PhD	1956–1960
JW Morrison, BSc, MSc, PhD	1960–1966
(to Research Station)	

Research Station
ED Putt, BSA, MSc, PhD	1966–1978
DK McBeath, BSA, MSc, PhD	1980–

WINNIPEG

Plant Pathology Laboratory
DL Bailey, BA, MS, PhD	1924–1928
JH Craigie, AB, MSc, PhD, DSc, LLD, FRSC, FAIC, FRS	1928–1945
WF Hanna, OC, CBE, OLM, BA, BSc, MSc, PhD, LLD, FRSC, FAIC	1945–1952
T Johnson, OC, BSc, BSA, MSc, PhD, DSc, FRSC, FAIC	1953–1957
(to Science Service Laboratory)	

CH Goulden, BSA, MSA, PhD, LLD, FRSC, FAIC, FASA   1925–1948
RF Peterson, BSA, MS, PhD, FAIC   1948–1957
(to Science Service Laboratory)

Forest Entomology Laboratory
    HA Richmond, BSF, MSc, PhD, FCIF   1937–1945
    RR Lejeune, BSA, MSc   1945–1951
    (to Forest Biology Laboratory)

Soil Survey
    WA Ehrlich, BSA, MSc, PhD   1939–1964
    HJ Hortie, BSc, MSc,   1964–1974
    RE Smith, BSc, MSc   1974–

Stored Products Insect Laboratory
    BN Smallman, BA, MA, PhD, FAAS   1946–1951
    FL Watters, BSc, MSc   1951–1957
    (to Science Service Laboratory)

Forest Biology Laboratory
    RR Lejeune, BSA, MSc   1951–1955
    WA Reeks, BSc, MSc   1955–1960
    (to Department of Forestry)

Science Service Laboratory
    T Johnson, OC, BSc, BSA, MSc, PhD, DSc, FRSC, FAIC   1957–1959

Research Station
    T Johnson, OC, BSc, BSA, MSc, PhD, DSc, FRSC FAIC   1959–1962
    JA Anderson, BSc, MSc, PhD, DSc, LLD, FCIC, FAIC, FRSC   1962–1963
    AE Hannah, BSA, MS, PhD   1963–1970
    WC McDonald, BSA, MSc, PhD   1970–1979
    DG Dorrell, BSA, MSc, PhD   1979–1983
    TG Atkinson, BSA, MSc, PhD   1983–

MELITA

Experimental Station
    JV Parker   1935–1959
    (closed)

PORTAGE LA PRAIRIE

Experimental Station
    EM MacKey, BSA   1944–1960
    WO Chubb, BSc, DSc   1960–1965
    TA Cheney   1965–1981
    G Loeppky   1981–

WABOWDEN

Experimental Station
    VW Bjarnarson, BSA   1955–1957

| P Braun, BSA | 1957–1965 |
| (closed) | |

SASKATCHEWAN

INDIAN HEAD

Experimental Farm
A Mackay, LLD	1888–1913
TJ Harrison, BSA	1913–1915
WH Gibson, BSA	1915–1919
ND MacKenzie, BSA	1919–1924
WH Gibson, BSA	1924–1949
JG Davidson, BA, BSA, MSA	1949–1952
JR Foster, BSA	1953–1972
RN McIver, BSA	1972–1978
WB Towill, BSA	1979–

Forest Entomology Laboratory
JJ de Gryse, PhCand	1923–1925
R Stewart	1927–1939
LOT Peterson, BSc, MSc	1939–1957
(closed, to Calgary and Winnipeg)	

Forest Nursery Station
(from Department of the Interior)
NM Ross, BSA, BF	1931–1941
J Walker, BSc, MSc	1942–1958
WM Cram, BSA, MS, PhD	1958–1969
(to Department of Regional and Economic Expansion)	

ROSTHERN

Experimental Station
WA Munro, BA, BSA	1909–1932
FV Hutton, BSA	1932–1940
(closed)	

SCOTT

Experimental Station
RE Everest, BSA	1911–1914
MJ Tinline, BSA	1914–1924
V Matthews, BSA	1924–1928
GD Matthews, BSA	1928–1959
RG Savage, BSA, MSc	1960–1964
CH Keys, BSA	1964–1977
KJ Kirkland, BSA, MSc	1978–

SASKATOON

Entomology Laboratory
| AE Cameron, MA, DSc, FES | 1917–1921 |

KM King, BSc, MSc, PhD   1922–1945
AP Arnason, BSc, MSc, PhD   1946–1952
H McDonald, BSA, MSc, PhD   1952–1957
(to Science Service Laboratory)

Plant Pathology Laboratory
WP Fraser, MA, MA, LLD   1919–1925
GB Sanford, BSA, MS, PhD   1925–1927
PM Simmonds, BSA, MS, PhD   1928–1957
(to Science Service Laboratory)

Soil Survey
HC Moss, BSA, MSc   1929–1959
JS Clayton, BSA, MSc   1959–1967
DF Acton, BSA, MSc, PhD   1967–

Dominion Forage Crop Laboratory
TM Stevenson, BSA, MSc, PhD, FAIC   1932–1938
WJ White, BSA, MSc, PhD   1939–1956
JL Bolton, BSA, MSc, PhD   1956–1957
(to Science Service Laboratory)

Forest Pathology Laboratory
CG Riley, BSA, MF, PhD   1948–1960
(to Department of Forestry)

Science Service Laboratory
MW Cormack, BSA, MSc, PhD, FRSC, FAIC   1957–1959

Research Station
MW Cormack, BSA, MSc, PhD, FRSC, FAIC   1959–1964
JER Greenshields, BSA, MSc, PhD, FAIC   1964–1979
JR Hay, BSA, MS, PhD   1981–

SWIFT CURRENT

Experimental Station
JG Taggart, CBE, BSA, DSc, FAIC   1921–1934
LB Thomson, OBE, BSc, FAIC   1935–1948
GN Denike, BSA   1948–1965
(to Research Station)

Soil Research Laboratory (financed by PFRA)
JL Doughty, BSA, MSc, PhD   1936–1957
(to Experimental Station)

Research Station
JE Andrews, BSA, MS, PhD, FAIC   1965–1969
AA Guitard, BSc, MSc, PhD, FAIC   1969–1978
WL Pelton, BSA, MSA, PhD   1978–1983
DM Bowden, BSc, MSc, PhD   1983–1985
                                                         1986–

REGINA

Experimental Station
    WS Chepil, BSA, MSc   1931–1936
    J Cameron, BSA   1936–1945
    JR Foster, BSA   1945–1953
    HW Leggett, BSc, BSA   1953–1962
    (to Research Station)

Research Station
    JR Hay, BSA, MS, PhD   1962–1980
    J Dueck, BSA, MSc, PhD   1981–

SUTHERLAND

Forest Nursery Station
    (from Department of the Interior)
    J McLean   1931–1941
    WL Kerr, BSA, MSc   1941–1962
    (closed)

MELFORT

Experimental Station
    MJ McPhail, BSA   1935–1948
    HE Wilson, BSA   1948–1960
    WN MacNaughton, BSc, MSc, PhD   1960–1966

Research Station
    SE Beacom BSc, MSc, PhD   1966–

ALBERTA

LETHBRIDGE

Experimental Station
    WH Fairfield, OBE, BSA, MS, LLD, FAIC   1906–1945
    AE Palmer, BSc, MSc, FAIC   1945–1953
    H Chester, BSA   1953–1959
    (to Research Station)

Entomology Laboratory
    EH Strickland, MS, DSc   1913–1921
    HL Seamans, BSc, MSc, PhD   1921–1944
    GF Mason, BSc, MSc   1944–1948
    CW Farstad, BSc, MSc, PhD   1948–1949
    (to Science Service Laboratory)

Animal Disease Research Institute (Western)
    (from Health of Animals Branch)
    LM Heath, BVSc, DVSc   1937–1938
    R Gwatkin, DVM, DVSc   1938–1945

RC Duthie, BVSc, DVSc   1946–1952
(to Production Service)

Livestock Entomology Laboratory
RH Painter, BSA, MSc   1946–1949
(to Science Service Laboratory)

Science Service Laboratory
WC Broadfoot, BS, MS, PhD   1949–1959
(to Research Station)

Research Station
TH Anstey, BSA, MSA, PhD, FAIC   1959–1969
JE Andrews, BSA, MS, PhD, FAIC   1969–1981
DG Dorrell, BSA, MSc, PhD   1983–

LACOMBE

389

Experimental Station
GGH Hutton, BSA   1907–1919
FH Reed, BSA   1920–1946
GE DeLong, BSA, MSc   1946–1955
JG Stothart, DSO, BSA, MSc, FAIC   1955–1959
(to Research Station)

Research Station
JG Stothart, DSO, BSA, MSc, FAIC   1959–1976
FJ Kristjansson, BSA, MSA, PhD   1976–1979
DE Waldern, BSA, MSA, PhD   1980

FORT VERMILION

Experimental Station
FS Lawrence   1907
R Jones   1908–1933
A Lawrence   1933–1944
VJ Low   1944–1956
CH Anderson, BSc, MSc   1956–1963
AG Kusch, BSc, BSA, MSc   1963–1965
B Siemans, BSA, MSc   1965–

BEAVERLODGE

Experimental Station
WD Albright, LLD   1915–1944
EC Stacey, BA, MSc   1944–1962
(to Research Station)

Research Station
AA Guitard, BSc, MSc, PhD, FAIC   1962–1969
LPS Spangelo, BSA, MSc, PhD   1969–1985
JD McElgunn, BS, MS, PhD   1985–

WAINWRIGHT

Cattalo Station
(in cooperation with the Department of the Interior)
J Wilson — 1919–1922
AG Smith — 1922–(1935)
AS McLellan — (1942)–1950
(to Manyberries)

EDMONTON

Plant Pathology Laboratory
GB Sanford, BSA, MS, PhD — 1927–1955
LE Tyner, BSc, MSc, PhD — 1955–1964
(to Lacombe and London)

Alberta Soil Survey
WE Bowser, BSA, MSc, PhD — 1936–1968
TW Peters, BSA, MSc — 1968–1974
WW Pettapiece, BSc, MSc, PhD — 1974–

MANYBERRIES

Range Experimental Station
LB Thomson, OBE, BSA, FAIC — 1927–1939
HJ Hargrave, BSA, FAIC — 1939–1947
HF Peters, BSA, MS, PhD — 1948–1964
(to Research Station, Lethbridge)

CALGARY

Forest Entomology Laboratory
GR Hopping, BScF, MSc — 1948–1952
(to Forest Biology Laboratory)

Forest Pathology Laboratory
VJ Nordin, BA, BScF, PhD — 1951–1952
(to Forest Biology Laboratory)

Forest Biology Laboratory
GR Hopping, BScF, MSc — 1952–1960
GP Thomas, BA, BScF, MF, PhD — 1960
(to Department of Forestry)

RALSTON

Entomology Laboratory
H Hurtig, BSc, PhD, FAIC — 1948–1956
(closed)

VAUXHALL

PFRA Drainage Division
    EA Olafson, BE, BSc, MSE, PhD   1949–1952
    CD Stewart, BSA, MSc, PhD   1952–1956
    EA Olafson, BE, BSc, MSc, PhD   1956–1960
    (to Research Station, Lethbridge)

Irrigation Station
    WL Jacobson, BSA   1953–1958
    LE Lutwick, BSc, MSc, PhD   1958–1959
    (to Research Station, Lethbridge)

VEGREVILLE

Solonetzic Soil Station
    RR Cairns, BSA, MSc, PhD   1957–1979
    MR Carter, BSA, MSA, PhD   1981–1982
    JR Pearen, BSA, MSA   1983–

YUKON AND NORTHWEST TERRITORIES

MILE 1019, Alaska Highway

Experimental Station
    JW Abbott   1945–1956
    WH Hough, BSA, MSc   1957–1959
    HJ Hortie, BSc, MSc   1960–1964
    JY Tsukamoto, BSc, MSc   1964–1969
    JRM Tait   1969–1970
    (closed)

FORT SIMPSON

Experimental Station
    JA Gilbey, BSA, MSc   1947–1960
    WA Russell, BSA   1960–1969
    AJ Tosh   1969–1970
    (closed)

WHITEHORSE

Soil Survey
    CAS Smith, BSc, MSc   1982–

BRITISH COLUMBIA

AGASSIZ

Experimental Farm
    TA Sharpe   1889–1911
    PH Moore, BSA   1911–1916
    WH Hicks, BSA   1916–1953

MF Clarke, BSA, MSA, PhD   1953–1959
(to Research Station)

Entomology Laboratory
  RC Treherne, BSA   1912–1916
  AB Baird, BSA, MS   1917–1921
  R Glendenning   1925–1953
  (to Chilliwack)

Plant Pathology Laboratory
  W Newton, BSA, MSA, PhD   1927
  (to Vancouver)

Research Station
  MF Clarke, BSA, MSA, PhD   1959–1972
  JE Miltimore, BSA, MSc, PhD 1973–1985
  JM Molnar, BSA, MSc, PhD   1985–

392   SAANICHTON

Experimental Station
  S Spencer   1912–1915
  L Stevenson, BSA, MS   1915–1921
  EM Straight, BSA   1921–1941
  JJ Woods, BSA, MSA   1941–1960
  (to Research Station)

Plant Pathology Laboratory
  W Newton, BSA, MSA, PhD   1929–1958
  WR Orchard, BA, MSA   1958–1960
  (to Research Station)

Animal Pathology Laboratory
  EA Bruce, VS   1937–1952
  (to Production Service)

Research Station
  H Andison, BSA   1961–1977
  JM Molnar, BSA, MSc, PhD   1977–1985
                             1985–

INVERMERE

Experimental Station
  GE Parham   1913–1919
  RG Newton, BSA   1919–1928
  (to Windermere)

SUMMERLAND

Experimental Station
  RH Helmer   1914–1923
  WT Hunter, BSA   1923–1931

RC Palmer, BSA, MSA, DSc, FAIC   1931–1951
TH Anstey, BSA, MSA, PhD, FAIC   1951–1958
CJ Bishop, BSc, AM, PhD, DSc, FRSC, FASHS, FAIC   1958–1959
(to Research Station)

Plant Pathology Laboratory
HR McLarty, BA, MA, PhD, FAIC   1921–1956
MF Welsh, BSA, PhD   1956–1959
(to Research Station)

Entomology Laboratory
J Marshall, BSA, MSc, PhD   1945–1959
(to Research Station)

Chemistry Laboratory
JM McCarthur, BA, MA, PhD   1948–1959
(to Research Station)

Research Station
CC Strachan, BSA, MSc, PhD   1959–1970
DV Fisher, BSA, MSA, PhD   1971–1975
GC Russell, BSc, MSc, PhD   1975–1985
DM Bowden, BSc, MSc, PhD   1985–

VERNON

Entomology Laboratory
RC Treherne, BSA   1915–1922
ER Buckell, BSc   1922–1939
J Marshall, BSA, MS, PhD   1939–1945
(to Summerland and Kamloops)

Forest Entomology Laboratory
R Hopping   1918–1939
GR Hopping, BScF, MSc   1940–1947
WG Mathers, BSA, MS   1947–1957
DA Ross, BSA, MSc, PhD   1957–1960
(to Department of Forestry)

VICTORIA

Fruit Insect Laboratory
W Downes   1919–1946
H Andison, BSA   1946–1959
(to Research Station, Saanichton)

Field Crop Insect Laboratory
KM King, BSc, MSc, PhD   1946–1957
(to Research Station, Vancouver)

Forest Entomology Laboratory
ML Prebble, BScF, MSc, PhD   1940–1945
HA Richmond, BSF, MSc, PhD, FCIF   1945–1955
(to Forest Biology Laboratory)

Forest Pathology Laboratory
    JE Bier, BScF, MA, PhD   1940–1947
    RE Foster, BA, BScF, PhD   1947–1955
    (to Forest Biology Laboratory)

Forest Biology Laboratory
    RR Lejeune, BSA, MSc   1955–1960
    (to Department of Forestry)

VANCOUVER

Forest Entomology Laboratory
    GR Hopping, BScF, MSc   1925–1934
    WG Mathers, BSA, MS   1934–1940
    (to Vernon)

Plant Pathology Laboratory
    W Newton, BSA, MSA, PhD   1928
    (to Saanichton)
    RE Fitzpatrick, BSA, PhD   1946–1959
    (to Research Station)

Soil Survey
    RH Spilsbury, BSA, MSA   1931–1938
    L Farstad, BSA, MSA   1939–1975
    T Lord, BSA   1975–1985
    DE Moon, BSc, PhD   1985–

Entomology Laboratory
    JH McLeod, BSc, MSc   1948–1955
    HR MacCarthy, BA, PhD, FESC 1955–1959
    (to Research Station)

Animal Pathology Laboratory
    IW Moynihan, DVM, MSc   1948–1952
    (to Production Service)

Chemistry Laboratory
    ME Reichmann, MSc, PhD   1955–1959
    (to Research Station)

Stored Products Laboratory
    JH Follwell, BSA, MSA   1949–1952
    P Zuk, BA   1952–1959
    (to Research Station)

Research Station
    REF Fitzpatrick, BSA, PhD   1959–1971
    M Weintraub, BA, PhD, FNYAS 1971–

WINDERERE

Experimental Station
    RG Newton, BSA   1928–1940
    (closed)

KAMLOOPS

Veterinary and Medical Insects Laboratory
    E Hearle, BSA, MSc   1928–1934
    GJ Spencer, BSA, MSc, FAAAS, FAAEE   1934–1936
    GA Mail, BSc, MSc   1937–1943
    JD Gregson, BA, MSc   1944–1954
    (to Entomology Laboratory)

Experimental Station
    EW Tisdale, BSc   1935–1939
    TG Willis, BSA, MSA, FAIC   1942–1962
    (to Research Station)

Field Crops Insect Laboratory
    ER Buckell, BA   1939–1949
    RH Handford, BSA, MSc, PhD   1949–1954
    (to Entomology Laboratory)

Entomology Laboratory
    RH Handford, BSA, MSc, PhD   1955–1962
    (to Research Station)

Research Station
    RH Handford, BSA, MSc, PhD   1962–1970
    JE Miltimore, BSA, MSA, PhD   1970–1973
    DE Waldern, BSA, MSA, PhD   1973–1978
    WK Dawley, BSc   1978
    JD McElgunn, BS, MS, PhD   1980–1985
    JA Robertson, BSA, MSc,PhD   1985–

SMITHERS

Experimental Station
    K MacBean, BSA   1937–1947
    WT Burns, BSA, MSc   1948–1954
    RG Savage, BSA, MSc   1955–1960
    (closed in 1969)

MILNER

Animal Pathology Laboratory
    JC Bankier, DVM   (1939)–1943
    (closed)

PRINCE GEORGE

Experimental Station
  RG Newton, BSA   1940–1945
  FV Hutton, BSA   1945–1954
  WT Burns, BSA, MSc   1954–1965
  WK Dawley, BSc   1966–1978
  WL Pringle, BSA, MSF   1979–

CRESTON

Experimental Station
  GR Thorpe, BSA   1940–1946
  FM Chapman, BSc, MSc   1946–1968
  (to Research Station, Summerland)

Plant Pathology Laboratory
  JM Wilkes, BS, PhD   1957–1967
  (to Research Station, Summerland)

Entomology Laboratory
  WHA Wilde, BA, MSc, PhD   1957–1967
  (to Research Station, Summerland)

CHILLIWACK

Entomology Laboratory
  HG Fulton, BSA   1953–1958
  (to Research Station, Vancouver)

APPENDIX III
EXTRACTS FROM LANDMARK DOCUMENTS

1. The Experimental Farm Station Act

Research Branch operates under the authority of the Department of Agriculture Act, assented to on 22 May 1868. The following is a copy of the Experimental Farm Station Act, assented to on 2 June 1886. Footnotes highlight some of the changes that have been made over the years, largely because of the development of Canada. The Act is substantially the same today as it was in 1886. The present Act may be found in the Revised Statutes of Canada, 1970, Volume III, Page 2887.

STATUTES OF CANADA
CHAPTER 57

An Act respecting Experimental Farm Stations. A.D. 1886.

Her Majesty by and with the advice and consent of the Senate and House of Commons of Canada, enacts as follows:—

1. This Act may be cited as *"The Experimental Farm Station Act."* 49 V.,c.23,s.1.

2. In this Act unless the context otherwise requires,—

(a.) The expression "the Minister" means the Minister of Agriculture;

(b.) The expression "farm station" means an experimental farm station established under the provisions of this Act. 49 V.,c.23,s.2.

3. The Governor in Council may establish, first, a farm station for the Provinces of Ontario and Quebec jointly; secondly, one for the Provinces of Nova Scotia, New Brunswick and Prince Edward Island jointly; thirdly, one for the Province of Manitoba; fourthly, one for the North-West Territories of Canada,[1] and fifthly, one for the Province of British Columbia;[2] and the farm station for the Provinces of Ontario and Quebec jointly shall be the principal or central station. 49 V.,c.23,s.3.

4. The Governor in Council may, for the purpose of establishing such farm stations, acquire by purchase an extent of land, not exceeding five hundred acres,[3] in the vicinity of the seat of Government, for the central farm station, and

---

[1]S.C. 1906,c.73 changed to: fourthly, one for the provinces of Saskatchewan and Alberta and the Northwest Territories jointly;

[2]S.C. 1949,c.6,s.12 add: the province of Newfoundland.

[3]S.C. 1906,c.73 add: Saskatchewan and Alberta.

an extent of land, not exceeding three hundred acres,[4] in either of the Provinces
of Nova Scotia, New Brunswick or Prince Edward Island, and a like extent of land
in the Province of British Columbia, for the farm stations secondly and fifthly
mentioned in the next preceding section; and the Governor in Council may, for
the like purpose, set apart in Manitoba and in the North-West Territories[2] of
Canada such tracts of unoccupied available public lands, which are the property
of Canada, as are necessary for the farm stations thirdly and fourthly mentioned
in the next preceding section; but the tract of public land so set apart shall not, in
each case, exceed one section:[3]

2. The Governor in Council may also set apart in the Province of Manitoba, and in that portion of the Province of British Columbia known as the
Railway Belt, in each a tract or tracts not exceeding ten sections, and in each of
the four provisional districts of the North-West Territories defined by order of the
Governor in Council, and known as Assiniboia, Alberta, Saskatchewan and
Athabasca, a tract or tracts not exceeding ten sections, for the purpose of tree-
planting and timber growing:

3. For the acquiring of lands for the purposes of this Act, all the powers
respecting the acquiring and taking possession of land conferred by "The
Expropriation Act," are hereby conferred upon the Minister; and all the provisions of the said Act respecting the compensation to be awarded for lands
acquired thereunder shall apply to lands acquired under the provisions of this
Act. 49 V.,c.,23,s.4.

5. The said farm stations shall be under the control and direction of the
Minister, subject to such regulations as are, from time to time, made by the
Governor in Council; and the Governor in Council may appoint[5] a director and
such officers and employees as are necessary for each farm station.
49 V.,c.23,s.5.

6. The Governor in Council may fix the rate of remuneration of the director and
officers and employees of each farm station, and such remuneration, and all
expenses incurred in carrying this Act into effect, shall be paid out of such
moneys as are provided by Parliament for that purpose. 49 V.,c.23,s.6.

7. Such officers of each farm station as are charged with such duty by the
Minister shall,—

(a.) Conduct researches and verify experiments designed to test the relative
value, for all purposes, of different breeds of stock, and their adaptability to the
varying climatic or other conditions which prevail in the several Provinces and in
the North-West Territories;

---

[4]S.C. 1928,c.25, S.1: remove: limitations on areas of land and add: acquire such other areas of land
as may be necessary for the establishment of such other Experimental Farm Stations,... as may be
considered advisable and in the public interest.
[5]R.S.C. 1927,c.61,s.7: To comply with the Civil Service Act, this was changed to: ...be appointed in
the manner authorized by law,...

(*b.*) Examine into the economic questions involved in the production of butter and cheese;

(*c.*) Test the merits, hardiness and adaptability of new or untried varieties of wheat or other cereals, and of field crops, grasses and forage-plants, fruits, vegetables, plants and trees, and disseminate among persons engaged in farming, gardening or fruit growing, upon such conditions as are prescribed by the Minister, samples of the surplus of such products as are considered to be specially worthy of introduction;

(*d.*) Analyze fertilizers, whether natural or artificial, and conduct experiments with such fertilizers, in order to test their comparative value as applied to crops of different kinds;

(*e.*) Examine into the composition and digestibility of foods for domestic animals:

(*f.*) Conduct experiments in the planting of trees for timber and for shelter;

(*g.*) Examine into the diseases to which cultivated plants and trees are subject, and also into the ravages of destructive insects, and ascertain and test the most useful preventives and remedies to be used in each case;

(*h.*) Investigate the diseases to which domestic animals are subject;

(*i.*) Ascertain the vitality and purity of agricultural seeds; and—

(*j.*) Conduct any other experiments and researches bearing upon the agricultural industry of Canada, which are approved by the Minister. 49 V.,c.23,s.7.

8.   The officer in charge, or such other officer at each farm station as the Minister designates, shall, for the purpose of making the results of the work done thereat immediately useful, prepare and transmit through the director to the Minister, for publication, at least once in every three months, a bulletin or report of progress. 49 V.,c.23,s.8.

9.   Such bulletins or reports, and all samples of grain, and of such plants and other products as are designated by the Minister, which are distributed for experiment and trial, may be transmitted in the mails of Canada subject to such regulations as to parcel postage as are prescribed by the Postmaster General. 49 V.,c.23,s.9.

10.   The officer in charge of each farm station shall prepare and transmit through the director to the Minister, on or before the thirty-first day of December in each year, a full and detailed report of the work accomplished, and of the revenue and expenditure at such farm station, which report shall be laid before both Houses of Parliament within the first twenty-one days of each session. 49 V.,c.23,s.10.

2.   Formation of a Research Branch

Dr. Robert Glen, Associate Director, Science Service, prepared the formal document submitted to the Minister of Agriculture, Treasury Board, and the Civil Service Commission recommending the establishment of a Research Branch. In his transmittal memorandum of 11 July 1958 to Dr. J.G. Taggart, Deputy Minister of the department, he suggested that "...this might well become a document of some historical significance ...."

The document reads:

RECOMMENDATION FOR THE ESTABLISHMENT OF A RESEARCH BRANCH[1] IN THE CANADA DEPARTMENT OF AGRICULTURE

The establishment of a single research organization within the Department of Agriculture has been a common subject of informal discussion for several years. More recently the question has received serious consideration by senior officers of Experimental Farms Service and Science Service. On June 26, 1958, a meeting convened by the Deputy Minister of Agriculture, with representatives of the Civil Service Commission and Treasury Board in attendance, reviewed the subject at length and approved the preparation of a formal submission as a means of reaching an official decision on the matter. (The minutes of this meeting are attached as Appendix A.) This statement has been prepared for this purpose and embodies a specific proposal and recommendation.

The Proposal

It is proposed:—

1.   That a Research Branch be established in the Canada Department of Agriculture effective April 1, 1959;

2.   that the Branch be formed initially by combining the present functions, personnel, and facilities of the Experimental Farms Service and the Science Service, but that consideration be given later to transferring to the Branch relevant work now being conducted in other parts of the Department, e.g., animal pathology from Production Service and agricultural economics from Marketing Service; and

3.   that the basic organization of the Research Branch be as outlined on the attached chart and explanatory statement (Appendix C).

Reasons for Reorganization

1.   The organization of research as it exists in the Department today is illogical. It does not fit the nature of the problems to be solved. For example, an understanding of soil fertility depends equally on a knowledge of soil chemistry and

---

[1]The name of the proposed organization and the departmental titles for senior personnel can only be given provisionally at this time.

soil microbiology as upon a knowledge of soil type and culture. But researches on the first two aspects are responsibilities of Science Service and on the last two of Experimental Farms Service. Likewise, the study of animal husbandry, animal diseases, and animal-infesting insects are currently the responsibility of three separate administrations, whereas the real problem is a unity: the economical production of high quality meat and milk. The farmer is concerned with the soil or the crop or the animal, but at present the research resources to deal with these "wholes" are scattered under separate authorities. It is imperative that responsible officers at regional laboratories and at headquarters be able to assemble under unified direction the requisite specialists to attack our major problems in comprehensive fashion.

2. The present organization of Departmental research is as inefficient administratively as it is illogical scientifically. Better use of existing Departmental talent and property could undoubtedly be achieved through a merging of the two administrations and through the greater sharing of research facilities that would follow. It is especially important that future laboratories and stations be designed for joint use.

3. The present proposal simply seeks official recognition for a union that is slowly taking place at the working level as a result of practical necessity. By voluntary action many research workers in the two Services are collaborating fruitfully. In the same way much needless duplication of facilities has been avoided. This trend has been more marked in recent years and integration of functions, staffs, and facilities, under unified direction, has already been achieved at a few points, e.g., St. John's West and Saskatoon. It is time that the Departmental organization for research reflected more nearly the basic requirements of the work and the convictions of the workers.

The historical background is reviewed briefly in Appendix B.

Objectives

The immediate objectives of the proposed reorganization are: to bring under unified direction the personnel of two distinct Services that are now responsible for separate but related aspects of the same problems; and to provide them with a common over-all authority and uniform administrative procedures. The long-range target, of course, is to improve the Department's research program mainly through achieving greater concentration on major problems.

401

Highlights of the Proposed Change

It is felt that the objectives can be achieved as follows:—
1. Unified direction of the research program would be ensured at different levels in the Branch through:
(a) a Director-General with over-all authority;
(b) a headquarters program directorate of senior scientists to develop and coordinate the whole research program and to assign the work to be done at each laboratory and station; and
(c) a research director at each field location (and at each headquarters laboratory) to be responsible for all departmental research personnel at that point and, wherever feasible, to have full authority for carrying out the assignments made from the program office.

The implementation of item (c) would immediately bring under unified direction the staffs of the two Services at more than a dozen locations, including such important research centres as Charlottetown, Kentville, Fredericton, Harrow, Winnipeg, Lethbridge, and Summerland.

2.   Unified administrative authority and procedures and the protection of scientific staff from these responsibilities would be achieved through:
(a)   an administration directorate headed by a senior officer with full responsibility for all business operations of the Branch; and
(b) assignment of administrative officers to each comprehensive field establishment.

3.   The vital relationship between the program directorate and the administration directorate would be maintained by:
(a)   the appointment of a Director of Administration who, in addition to having an aptitude for administration, has broad training and experience in research;
(b) organizing the headquarters offices to facilitate day to day liaison between the Director of Administration and the Director of Program; and
(c)   providing the Director of Administration with adequate staff to relieve him of the necessity of personal participation in the routine operations of his directorate.

This question is dealt with at greater length in Appendix C, part III.

4.   Greater concentration of effort on major agricultural problems would be sought through:
(a)   combining of staffs as referred to in 1(c) above and the development of comprehensive regional laboratories and stations designed to study in depth the important problems of their respective regions;
(b) specific assignments of national projects to selected regional laboratories; and
(c)   gradual development of National Research Institutes to work primarily on basic researches of wide application to major practical problems.

In the proposed new Branch there is no equivalent of the present research divisions of Experimental Farms Service and of Science Service. When originally established these semi-autonomous divisions were equally responsible, within their own subject fields, for several major functions such as program planning and coordination, research direction, and business administration. Modifications have been made in recent years. But in the proposed Branch these major functions are separated operationally wherever feasible. At headquarters, major emphasis has been shifted from line to staff relationships; and in the field greater authority and responsibility has been given for carrying out the actual research.

A detailed description of the proposed organization is given in Appendix C.

Costs

It is not possible at this time to estimate the monetary savings, if any, that might accrue from the proposed amalgamation. However, the only foreseeable annual increase in expenditures relates to some 20 senior positions for which higher salaries should be approved because of increased responsibilities. It is unlikely that the extra annual cost will exceed $20,000 to $25,000. Since some officers are now receiving salaries commensurate with their expected new duties while others will be assigned greater responsibilities than they now have, it is not possible until the actual appointments are made to make a more accurate estimate.

For the most part, existing classifications could be used in establishing suitable pay scales. The following are suggested:

Provisional Departmental	Titles Pay Classification
Director-General	?
Associate Director-General	Senior Officer 3
Director of Administration	Senior Officer 2
Director of Program	Senior Officer 2
Associate Directors of Program(5)	Senior Officer 1
Assistant Directors of Program(13)	Chief Res. Div. 2

Recommended Action

If this proposal is accepted in principle, arrangements should be made immediately thereafter for the following:

1. Clearance in principle with the Civil Service Commission and the Treasury Board.

2. A final decision on the name of the new organization and on departmental titles for the senior personnel.

3. The appointment, on an acting basis if necessary, of the Director-General and of other key personnel on whom will fall much of the responsibility for developing the plan beyond its present preliminary phase, e.g., the Associate Director-General, the Director of Program, and the Director of Administration.

4.   Agreement on an appropriate temporary change in the financial vote structure to facilitate the preparation of estimates for 1959–60.

5.   An official announcement of the change to be made in Departmental
organization.

Appendices attached:
A – Report of Meeting of June 26, 1958.
B – Notes on the Development of Research Organization in the Department.
C – Description of the proposed new Branch.

July 10, 1958.

(The appendices have not been included here because of their length.)

404

APPENDIX IV
SUMMARY OF EXPENDITURES AND STAFF—EXPERIMENTAL FARMS AND SUCCESSORS

Year	Consumer price index[1]	$Expenditures Current[2]	Adjusted[3]	Number of staff[4] Professional	Total
		(thousands)	(thousands)		
1890	9	112	1 243	10	75
1895	10	83	830	9	103
1900	10	90	900	12	113
1905	11	100	909	16	239
1910	11	231	2 100	19	360
1915	10	697	6 970	74	682a
1920	19	1 672	8 800	82	690
1925	15	1 698	11 320	118	749
1930	15	2 815	18 767	375	1 290
1935	12	2 337	19 475	295	1 390
1940	13	3 019	23 223	326	1 300b
1945	15	3 923	26 153	370	1 570c
1950	26	15 297	58 835	971	4 276
1955	23	18 935	82 326	1 249	4 259d
1960	26	21 885	84 174	1 023	3 835e,f,g
1965	28	32 593	116 403	964	3 583
1970	34	47 753	140 451	899	3 293
1975	48	95 633	199 232	864	3 222
1980	73	138 714	190 024	923	3 650
1985		263 267[5]	263 267	948	3 638
Total[6]		3 254 270	6 277 010		

[1]Consumer Price Index (CPI 1984 = 100) is from Statistics Canada. For 1890 to 1910 an estimate was calculated by averaging the salaries of the Director; Chief, Plant Pathology; Superintendent, Experimental Farm; Farm Foreman; and Teamster.

[2]Current dollars are the sum of audited expenditures for Experimental Farms, Entomological Branch, Science Service, and Research Branch, as appropriate.

[3]Dollars adjusted to 1984 dollars.

[4]Staff is in Person Years calculated from the reports of the Auditor General and the Estimates. Whenever possible, audited data were used. However, methods of presentation by the Auditor General and in the Estimates varied throughout the 100-year history and therefore data from year to year are not strictly comparable. Data for 1980 and 1985 are from branch records.

[5]Estimate, not expenditure.

[6]Totals are five times the arithmetic sums and therefore estimates.

a. Includes Entomological Branch and those employed under the Destructive Insects and Pests Act effective in 1914.
b. Includes Science Service effective 1937, of which Entomology and Animal Pathology were a part.
c. Includes Plant Protection, which became part of Science Service in 1942.
d. Excludes Animal Pathology, which transferred to Production Service in 1953.
e. Excludes Plant Protection, which transferred to Production Service in 1957.
f. Experimental Farms and Science Service amalgamated to form the Research Branch in 1959.
g. Forest Biology moved to the Department of Forestry in 1960.

406

INDEXES

Research Branch
Agriculture Canada
1886–1986

The names of those individuals only appearing in the lists of referenced material or in the appendixes are omitted from this index.

410

411

Masson, A.B., 215, 226
Mathews, A.E., 231
McArthur, J.M., 183
McBean, D.S., 222
McCubbin, W.A., 286
McDonald, H., 53
McDonald, W.C., 108, 222
McDunnough, J.H., **57**
McEachran, D., 4
McElroy, W.R., 240
McFadden, A.D., 214, 218, 222
McGillivray, C.S., 330
McGinnis, A.J., 108
McKenzie, R.I.H., 218
McLaine, L.S., 56, 69
McLarty, H.R., 286
McLennan, H.A., 235
McMechan, A.D., 319
McMullen, R.D., 287
McNaughton, A.G.L., 63
McRostie, G.P., 83, 235, 243
Mellor, F.C., 289, 291
Metcalfe, D.R., 222
Michaud, R., 235
Migicovsky, B.B., **97**, 101–103, 106, 107, 227, 355
Miller, H., 223
Miltimore, J.E., 183, 186
Minshall, W.H., 295
Moberley, F., 36, 261
Molnar, J.M., 164, 252, 320
Moore, R.J., 294
Morrison, J.W., 100, 102, 103, **104**, 105, 107, 109, 111
Mortimore, C.G., 232
Motherwell, Hon. R.W., 47, 352
Mountain, W.B., **106**, 107, 109
Moyls, A.W., 333
Moynan, J.C., **79**
Muir, G.W., 174
Mündel, H.-H., 230
Munro, W.A., 35
Munroe, S.S., 197
Murphy, P.A., 46, 286
Murray, B.E., 236
Murray, W.A., 235
Murwin, H.F., 245

Nass, H.G., 216, 217
Neal, J.L., 169
Neatby, K.W., 43, 64, 67, **70**, 71–76, 89–92, 95, 213, 351
Neilsen, J.J., 214
Newman, J.A., 179
Newman, L.H., **44**, 47, 219
Newton, D., 46
Newton, J., 46
Newton, M., 43, 46
Newton, R., 46

Newton, W., **46**, 286, 333
Nicholaichuk, W., 155
Nobel, C.S., 125
Nowosad, F.S., 242
Nyborg, M., 137

Ogilvie, I.S., 233
O'Halloran, G.F., 39, 42, 259
Oakland, G.B., 74, 309
Olsen, O.A., 144, 283
Olson, Hon. H.A., 103
Owen, C.W., 228

Painter, R.H., 73, 279
Paliwal, Y.C., 291
Palmer, A.E., 125, 152
Palmer, R.C., 249, 253
Pankiw, P., 239
Parent, R.C., 135
Parks, N.M., 246
Partridge, G., 46
Pawlowski, S.H., 227
Peake, R.W., 241
Pearson, Rt. Hon. L.B., 70
Pelletier, M., 318
Pelton, W.L., 110, 357
Penhallow, Prof., 5
Penney, D.C., 137
Petch, C.E., 53, 271
Peters, H.F., 177, 195
Peterson, D.G., 102
Peterson, R.F., 213, 221
Phillips, W.R., 327
Pickett, A.D., 68, 271
Pigden, W.J., 81, 105, 183
Pilote, Rev. A., 7
Platt, A.W., 73, 207
Poly, J., 355
Pope, Hon. J.H., 4, 19, 35
Porritt, S.W., 328
Porter, S.G., 152
Prebble, M.L., **75**
Prentice, R.M., 353
Preston, I., 252, 253, 255
Proudfoot, F.G., 202
Proudfoot, K.G., 283
Provancher, L., 271
Proverbs, M.D., 274
Putman, W.L., 271

Quamme, H.A., 299

Rahnefeld, G.W., 179
Rajhathy, T., 111, 220
Rasmussen, H.K.C.A., **80**, 92, 96, 97, 99, 102, 175
Reid, J., 12
Reinbergs, E., 222
Rempel, J.G., 281
Rennie, R.J., 169

413

Subject Index

The names of plant varieties, species, provinces, states, and countries are omitted from this index.

415

416

419

420

423

hog(s), 191, 192
Holstein. See breed.
honey, 40
hookworm(s), 204
hop(s), 46
hormone(s), 67
horse(s). See also breed. 12, 14, 30, 36,
    69, 217, 281
horsebean. See vegetable, bean.
horticulture. See also division. 16, 62, 85,
    153, 316, 351
horticulturist(s), 19, 30, 44, 66, 78, 83, 87,
    170, 212, 244, 246–248, 253, 255,
    256, 317
Hull, 63, 69, 204
Humboldt, 355
humus, 125, 235
hybrid, 176–180, 190, 191, 194, 197, 231,
    232, 234, 248, 254–256
hydrogen, 169
hydroponics, 85

424

IAEA. See agency, International Atomic
    Energy.
ice, **167**
icehouse, 167
IDRC. See centre, International Development
    Research.
illustration station(s). See also division. 77,
    85, 93, 130, 135, 139, 205
immigrant(s), 3
incubator(s), 26
Indian Head, 11, 13, 14, 17,18, 37, 38, 43,
    48, 55, 83, 83, 99, 124, 130, 133,
    179, 191, 192, 212, 247, 285, 293,
    297, 310
industry (industries), 11, 18, 34, 65, 87, 90,
    101, 109, 170, 258, 289, 313, 314,
    331, 339
  animal, 34, 234
  apple, 135
  barley, 221
  beef, 143, 243, 337
  canola, 228
  dairy, 143, 340
  fiber flax, 41, 257
  flower, 164
  forest, 51, 74
  greenhouse, 164
  leafcutter bee, 239
  livestock, 84, 173, 278
  ranching, 234
  seed corn, 232
  seed potato, 289
  soybean, 228
  tobacco, 259, 260
  tree fruit, 73, 287
  vegetable, 73, 164
  wheat, 213
infection, 278, 285–288

inoculum, 185
insect(s). See also pest, insect. 5, 11, 17, 19,
    20, 51, 68, 71, 75, 101, 267,
    270–272, 274, 275, 277, 280, 295,
    296, 299, 329, 352
  livestock, **278**
insecticide(s), 25, 40, 45, 68, 268, 271,
    274, 276, 277, 279–281, 295, 319
insectory, 36
insemination, **174**, 205
inspection, 64, 69, **336**, 355
inspector(s), 38, 338
  cannery, 330
  nursery, 36, 52
  potato, 87
institute(s),
  Agricultural Institute of Canada (AIC),
    89, 349
  Agricultural Machinery Institute, 355
  Animal Disease Research Institute
    (ADRI), 109, 198, 199
  Animal Parasitology Institute, 46
  Animal Research Centre (ARC), 110, 111,
    189, 192, 198
  Animal Research Institute (ARI), 103,
    110, 195, 201, 202, 341
  Biosystematics Research Institute (BRI),
    38, 58, 106, 189, 275, 294
  California Institute of Technology, 313
  Canadian Institute of Food Science and
    Technology, 336
  Cell Biology Research Institute (CBRI),
    102, 169, 291
  Chemistry and Biology Research Institute
    (CBRI), 104, 107, 111, 166, 189,
    291
  Commonwealth Institute for Biological
    Control (CIBC), 275, 353
  Commonwealth Institute of Entomology,
    353
  Engineering and Statistical Research
    Institute (ESRI), 108, 166, 167, 317,
    318, 323, 335–337
  Entomology Research Institute (ERI), 106
  Fresh Water, 282
  Food Research Institute, 64, 327, 339,
    341
  Genetics and Plant Breeding Research
    Institute, 329
  Institut national de la recherche
    agronomique (INRA), 355
  Institute for Biological Control, 104, 295
  Institute of Agricultural Technology, 107
  International Institute of Agriculture, 355,
    356
  Land Resource Research Institute (LRRI),
    111, 123, 138, 170, 320
  Microbiological Research Institute, 285
  Pesticide Research Institute, 99, 110

425

426

431

1886–1986